JUSTIFYING NEW MANUFACTURING TECHNOLOGY

JUSTIFYING NEW MANUFACTURING TECHNOLOGY

EDITED BY JACK R. MEREDITH

Published by
Industrial Engineering and Management Press
Institute of Industrial Engineers

Industrial Engineering and Management Press, 25 Technology Park/Atlanta
Norcross, Georgia 30092 404/449-0460

Published 1986
Printed in the United States of America

ISBN 0-89806-084-2

TABLE OF CONTENTS

PREFACE

The problem of justifying the acquisition of automation equipment and other advanced production technology in our manufacturing industries has become almost a national crisis. Survey after survey and conference after conference reiterate this fact: we cannot seem to get these systems, which may be an individual firm's only hope of surviving another half-decade, approved by top management. Historically outgunned by heavyweights in finance, accounting, and marketing, many manufacturing managers have simply given up. In all too many firms, top management doesn't understand manufacturing, or even want to, much less the non-cost and qualitative benefits such automation would bring.

This collection of carefully selected papers is being published to help managers in manufacturing, engineering, and even top management better understand the issues and ways that other firms have attacked, and in many cases solved, this problem. This is not the first collection of ideas and approaches for addressing the justification problem—indeed, such collections seem to appear regularly in the manufacturing literature. For example, *American Machinist* published a guidebook for machine-tool investments in 1979, although it was directed to somewhat different issues and is dated by now. Also, the Society of Manufacturing Engineers published an equipment justification monograph in 1977 that included forms and procedures for the standard justification process.

These papers were selectively gathered to address comprehensively the justification issue. Thus, these are not "engineering" papers, or "economic" papers, but a collection from a number of sources whose purpose is to give managers guidance on the variety of difficulties they face in justifying new systems. The book has been arranged to read like a textbook, with a logical flow of ideas and concepts, rather than like a book of random readings. Also, the papers were selected with a manager's viewpoint in mind, not an engineer's or a technician's. They are thus eminently readable and include a large number of case studies from well-known manufacturing firms.

However, this does not mean that the works are all philosophical and abstract. The point is to provide those facing the justification issue with some solid ammunition—cost analysis techniques as well as analytical techniques such as risk analysis, value analysis, simulation, scoring models, and so on. Nor does this collection stop there, because advice is given also on the best way to conduct the justification process. And numerous papers address the hard realities of the new technology's strategic value, the real reason for obtaining it.

In Part I: The Justification Problem, we look at the variety of difficulties firms experience with justification and what causes them. In Part II: Justification Processes, others share with us their thoughts on the justification process and how it should be conducted. Part III: Accounting Methodologies initiates our analysis of specific justification techniques starting with basic economic, or financial, approaches. Part IV: Analytic Approaches continues the thrust of Part III and illustrates approaches for handling risk and non-monetary costs and benefits, including qualitive aspects. Part V: Strategic Considerations looks at the bigger picture and the potential competitive gains for the firm, offering advice for justifying the technology along these lines. Part VI: Cases in Justification includes a number of cases where these approaches were applied and what the results were.

We hope that this collection of readings will help managers and engineers become better prepared to justify new technology and thus make our firms more competitive in a highly competitive world that is growing more so by the month.

Jack R. Meredith
University of Cincinnati

THE JUSTIFICATION PROBLEM

This part of the book gives an overview of the justification problem, its history, and what users and others think about it. The papers will illuminate the many facets of the difficulties and, hopefully, illustrate how these difficulties arise. It is not unusual for different firms to have different difficulties with the same issue, and simply because a firm may have solved one problem doesn't mean that another, even bigger problem isn't on the horizon.

The first paper, by Stevenson, puts you in the shoes of a supervisor who is facing the justification dilemma. The paper by Graham is one of the oldest in the collection and shows how little things have changed in sixteen years. Some of the material is naturally dated, but Graham forces us to put the justification problem in a wider perspective and see that there are larger issues to consider as well. The third paper, by Rosenthal, illustrates how pervasive the justification problem actually is. Rosenthal reports on his survey of users, vendors, and experts who offered their opinions.

The next paper, by Kaplan, is a recent classic on the weaknesses of corporate accounting and its dangers for the firm, particularly for manufacturing and the acquisition of new technology. In the fifth paper, Gold continues Kaplan's theme by summarizing the conclusions of his twenty years of research on the acquisition of advanced manufacturing technology and the errors that occur in justifying and implementing it.

Next, Skinner gives us a historical overview of the situation manufacturing managers have faced for the last forty years, how these manufacturing problems developed, and automation's potential for solving them. Finally, Hunter reviews the main themes of justification research and what the researchers feel are the primary difficulties and best hopes for solutions.

Reprinted from ***Industrial Engineering,*** *November 1984.*

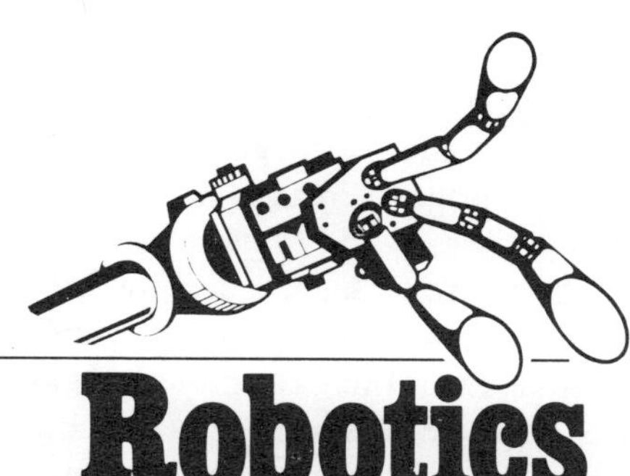

Robotics

Decision-Making On Investments Fuels Productivity Decline

By David O. Stevenson

The purpose of this article is to discuss a problem which has served to fuel the decline of manufacturing productivity in U.S. industry: the frequently prevailing approaches to investment decision-making. Responsibility for this should be shared by industrial management, business schools, finance professors, MBAs and financial managers.

This topic is by no means a new one. But unfortunately, the problem is still a very real concern to many U.S. firms. My intention is to look at the problem from a different angle. That is, I want to address the problem from the viewpoints of project engineers and manufacturing managers, not the viewpoints of top management.

Of course, changing top management's mindset on investment decision making is of the utmost importance. After all, they analyze the competitive environment and develop the strategic direction for the company. But lower down the decision-making ladder, where many projects originate which may serve to promote that very direction, the same problems can be found, the investment philosophies of the company leaders tend to become the philosophies of the operating management.

An investment opportunity

I believe the most effective way to present the problem is in the form of a hypothetical investment opportunity. The presentation is limited in data and is not meant to be considered an academic treatment of an actual occurrence.

The investment opportunity originated at the plant level: a first-line maintenance supervisor made some good observations and realized that there was some potential for a robotic application in the packing area, where there has been some turnover problems.

The packing job consisted of repetitive lifting of heavy packages and placing them into boxes. Based on what he had read in a trade journal, he thought that possibly robots could be used to do the same job.

This supervisor developed a brief description of the proposal and submitted it to the engineering department for a technical and economical feasibility study. Subsequently, a project engineer was assigned to examine the project.

It appeared to be a relatively small-scale automation project which would reduce the direct labor associated with the packing job and would probably provide some other benefits to the manufacturing operation.

Based on his cycle time calculations, the project engineer determined he would need one and a half fairly sophisticated jointed-arm robots. So for the heart of the project, he decided on two jointed-arm robots. After a number of phone calls and a lengthy number crunching session, he decided on the additional equipment he would need to interface the robots with the existing manufacturing operation. This equipment included in-going and out-going conveyor systems and a programmable controller to enable pieces of automation equipment to talk to each other and to the existing process machinery. He also figured in a sum for technical labor to include engineering design and system software development. Finally, he lumped in a sum for installation charges and debugging costs.

Then, he ran into something he had not expected: he had to outfit the plant with entirely new product carriers so that the robots could work with the product. The old carriers, 10-year-old hand trucks, were so non-uniform that the robots would not be able to work with them.

After he felt he had the investment numbers deemed sufficient for his preliminary analysis, he tackled the calculation of the savings the project would generate. He started with the simple savings numbers and calculated the direct labor savings plus direct labor fringes. He then calculated the net present value of the cash flows the project would generate. The value ended up on the low side, but it was positive.

Next, realizing that he must be able to talk about the project using the accepted company financial language, he calculated the payback.

Disappointment quickly set in: 5.7 years! He knew he would really have to dig to make this project look financially pleasing.

Productivity savings

The packing operation had always been a bottle-neck in the operation. It made sense that productivity would be increased by going to the two robots. But *how much* was difficult for him to quantify. He quantified a very conservative level of productivity gain and was able to reduce the amount of indirect labor and overhead allocated to the operation.

But could the automation of this process serve to increase the quality of the product? It made sense to him that some level of increase in product quality would be generated by automating the operation. He had heard many examples of customer complaints about poorly packed product; in transit the product was damaged or at times the customer received something he did not pay for.

Automation would take out all of the human variables and ensure consistent product packing. He wrote "increased quality" down as a desirable result of the project but felt uncertain about quantifying it.

He listed additional savings items. He knew he would be able to track the product through the manufacturing process much more accurately; automation should be very predictable. This accurate tracking ability should help to bring about the increase in quality but could also reduce some paperwork and possibly a portion of the labor associated with some of the accounting functions.

But how to quantify? He ended up making an educated guess at some numbers and lumped that figure in with his savings. He sat down and re-figured the payback at 4.7 years.

The big day came to discuss plant investment opportunities with management. The project engineer had always felt good about the robot packing project except, of course, for the payback. In his presentation, he described the project in such terms as: the "very possibly large increases in productivity," the "very possibly large increases in quality," the "very possibly large decreases in product tracking costs" through the "very possibly large increases in predictability."

The controller knew a decision had to be made soon, but he felt uncomfortable. Everything the project engineer had been saying made good sense, but it all seemed so nebulous. The air thickened as he uttered his all-important words, "Uh, yes, but what's the payback on this?"

Some may think that when something absolutely has to be done for survival, who cares about the dollars involved? Or, who cares how the financial tools measure an investment project if implementing that project could allow the company to successfully address a shifting market, meet the competition and survive long term? Of course, the question at hand is will that project provide benefits to the company—especially when the financial tool says it will not?

The truth is, and this should not sound surprising, most financial analysis tools are not just good tools but great tools. Though some of them are better than others, basically they all can significantly enhance our ability to make decisions. They add objectivity into frequently too-subjective investment decisions, they supply a means of comparison for different investment projects (an absolute necessity when investment capital is limited; and they help to create a uniform atmosphere within a company for making a decision.

So what is the problem with the tools? The real problem is not the tools, but the relative importance they have been given in investment decision making. Actually, the tools have advanced to the same position in many managers' minds as the *actual decision.*

Once we quantify everything, run all those beautiful numbers through our company routine, it's "cut and dried." All we have to do is "invest in those projects that are above this line. . ."

But note something I find very important here: it is difficult to assign the blame to any one particular group associated with this chain of events. Each one of these groups should shoulder a portion of the responsibility because in many ways, the problem is the result of an unfortunate, yet convenient, lack of communication.

Somewhere along the communication lines, we started to forget that in order for an investment decision to be clear-cut, *complete and accurate* numbers must be utilized. Otherwise, the investment decision is anything but a clear-cut selection of those projects which make it over a line drawn by the company leaders.

The real manufacturing world

It is unfortunate, but most of us do not have the time, money or frequently, even the ability to search, analyze and rework the numbers to the point that we can say this is *exactly* what an investment opportunity is worth to the company. Also, quantifying some things, such as the strategic merits of a project, is downright impossible.

Surely, we all work diligently to quantify as much as we can, and we absolutely desire to get the numbers down as accurately as possible. But let's not kid ourselves. The numbers will never be the kinds of numbers we like to think they are when we bravely make an investment decision.

I think most project engineers or managers can do a pretty good job making cost predictions, but they are not mediums. And now there is an additional variable thrown into the financial analysis: lack of experience in a new strategic direction. For many companies, the automation of manufacturing processes is still a very new thing. The investment opportunities associated with automation will contain new, state-of-the-art equipment that many U.S. manufacturing firms have very little experience with.

Proof of the problem

The next time you see a list of the possible investment projects your department or company may undertake, take note of the other pieces of information which are presented on that list. My bet is that each project's financial analysis score will hold a very prominent position on that list. When those individuals who do not fully understand the strategic implications of each project go over that list, which projects will be considered more favorably?

Unfortunately, they probably will not be the robotic and automation projects. For many industries, especially those with lower wage rates, sophisticated automation is very expensive. Attempting to obtain management acceptance of these projects by using the old company financial tools probably will relegate them to a remote place in the filing cabinet.

Of course, just because robotics and automation happen to be today's industry buzzwords, I am not advocating the blind automation of manufacturing processes. But I am also not for the ridiculous shelving of sound, strategically important robotic and automation projects because they do not measure up on the company's financial scoreboard.

The financial tool should not be the decision. A thorough, strategic analysis which examines all of the implications of an investment opportunity should provide the answer to the investment question.

In summary, I would like to offer the following suggestions:

☐ Recognize the problem: poor investment decision making can occur at many levels. At the lower levels, projects which could serve to promote the strategic direction set by upper management could be snuffed out.

☐ Realize the merits of financial analysis tools but recognize their limitations: the tool is only as accurate as the numbers which feed it. The numbers which feed the analysis come from projections and predictions and cannot possibly contain all of the characteristics of a project.

☐ Relegate these tools to their proper position: financial analysis tools should be only a part of a greater strategic analysis.

☐ Recognize the strategic merits of investment opportunities: the strategic implications of a project are frequently the most difficult to quantify but are often the most important characteristics of the project.

☐ Try to add other decision analysis tools into the investment decision making toolbox: quantitative decision analysis tools, such as a weighted average analysis of the various characteristics of a project, may help to reduce the subjectivity involved in decision making and could help to provide uniform decision making within a company.

☐ Accept the fact that investment decision making will contain a certain level of subjectivity. Investment decision making is rarely an easy or comfortable thing to do. It is usually a difficult and imprecise art of weighing out all of the implications of a project. Also, do not leave intuition completely out of the picture.

For further reading:

Dickinson, Roger, Herbst, Anthony, and O'Shaughnessy, John, "What Are Business Schools Doing for Business?" *Business Horizons,* November-December 1983, pp. 46-51.

Ellsworth, Richard R., "Subordinate Financial Policy to Corporate Strategy," *Harvard Business Review,* November-December 1983, pp. 170-182.

Fraker, Susan, "Tough Times for MBAs," *Fortune,* December 12, 1983, pp. 65-72.

Hayes, Robert H., and Abernathy, William J., "Managing Our Way to Economic Decline," *Harvard Business Review,* July-August 1980, pp. 67-77.

Hayes, Robert H., and Garvin, David A., "Managing As If Tomorrow Mattered," *Harvard Business Review,* May-June 1982, pp. 71-79.

"High Tech Productivity Potential Blindfolded By Financial Tools," *Automation News,* May 14, 1984, pp. 19 and 30.

Isaacs III, McAllister, "Robot Packs Yarn From Computerized Processing," *Textile World,* March 1984, pp. 77-78.

Michaels, Lawrence T., Muir, William T., and Eiler, Robert G., "Improving Technology Cost-Benefit Analysis," *Material Handling Engineering,* February 1984, pp. 49-54.

Naidish, Norman, L., "Return on Robots," presented at Robot VI Conference, March 1982, pp. 1-9.

Quinn, James Brian, "U.S. Industrial Strategy: What Directions Should It Take?" *Sloan Management Review,* Summer 1983, pp. 3-21.

Rehder, Robert R. and Porter, James L., "The Creative MBA: A New Proposal for Balancing the Science and the Art of Management," *Business Horizons,* November-December 1983, pp. 52-54.

Rudden, Eileen M., "The Misuse of a Sound Investment Tool," *Wall Street Journal,* November 1, 1982, p. 30.

Van Blois, John P., "Strategic Robot Justification: A Fresh Approach," *Robotics Today,* April 1983, pp. 44-48.

David O. Stevenson is a project engineer in the automated factory systems group with a Fortune 200 firm. His present job responsibilities include plant surveying, justification analysis and implementation support of robotic and automation projects. He holds a BS degree in mechanical engineering from Virginia Polytechnic and State University and an MS in industrial administration from Purdue University's Krannert Graduate School of Management. Stevenson is a member of IIE. He previously worked for DuPont in the Naval Sea Systems Command.

*Reprinted from **Automation,** May 1970.*

cost justification
of capital expenditures

By PEARSON GRAHAM, Director, Facility Planning and Control
TRW Corporate Staff Manufacturing Services, Cleveland, Ohio

As American business enters the seventies, operating and technical people in many companies find themselves attempting to justify expenditures for capital equipment as they did in the fifties and sixties. Meanwhile, managers are growing more knowledgeable about the broad scope of their business, are more thoughtful about the effects of their actions, and are under greater pressure to make the most of each dollar. Many are dissatisfied with their capital programs. As a result, conventional techniques for capital expenditure analysis are coming under fire.

Three basic conventional techniques for capital expenditure analysis are presently in widespread use. Payback period is the oldest and is out of academic favor. Payback period is equal to the time span required for an investment to self-liquidate. It is a measure of capital liquidity and, by inference, is a measure of short-term profitability.

Another technique, the well-known MAPI method of justification, was developed by George Terborgh of the Machinery and Allied Products Institute in the late 1940s as a means of ranking machinery replacement candidates. It is a return-on-investment type of analysis in which operating savings, next year capital consumption incurred, and next year capital consumption avoided are related to net investment. The capital consumption concept is taken from the field of economics. In practice, the applier uses preprinted charts supplied by MAPI and data derived from averages in order to estimate capital consumption for the specific machines in question.

Discounted cash flow is the technique most widely favored. It was introduced as a means for capital expenditure analysis by Ray Ruel and Joel Dean in the early fifties. Future cash flows are discounted to present value using some discount factor. Two variations are popular. Using "Internal Rate of Return," cash flows (including the outward cash flow for the initial investment) are discounted at a rate such that present values of inward and

. . . what is needed is less effort taking a microscopic look at the details of the answer and more effort looking through wide-range field glasses at the entire area of cost justification . . .

outward cash flows are equal, for a net present value of zero. Comparing alternatives under these rules, the highest discount factor wins.

Alternatively, using the "Present Value" method, it is possible to employ an established discount rate which is often referred to as the cost of capital, probably erroneously. Future cash flows are discounted at the established rate, and the present value of such flows is compared to the investment required. Under these rules, the alternative with the highest ratio of present value to net investment is the winner.

Since these basic techniques were introduced, countless refinements have been developed. These include the discounted payback period, a number of improvements to the MAPI approach, and "two rate" discounting, to name a few. In general, each refinement reflects considerable careful thought and a valid philosophy of capital expenditure cost justification. No criticism is intended, because it is through careful study that a better understanding of capital is gained. Still, as the scopes of the studies become narrower and narrower, operating managers tend to feel less and less comfortable with the results. They feel intuitively that the answers, while valid, are to the wrong questions.

An operating man can make, at best, only an educated guess as to what his product demand will be several years in the future. He must mentally cope with changes in technology, foreign competition, and shifts in demand, for example. Under these circumstances, he has little interest in a refinement in technique that improves the accuracy of his computation by a percent or two. In the view of most operating managers, what is needed is less effort taking a microscopic look at the details of the answer and more effort looking through wide-range field glasses at the entire area of cost justification, questioning what is really important.

A broad investigation of cost justification is now under way. Older techniques are being questioned and re-evaluated. Very basic questions are being asked:

- Why do we want to invest in capital equipment?
- What could go wrong with the investment?
- How bad off could we be?
- How well off could we be?

Furthermore, the idea of summarizing a whole investment in one number is being questioned. The decision to invest is a complex one, and it is possible that we have gone too far and oversimplified. It seems about time to consider the importance of judgment when justifying capital investments. Furthermore, it seems about time to start giving the operating manager all the information he needs to make a good decision. A new approach to cost justification is to present all the important facts to a manager in a way that will make it easy for him to exercise his best judgment. Simple as it sounds, it is not always easy. It requires going right back to the basic questions.

Why Invest In Capital Equipment?

Probably the most basic question of all is, "why do we want to invest in capital equipment?" It is usually answered "to make money." This is a fair answer—most investments are for the purpose of making money, but not always directly. Thumbing through a few actual capital projects; the following ways of "making money" appear:

- Increase capacity to supply growing market
- Add locker facilities to aid employee relations
- Purchase pollution control equipment
- Replace machine for cost reduction
- Replace roof

- Acquire R&D facilities for new product development
- Establish new branch warehouse to give better service

All of these projects are related to earnings, but in a wide variety of ways. Added profitability of a new plant serving a growing market is easy to understand, but what about the new locker room? Good employee relations are certainly an important asset, but can profitability be related to a locker room? Or pollution control equipment? Or even a new roof? Often the relationship is so complex that it is difficult to see. Thus, a more explicit description of the reason for investment is essential for a manager to make a good decision. Also, there is the clear danger that in a zeal to be "scientific" in all justification studies, answers will be forced to fit questions that shouldn't really apply to a situation.

Consider this example: One day last year, a $25,000 capital expenditure project arrived for a new roof. Sales of the operation were $10 million annually, and an engineering economy study had been carefully prepared on the basis that if the roof were not repaired, the plant would shut down and $10 million in sales would be lost. The return on investment was, obviously, exceptionally good. It was not nearly as good as the project which arrived a month or so later, however. This time the entire $10 million in sales hinged on replacement of a boiler costing only $15,000—resulting in an even higher return on investment!

Obviously the problem lies not in the technique but in the way it is used. The alternatives chosen are extreme and the calculation represents an attempt to force the answers to fit the question, "what is the return on investment?" How much more reasonable it would have been to say: "We need a new boiler because every year the cost of maintaining the boiler goes up; the boiler is now 45 years old; and we are told that we only have a 30 percent chance of making it through the winter without replacing it." These were the actual reasons for needing the boiler.

Not all projects are this simple. For example, a new automotive product had been developed and the firm was engaged in limited production. The initial year's requirement was 50,000 pieces, growing to 70,000 the next, followed by 100,000 and 200,000 units in succeeding years. When the next year's forecast was established at 650,000 pieces, additional capital equipment was requested, amounting to nearly ½ million dollars. Management was understandably reluctant to make the investment if it looked like the product

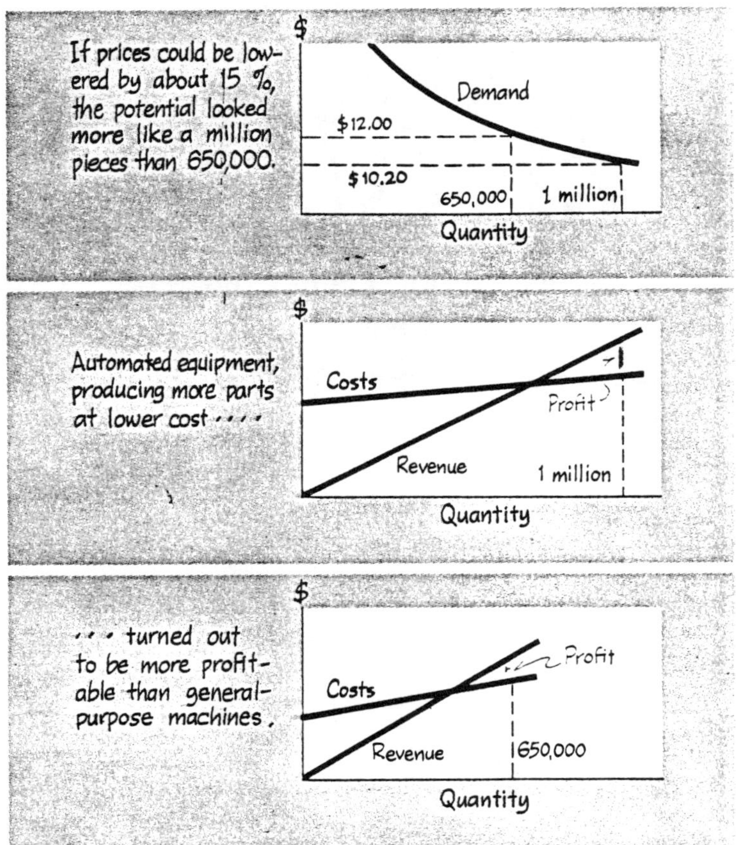

would be a "flash in the pan." On the other hand, at first glance, the potential looked good. Margins were satisfactory, and the technology was closely related to other successful products. Attempts to determine return on investment floundered, however, as information on prices and costs was elusive and seemed to change regularly.

Because of the size of the investment and its apparent potential, a team was assembled to study the project further. It was found that changes in technology favored market growth for the product, although there were alternatives for most applications. A limited market could be supplied at existing prices, but if prices could be lowered by about 15 percent, the potential looked more like a million pieces rather than 650,000 although it would take two or three years for the demand to develop.

As previously mentioned, margins were good but not great enough to allow a 15 percent reduction in price without a corresponding cost reduction. A feasibility study was therefore run by the project team to determine if cost reductions could be made at forecast volumes. The recommendation of the team was that reductions could be made, provided certain changes in equipment were made, such as substituting a

TABLE 1—Types of Uncertainty In Capital Decisions

Risk	Factors
Tangible	
Size of market	Aggregate consumer demand Alternative products New products
Share of market	New competitors Effectiveness of marketing efforts
Market growth rate	Growth of economy Structural changes in economy
Selling price	Time decay of prices Competition
Required investment	Price changes Quality of estimates Machine loadings
Residual value	Demand for used machinery Demand for scrap
Useful life	Deterioration Obsolescence
Operating costs	Learning curve Methods improvements Labor rates
Fixed costs	Overhead rates Administrative expenses
Product developments	Technical Breakthroughs Substitute products Substitute materials Process technology
Intangible	
Employee reactions Customer reactions Public reactions	

transfer machine for general-purpose machine tools. Lead times were such that the equipment would be under-utilized for only a relatively short time, if forecast demand turned out to be correct.

The study therefore recommended buying automated equipment, spending slightly more than the amount specified in the original request. The reason for investing was to produce increased quantities of a new product at reduced costs, resulting in greater earnings.

With an investment such as this, the traditional reasons for investing—such as obtaining increased capacity, reducing costs, and producing a new product—are so tightly tied together that an attempt to separate them can not tell the whole story. And the whole story needs to be known in order to make a good decision. This is especially true if risk is to be evaluated.

What Could Go Wrong?

Any capital investment depends on future events. Consequently, there is uncertainty involved in every capital decision. Depending on a person's point of view, it would be equally correct to ask "what could go well with the investment?" as "what could go wrong?". For nearly every situation there is a range of possibilities from very good to very bad. For every price-cutting competitor who might appear on the scene, there is another who might go out of business. For every user who may decide to manufacture his own requirements, there is another who may decide to start purchasing. Before we can really evaluate a capital investment, we must have some idea of the significant types of uncertainty which are associated with it.

A list of typical risks is shown by TABLE 1. This is merely representative, of course. A list of everything that could possibly go wrong with a project would be extremely long. If a thorough evaluation has been made of the reasons for investing, however, the major areas of uncertainty are usually few in number, and relatively easy to define.

For example, there are only a few risks in the example of the project to replace the boiler. The example of the automotive product is considerably more complex, and it is with a project such as this that definition of risk plays an important role. It is rarely possible for one man to sit at a desk and determine what might go wrong. Defining major risks requires the knowledge and skills of a number of people, generally including representatives from marketing, finance, manufacturing, and engineering. In this particular project, it was determined that the most important areas of uncertainty were:

- Continuation of automotive technological development trends
- Ability to maintain technical and manufacturing edge over competitors
- Ability to reduce labor costs
- Availability of labor
- Ability to reduce material costs
- Continuation of automotive overall market growth
- Timing of customer requirements
- Possibility of competitors selling identical product for less
- Ability to purchase facilities for estimated costs
- Ability to get machines in time
- Ability to get machines working properly

Having evaluated these factors, it is possible

to pick the most likely income, costs, and investment and, using these, compute a MAPI or discounted cash flow return on investment or payback period. Uncertainties would be merely recognized, and a judgment made as to the extent and likelihood of deviation from most likely values. Until recently, this was all that could be done. In some situations, and in some companies, it is all that needs to be done. The situations are familiar enough that it is not difficult to pass judgment. In most situations, this is more than is being done at present.

Usually, the most likely figure is presented to top management as if the future were certain. No indication is given that anything other than the most likely events could take place. Little wonder that managers search through the backup papers looking for a clue where things might go wrong, and little wonder that they feel investment theory is academic and irrelevant. Of what importance is a half percentage point if it turns out that the customers aren't there?

What Are the Odds?

A few companies, which are heavily dependent on major capital investments, are going a further step beyond just defining the areas where risk exists. If their experiments work out, and they seem to be doing so, it is likely that we will all soon be quantifying risk—that is, determining what is uncertain, how good or how bad things might be, and what the odds are. This amounts to sort of a statistical quality control for capital investments, establishing probability curves for return on investment. This has actually been done in the chemical industry where a capital decision can make or break a company. Risks have been defined, and probabilities of each type of risk determined and combined, using a computer to establish a composite risk curve. This showed not only the most likely return, but its probability, and the three-sigma limits, indicating how good or how bad the return might be.

There is one big difference between the probability of risk and the kind of probability used in quality control, however. The classical concept of probability depends on the frequency of various outcomes of something which happens numerous times. The more times it is repeated, the more accurately we predict the frequency of the various outcomes. With capital investments, however, the event is not repeated many times. Usually it is done only once. When the investment is made, the experience is usually of little or no benefit in evaluating the next project because each investment is unique. As a result, when assigning probability values to risk, it is necessary to use personal or subjective probabilities. That is to say, the probability is the best opinion of a knowledgeable and unbiased person (or group) concerning the various outcomes of a one-time event.

In a very real sense, we are gambling on future events every time we make a capital investment. It is only good sense to try to establish the odds before the money is put on the line. How to do it is the problem.

Basically, the answer is simple: ask someone. It is not an easy question to ask nor is it easy to answer, but it can be done. The answer is not likely to turn out in terms of modes, medians, and three-sigma limits. It is often possible, however, to get an answer for the most likely value, how good or how bad things might be, and what the chances are. The marketing man, for example, might say that by 1975 a specific market would most probably amount to $15 million per year. It might go as low as $12 million, and could even go as high as $20 million, but would be extremely unlikely—less than one chance in a hundred—to go beyond those limits. From this information, it is possible to plot a probability distribution.

The familiarity of many specialists with statistical quality control techniques can help, too. Some of the ways that have been used to establish probability distributions include:

- Asking a person to draw his best idea of the distribution.
- Showing a group of curves and asking a person to pick the one that looks best.
- Relating uncertain outcomes to hypothetical gambles involving a known probability. For example, a jar contains 8 black balls and 2 red balls. Would you rather bet $10 on getting a 50 percent market share or getting a red ball?

From these it is possible to establish a mathematical probability distribution for each important area of risk. Computer programs have been developed using Monte Carlo simulation techniques, which randomly select sets of factors and compute an overall probability distribution curve for the return on investment.

Is this advanced? Of course it is. But not too advanced for the seventies when American business will be under mounting pressure to increase productivity and output, making every dollar of capital count. MAPI, Payback, and Discounted Cash Flow techniques all answer the questions they were intended to be asked, but they are no longer enough. The next step forward in cost justification of capital expenditures may well be evaluation of the risks inherent in the numbers and assumptions used in viewing future events.

Reprinted from *Operations Management Review*, Winter 1984.

A Survey of Factory Automation in the U.S.

by
Stephen R. Rosenthal

This paper summarizes the results of a survey of users, vendors, and "experts" in factory automation. The survey clearly identifies the dilemma of users of factory automation and the difficulties they face in implementing these new technologies, as well as the problems vendors face in the same task. Stephen Rosenthal is an Associate Professor of Operations Management at Boston University and teaches their Management of Technology course. He recently concluded a trip to Japan in his continuing investigation of factory automation and its implementation.

Despite the considerable attention being placed on the potential payoffs from programmable manufacturing technologies, remarkably little systematic research has been conducted on organizational plans, decisions, and associated actions. The 1982-83 factory automation surveys project of the Boston University Manufacturing Roundtable reported on here was designed to develop a broad managerial view of computer-aided manufacturing processes in the U.S. Our purpose was to create a body of information, drawn from the recent experience of leading-edge practitioners, that would begin to describe the range of behaviors accompanying the adoption and implementation of factory automation technologies.

We sought not to limit our research to a single technology, but instead to explore the full range of new technologies, including CNC (computer numerical control), DNC (direct numerical control), PC (programmable controllers), AS/RS (automatic storage and retrieval systems), Robotics, CAD/CAM, and FMS (flexible manufacturing systems). Throughout our surveys we used the all-encompassing term computer-aided manufacturing processes, or CAMP for short, to refer to any of these technologies.

In each of our three surveys we had a specific target population in mind. The "user" survey was aimed at manufacturing organizations that were leaders in adopting factory automation technologies. This survey was designed to address "best practice" at the level of a single strategic business unit rather than across an entire corporation. Furthermore, we asked user respondents to describe individual automation projects, rather than to attempt to generalize from a broad base of project experiences. Our "supplier" survey was aimed at the leading producers of each of the factory automation technologies included in our definition of CAMP. Here we sought the observations and opinions of those firms with the most success (and therefore experience) in marketing computer-based manufacturing technologies. In these organizations, we sought responses from marketing executives who were in the best position to comment on their customers' behavior and on their own roles as suppliers of new technology. Finally, our "expert" survey was designed to test some of our preliminary findings from the users and suppliers, including insights gathered from in-depth follow-up telephone interviews with respondents to those two surveys. We intentionally tried to identify a broad sample of individuals whose views would be based on extensive factory automation experience in different capacities.

The potential benefits are highly uncertain (and) considerable risk is perceived to exist.

SURVEY OF USERS

The 57 business units included in this survey are all discrete-part manufacturers and are concentrated in machinery, electrical equipment, fabricated metals, and transportation equipment industry groups. They all generally claimed to do their own strategic planning for manufacturing technology, to automate particular manufacturing functions in accordance with an overall plan, and to examine this manufacturing systems plan in terms of its consistency with a long-range business plan. It is worth noting, however, that in each of these areas about a third of our sample acknowledged that they had not yet developed the associated formal planning capacity for manufacturing technology. Nevertheless, almost 90 percent of the participants reported that their business unit has made a strong commitment to implement additional "factory of the future" capabilities in the next few years.

Almost all . . . reported that their business units have made formal commitments to improve manufacturing productivity.

Many of the 57 projects reported on represent the most recent step in the use of factory automation by that business unit. Frequently, a project could be seen as a natural evolution from prior manufacturing capabilities or as one element in a broader automation initiative. Almost 90 percent of participating business units have adopted computerized process planning procedures to some extent. By doing so, these manufacturing organizations have notably improved (in varying degrees) such areas as communication with the shop floor, response to engineering changes by operating departments, and more detailed and uniform process plans. Another distinguishing characteristic of this sample is the extent of effort expended on automating production planning and control activities within the business unit.

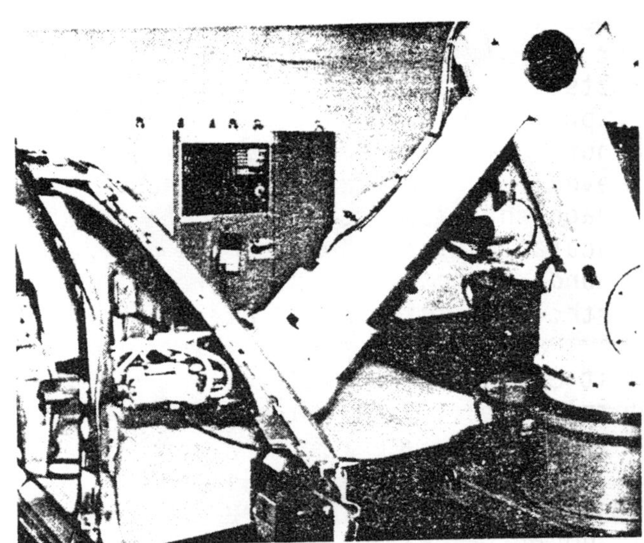

(Photo courtesy of GMF Corp.)

The projects in our survey were equally split between modifications to existing production systems and parts of new production systems, with only a few being labeled as off-line experiments. CAD/CAM projects were mostly modifications to existing production systems, while over half of the projects that developed factory management capabilities were associated with new production systems.

To achieve the desired benefits from their investment in factory automation technologies, users frequently had to take simultaneous and related steps in modifying production activities. Areas most commonly modified were work measurement/standards, quality control/inspection, routing, maintenance, and production control. CAD/CAM projects reported making relatively small adjustments, while the introduction of factory management systems was usually accompanied by substantial modifications in other areas as well.

Stimulus to Adoption

Almost all of our participants reported that their business units have made formal commitments to improve manufacturing productivity. Even if there was no formal stimulus, manufacturing management and staff felt that top management was generally supportive of opportunities for technological innovation. Among the several possible strategic reasons for adopting these technologies, the desire for new processes to produce new products was cited most frequently. Those responsible for developing advanced manufacturing technologies also believed that they could increase top management support by continuing to educate them, at least to the point of understanding the significance of forthcoming proposals.

Most of the business units in our sample used traditional return on investment or payback period calculations in their decisions to adopt automated manufacturing processes. Although these users seemed unable to escape traditional investment techniques, they usually cited strategic and qualitative factors as being equally important. Users were satisfied with traditional techniques when applied to stand-alone projects, such as an isolated CAD installation, where impacts are easily measured and the payoff is calculated in terms of either direct labor savings or reduced cycle time.

In order to learn more about potential applications, users depended heavily on presentations by suppliers and visits to other manufacturers who had already procured the technology. Other readily available sources--such as popular press, trade literature, academic experts, professional meetings, and manufacturing consultants--were also tapped routinely.

The ability to react more quickly to product design changes was consistently identified as the capability that was most enhanced.

Factors Considered in Selecting Among Options

The selection of a vendor is a key decision in adopting advanced manufacturing

Factor	Mean Score*
Reliability Already Proven in the Field	2.6
Maintenance Service Available	2.6
Price	2.3
Delivery Lead Time	2.0
Willingness to Upgrade Software and Hardware on Attractive Terms	2.0
Compatibility with Equipment supplied by Other Vendors	2.0
Prior Experience with Vendor	2.0

*Key: 3 = Very Important

2 = Somewhat Important

1 = Not Important

Exhibit 1: Factors in Selecting a Vendor

technologies. Users consider a number of factors at the point of selecting a vendor. As shown in Exhibit 1, users generally cited reliability and maintenance service as the most important selection criteria, followed by price. For a specific number of projects, other factors--compatibility, prior experience, upgrading potential, and delivery lead time--were stated to be important.

. . . the desire for new processes to produce new products was cited most frequently.

Waiting until a better system becomes available generally is not a consideration when leading-edge users are thinking about adopting automation technologies. They perceive their needs to be urgent, and they want "hands-on" experience with these computer-based technologies as soon as possible. In our sample, the most significant factors considered at the selection stage were ease of integration (interface requirements), expansion (future add-on capabilities), flexibility (degree of modularity), security (recover from system failure), and feasibility (special software and control requirements). Users were especially concerned about potential impacts on operations resulting from production downtime (due to computer malfunction) and subsequent upgrading of the technology.

When implementing computer-based applications, users often relied heavily on suppliers for consulting, training, trouble shooting, system design, and software preparation. Variations across the different technologies can be noted. For CAD/CAM projects, operator training by the supplier was always important, while system design was rarely provided. Typically, suppliers were least involved during projects, such as factory management systems, in which the user took responsibility for coordinating implementation. When a complex project was to be implemented incrementally, the suppliers also tended to play limited roles. This was particularly true when several suppliers were simultaneously participating in a single project.

Manual Operations and Labor

For most of the projects in this survey, coordination of automated processes with manually paced operations was not extensive. However, in over three-quarters of the projects, users took some steps to minimize the impact of the new process on ongoing operations. Particularly in job-shop environments, rigorous coordination was not required when buffers separated highly automated processes from more conventional equipment. Coordination was important for projects in material handling and CAM, since they required close synchronization between job orders, tools, raw materials, and programmed instructions.

Computer-aided manufacturing technologies with direct impact on production activities (robots, CNC, DNC, FMS) brought about a reduction in touch-labor. Whenever only a portion of an existing manufacturing facility was automated, participants reported that the affected employees were retrained and/or transferred to other areas. For factory management projects, the enhancement of workers' responsibilities, rather than their replacement, was the prevailing approach.

Evaluation of Results

Comprehensive evaluations often turned out to be infeasible, since available data were limited prior to implementation of the project, thereby limiting the possibility for before/after comparisons. Informal evaluations sufficed whenever a project was originally justified in terms of direct savings. In such situations, managers tended simply to ask if the equipment was functioning as expected. In contrast, factory management projects were more difficult to justify initially, and management was more likely to rely on an evaluation after implementation, to confirm accurate and timely flow of information.

Regardless of the forms these evaluations took, participants seemed to know the ways in which productivity had been improved. Exhibit 2 presents a list of the impact areas that were generally cited as being most significant, along with a breakdown by type of technologi-

	Type of CAMP			
	CAD	CAD/CAM	CAM	Factory Management & Control Systems
Greater Labor Productivity	H	H	H	H
Reduction in Cycle Time	M	H	H	H
Reduction in Set-Up and Run Time Plus Better Quality	L	M	M	M
Improved Communication Between Engineering and Manufacturing	M	H	L	M
Less Rework and Shape	L	M	M	H
Reduction in Work-in-process Inventory	L	L	M	H

Key: H = More than 80% of projects surveyed

M = From 60% to 80% of projects surveyed

L = Less than 60% of projects surveyed

Exhibit 2: Most Significant Improvements in Productivity Resulting from Installation of CAMP

cal innovation. The most frequently cited improvements were: the ability to achieve greater production without corresponding increases in labor or salaried staff; and reduced cycle, setup, and run time. For CAD/CAM projects a major reported impact was improved communication between engineering and manufacturing.

Participants reported a number of strategic benefits from their experience with computer-aided manufacturing technologies. Although all of these benefits were indirect and hard to measure, they were clearly felt. The ability to react more quickly to product design changes was consistently identified as the capability that was most enhanced.

Pricing, quality, response to changes in volume, and timely customer service were the other strategic factors that were seriously affected in a significant proportion of the projects. For example, more than half of the factory management systems enhanced the users' ability to react more quickly to changes in volume. For almost half of the same set of projects, participants noted improved ability to offer competitive prices by keeping costs down. Because our survey covered such a wide range of technologies, it is inappropriate to generalize about their strategic impacts.

SURVEY OF SUPPLIERS

The 38 suppliers in our sample serve a wide range of potential customers, including those with extensive experience with factory automation technology that are considering integrated applications and those that are just getting started. With this full spectrum in mind, suppliers reported that most potential users lack a good understanding of their needs regarding the implementation of factory automation technologies. Furthermore, suppliers observed that this shortcoming is especially pronounced for integrated applications in comparison to those that are incremental. They held similar views of their potential customers' understanding of the capabilities of the technologies being offered. Once again, this concern was notably stronger for integrated applications than for incremental ones. These findings are summarized in Exhibit 3.

Clearly, suppliers believe they are usually approached by potential customers who do not understand their own needs or the capabilities of the technology. Accordingly, suppliers offered to fill this gap through a range of services such as referrals to current users, general education in technology, and assistance in implementation planning. Most suppliers noted that customers generally do not have a realistic sense of the time and resources required for implementing their technologies. To assist in the implementation process, suppliers are very active in providing ongoing services including training, maintenance, and product updating. Most suppliers emphasized that involving factory floor personnel early on in procurement decisions would result in smoother implementation. This was viewed as particularly important in successfully introducing robots and machine tool control systems.

Suppliers reported frequent contacts with a wide range of people in each organization. The type of technology being considered affected the extent of participation within a user organization. The primary actors in factory automation decisions, according to suppliers, were manufacturing management, followed by corporate management and automation task forces. Manufacturing management was particularly dominant in the acquisition of robots and factory management systems. Decisions concerning machine tool hardware involved corporate as well as production management. Finally, automation task forces tended to play the strongest role in decisions regarding CAD/CAM systems.

Suppliers reported that only a small proportion of their customers have installed their product as an experiment rather than as an ongoing operational

	Perceived Level of Understanding by "Typical Customer" (% of Suppliers in Sample)		
	Poor	Some	Good
How well do potential customers generally understanding their needs for CAMP?			
Incremental Applications	16%	58%	26%
Integrated Applications	53%	42%	5%
How well do potential customers understand your technology at point of initial contact?			
Incremental Applications	16%	60%	24%
Integrated Applications	47%	50%	3%

Exhibit 3 Suppliers Views of Users' Sophistication

	Mean Score*
Does not meet investment criteria	2.1
Unable to adequately quantify returns	2.0
Wrong people involved in making decisions	1.9
Incomplete understanding of technology	1.9
Skills not in place	1.8
General risk too great	1.8
Labor or organizational problems	1.5

*Key: 3 = very significant factor; 2 = significant factor
1 = insignificant factor

Exhibit 4 Reasons that Potential Users Decide Not to Purchase Factory Automation Technology (Suppliers' Perceptions)

innovation. They also noted variations in the extent of evaluation being performed; the most common issue users study, according to suppliers, is the compatibility of the new technology with the rest of the manufacturing process. Users sometimes compare costs and savings with original expectations, and, occasionally, also try to determine effects on other functions of the business.

Suppliers unanimously reported that they know why potential users decide not to purchase factory automation technology. As shown in Exhibit 4, the reasons given most frequently were that the proposed project did not meet investment criteria or that the expected returns could not be quantified adequately. The identified barriers to adoption were incomplete technical understanding or "inhouse" skills and the degree of perceived risk. If suppliers are correct in their perceptions, users face two difficult dilemmas:

- They need to use traditional investment criteria to justify factory automation purchases, although they know that the implied quantification often does not reflect the potential returns;
- They lack sufficient understanding of these technologies but do not have adequate "internal" skills to improve their knowledge.

Suppliers present a necessarily broad view in this survey, since participants were asked to formulate generalizations by considering their full range of actual and potential customers. Nevertheless, these generalizations complement the findings from the survey of users. Most manufacturers, it seems, have a long way to go before they are sufficiently skilled to implement long-range commitments to these technologies. The prerequisite for such commitments, however, would be the development of a technological base on which to implement. Suppliers clearly have an incentive to educate users on appropriate criteria for evaluating specific factory automation options, as well as to help them understand the associated technological capabilities.

. . . the most difficult problems in achieving CIM are primarily managerial rather than technical.

SURVEY OF EXPERTS

Our panel of 64 experts agreed with each other on some points but disagreed on others. Each of these two kinds of findings is equally important. When a sample as large and as varied as this one exhibits strong agreement, the policy implications of the associated conclusion should be carefully examined. A basic split in opinion, in turn, calls for further investigation. Is current knowledge simply inadequate to resolve major ambiguities? Or does "the truth" depend on the perspective taken?

On questions relating to the adoption of computer-aided manufacturing processes the following points emerged (with varying degrees of concensus):

- To many users, the potential benefits of computer-aided manufacturing technologies are highly uncertain, while the associated costs are often easier to predict. For complex factory automation systems, considerable risk is perceived to exist.
- Manufacturers are tied to quantitative investment criteria as the basis for making decisions regarding the automation of the manufacturing processes.

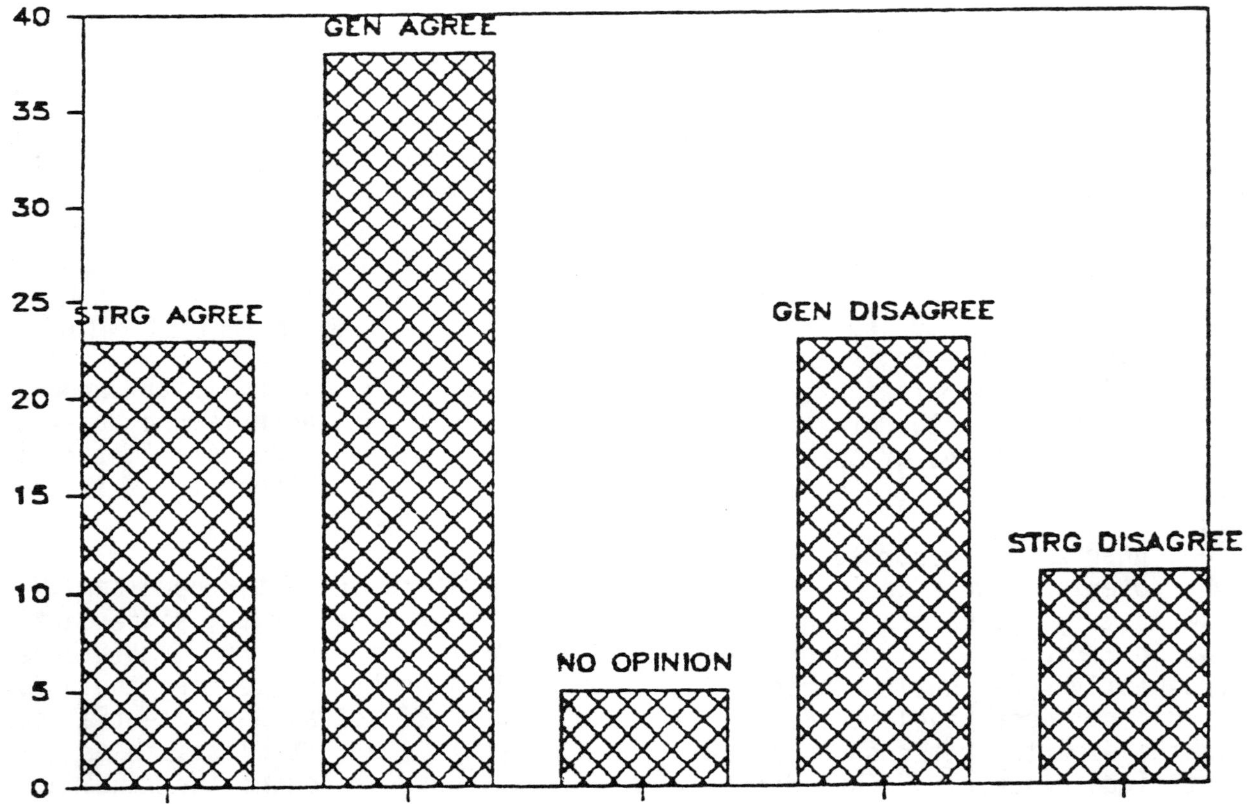

Ultimately, plant managers are likely to be a barrier to the adoption of sophisticated programmable manufacturing systems, since their own rewards are not likely to reflect the risks they must take.

Exhibit 5 Experts' Views of Plant Managers' Risk/Reward Profile

- Top management judgment is weak (i.e., inexperienced) in these matters and should not be used as a substitute for careful analysis.
- Plant managers are often not given adequate incentives to take risks in the adoption of new manufacturing technologies.
- Despite the current uncertainties and lack of experience, users would not significantly reduce their risks by postponing commitments to adoption of computer integrated technologies.

The implications of this set of strongly held views are worth noting. First, although U.S. manufacturers routinely calculate the return on investment from adopting computer-aided manufacturing technologies, this technique tends to hide, rather than highlight, the potential longer-term benefits of these technologies. At the same time, as shown in Exhibit 5, should the potential benefits of the proposed venture become apparent, there is likely to be some resistance by the plant manager because of the perceived level of risks. Even if top management supports automation in general, their specific judgments on any particular technology need to be carefully balanced with sound technical analysis. In short, for anything but the most straightforward initiatives toward factory automation, the adoption process is currently hampered by inappropriate decision criteria, low incentives and a weak base of knowledge.

On the issues of implementation, as shown in Exhibit 6, the panel of experts was almost unanimous on one conclusion: the most difficult problems in achieving computer integrated manufacturing are primarily managerial rather than technical. Involving workers in upcoming automation decisions and continuing to invest in people who have appropriate engineering and computer skills were noted as ways of strengthening a manu-

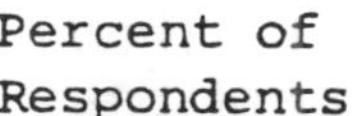

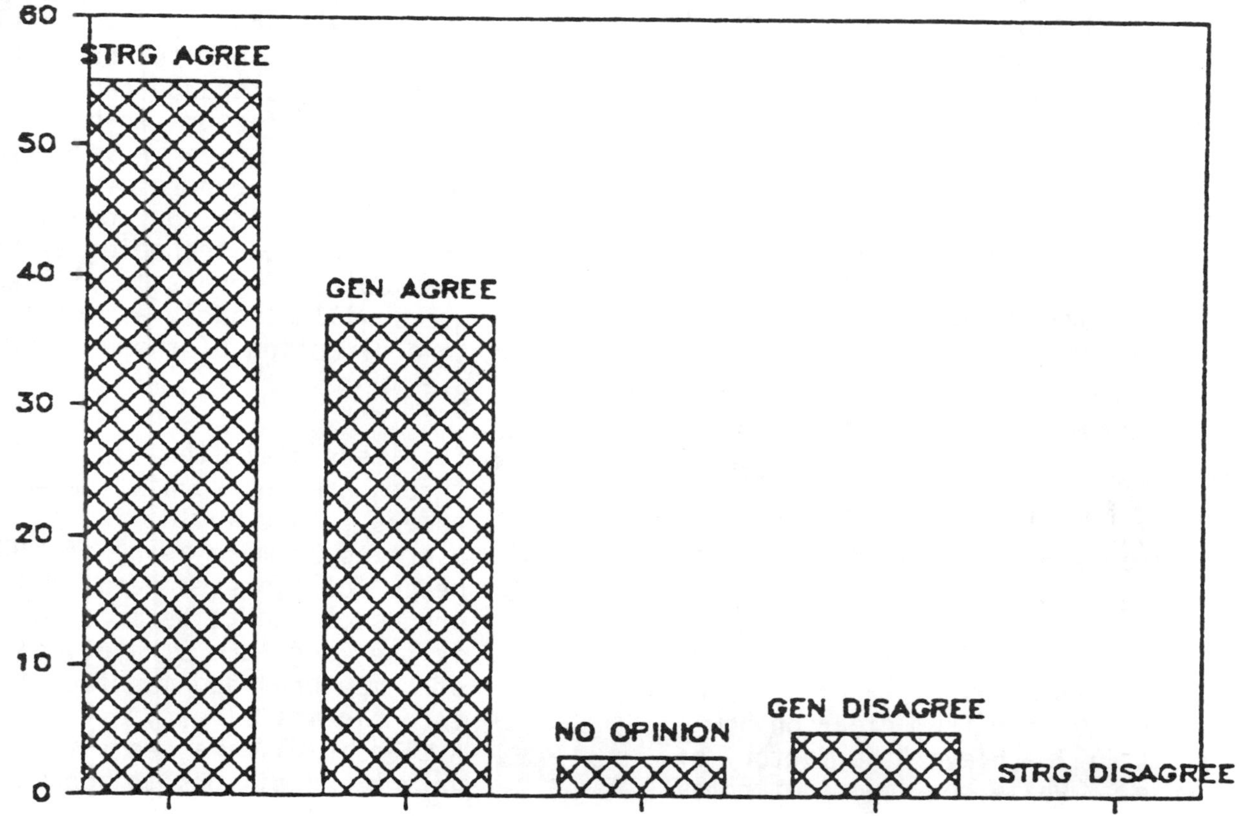

There are technical challenges in the achievement of computer-integrated manufacturing but the toughest problems at this time are managerial.

Exhibit 6 Experts' Views of Nature of the Challenge

facturer's ability to successfully design and develop computer integrated manufacturing. At the same time, three-quarters of the experts strongly emphasized the experience and capabilities of the supplier organizations as determinants of the widespread future adoption of computer integrated manufacturing.

Regarding the future of factory automation in the U.S., respondents offered the following opinions:

- A "split-vote" on whether the technology currently exists to allow computer integrated manufacturing to become a widespread reality.

- Almost a unanimous view that the leading-edge users of computer integrated manufacturing would strongly affect the future direction of these technologies.

- A majority view that a modular strategy for achieving computer integrated manufacturing can be viable if attention is paid to the subsequent linkage of the "islands of automation."

- A majority view that the U.S. manufacturers are less strategic in their view of the benefits of computer integrated manufacturing than are their counterparts in other leading countries.

Our purpose was to create a body of information . . . to describe the range of behaviors accompanying the adoption and implementation of factory automation technologies

In summary, our panel of experts, taken as a group, was far from clear on how U.S. firms will achieve "the factory of the future." While the important roles of leading users and suppliers was acknowledged, today's gaps in human resources, strategic orientation, and technological feasibility seemed to loom large. The experts cautioned us that there are no easy answers and that considerable learning through experience is still required before computer integrated manufacturing will be a widespread reality in this country.

CONCLUSION

Initial findings from each of the three surveys presented in this report form certain patterns that transcend the isolated facts commonly presented in the literature. The dominant impressions that emerged from the data were:

- Leading-edge users of computer-aided manufacturing processes believe in learning by doing. Most have proceeded in an incremental fashion to develop an internal base of experience with factory automation technologies. They tend to have supportive management and a sense of where they are heading. Their current measurement capabilities, however, restrict the basis for making adoption decisions as well as the subsequent evaluation of impacts from recent technological innovations. These users usually rely heavily on outside suppliers for critical technical assistance and consider reliability factors to be more important than price in selecting a vendor. They generally feel they cannot afford to postpone decisions until improved technologies become available. Current leading-edge users will strongly affect the future direction of these technologies.

- Suppliers claim that most manufacturers are not sophisticated customers. They would like potential users to be more aware of their needs for improved manufacturing processes and to be more interested in the long-term strategic benefits of computer-aided manufacturing technologies already on the market. A classic dilemma seems to have arisen: decisions to adopt expensive factory automation technologies are often made by managers who lack the background to assess technological options, while staff familiar with the new technologies are less able to appreciate associated strategic dimensions. A likely outcome is a decision that is either short-sighted or misguided. Suppliers with limited direct experience in new applications of their technologies and little in-depth knowledge of the business situations of their customers cannot be expected to help users avoid such errors.

- The most difficult problems in achieving computer integrated manufacturing are managerial rather than technical. Manufacturers contemplating factory automation face different issues depending on whether they adopt a retrofit strategy using existing production systems or whether they attempt to make a fresh start as new facilities are brought on stream. In either case, they often have inadequate internal technical resources, strong barriers to communication across traditional functional lines, and a reward structure that does not encourage risk-taking in the interest of long-term strategic gains. Manufacturing organizations that do not deal directly with these kinds of managerial problems are not likely to succeed in factory automation, even if appropriate technological options exist.

If these impressions are correct, then manufacturing managers ought to reflect on their organization's current strengths and weaknesses as they formulate a strategy for factory automation. In particular, they should begin with a healthy concern for the adequacy of their current base of knowledge, mix of human resources and formal measurement systems. They should also refine their expectations for integrated factory automation applications, to be sure that general goals such as "enhanced flexibility" have explicit operational meanings. Finally, they should improve their organization's ability to identify types of indirect costs, as well as benefits, that are implicit in such goals. OMA

Efforts to revitalize manufacturing industries cannot succeed if outdated accounting and control systems remain unchanged

Yesterday's accounting undermines production

Robert S. Kaplan

Many U.S. companies are now exploiting new process technologies, new inventory and materials handling systems, new computer-based abilities in design, engineering, and production, and new approaches to work force management. But these developments, promising as they are, rest on a foundation that is obsolete and in need of repair.

As the author makes clear, most accounting and control systems have major problems: they distort product costs; they do not produce the key nonfinancial data required for effective and efficient operations; and the data they do produce reflect external reporting requirements far more than they do the reality of the new manufacturing environment. Only when management accounting systems are brought in line with the new competitive environment will efforts to upgrade production prove genuinely successful – and permanent.

Mr. Kaplan, formerly dean of the business school at Carnegie-Mellon University, holds a joint appointment as professor of industrial administration at Carnegie-Mellon University and as the Arthur Lowes Dickenson Professor of Accounting at the Harvard Business School. This article grows out of his research on the new challenges for accounting systems in manufacturing industries.

Illustrations by John Devaney.

The present era of intense global competition is leading U.S. companies toward a renewed commitment to excellence in manufacturing. Attention to the quality of products and processes, the level of inventories, and the improvement of work-force policies has made manufacturing once again a key element in the strategies of companies intending to be world-class competitors. There remains, however, a major – and largely unnoticed – obstacle to the lasting success of this revolution in the organization and technology of manufacturing operations. Most companies still use the same cost accounting and management control systems that were developed decades ago for a competitive environment drastically different from that of today. Consider, for example, the following cases drawn from actual company experiences.

During Richard Thompson's two years as manager of the Industrial Products Division of the Acme Corporation, the division enjoyed such greatly improved profitability that he was promoted to more senior corporate responsibility. Thompson's replacement, however, found the division's manufacturing capability greatly eroded and a plunge in profitability inevitable.

Careful analysis of operations during Thompson's tenure revealed that:

- ☐ Increased profitability had been largely caused by an unexpected jump in demand that permitted the division's facilities to operate near capacity.
- ☐ Despite this expansion, the division's market share had decreased.
- ☐ Costs had been reduced by not maintaining equipment, by operating it beyond rated capac-

ity, by not investing in new equipment or product development, and by imposing stress on workers to the point of alienating them.

- ☐ Many costs had been absorbed into a bloated inventory position.
- ☐ Unit productivity had actually fallen.

By this time, however, Thompson was secure in his senior position and was still receiving credit for the high profits Industrial Products had earned under his direction.

The Carmel Corporation had made significant investments in labor-saving equipment. Yet with total costs, particularly overhead, still increasing, it was hard pressed to maintain market share with prices that fully recovered all of its costs. Carmel used a standard cost accounting system that allocated all nondirect costs on the basis of direct labor hours. The company had installed this procedure many years ago when direct labor accounted for more than 60% of total costs and machinery was both simple and inexpensive. Over the years, however, investment in sophisticated new machinery had greatly reduced the direct labor content of the company's products.

Staff costs rose as Carmel expanded its design and engineering staffs to develop specialized high-margin products. Because these new products required advanced materials, Carmel also expanded its purchase of semifinished components from suppliers. With direct labor (at an average wage of $12 per hour) plunging as a fraction of total costs, the accounting system was allocating the growing capital and overhead costs to a shrinking pool of direct labor hours.

The predictable result: a total cost per direct labor hour in excess of $60 – and projections that it would soon rise to $80. Worse, efforts to offset these higher hourly rates by substituting capital and purchased materials for in-house production only compounded the problem. The accounting system was distracting management attention from the expansion of indirect costs.

Both these examples offer a pointed reminder that poorly designed or outdated accounting and control systems can distort the realities of manufacturing performance. Equally important, such systems can place out of reach most of the promised benefits from new CIM (computer-integrated manufacturing) processes. As information workers like design engineers and systems analysts replace traditional blue-collar workers in factories, accounting conventions that allocate overhead to direct labor hours will be at best irrelevant and more likely counterproductive to a company's manufacturing operations. And with the new manufacturing technology now available, variable costs will disappear except for purchases of materials and the energy required to operate equipment.

Not only will labor costs be mostly fixed; many of them will become sunk costs. The investment in software to operate and maintain computer-based manufacturing equipment must take place before any production starts, and of course that investment will be independent of the number of items produced using the software program. With the decreasing importance of variable labor costs, companies that allocate the fixed, sunk costs of equipment and information systems according to anticipated production volumes will distort the underlying economics of the new manufacturing environment.

In this environment, companies will need to concentrate on obtaining maximum effectiveness from their equipment and from their increasing investment in information workers and in what they produce. Controlling variable labor costs will become a lower priority. This major change in emphasis requires that managers learn new ways to think about and measure both product costs and product profitability.

Nonfinancial aspects of manufacturing performance

It is unlikely, however, that any cost accounting system can adequately summarize a company's manufacturing operations. Today's accounting systems evolved from the scientific management movement in the early part of the twentieth century. They were instrumental in promoting the efficiency of mass production enterprises, particularly those producing relatively few standard products with a high direct labor content. Reliance on these systems in today's competitive environment, which is characterized by products with much lower direct labor content, will provide an inadequate picture of manufacturing efficiency and effectiveness.

Measurement systems for today's manufacturing operations must consider:

Quality. To excel as a world-class manufacturer, a company must be totally committed to quality – that is, each component, subassembly, and finished good should be produced in conformity to specifications. Such a commitment to quality entails

1 See, for example, Jinichiro Nakane and Robert W. Hall, "Management Specs for Stockless Production," HBR May-June 1983, p. 84; Larry P. Ritzman, Barry E. King, and Lee J. Krajewski, "Manufacturing Performance – Pulling the Right Levers," HBR March-April 1984, p. 143; and Hal F. Mather, "The Case for Skimpy Inventories," HBR January-February 1984, p. 40.

major changes in the way companies design products, work with suppliers, train employees, and operate and maintain equipment. But this commitment must also extend to a company's measurement systems. Data on the percentage of defects, frequency of breakdowns, percentage of finished goods completed without any rework required, and on the incidence and frequency of defects discovered by customers should be a vital part of any company's quality-enhancement program. Otherwise, the impact of variations in quality will show up in cost and market share data too late and at too aggregate a level to be of help to management. Direct quality indicators should be reported frequently at all levels of a manufacturing organization.

Inventory. A second nonfinancial indicator of manufacturing performance is inventory. American managers are well versed in optimizing inventory levels according to the economic order quantity (EOQ) model, which balances the cost of additional setup time with the cost of carrying inventory. They are less familiar with the effort, common among Japanese producers, to eliminate setup times and to implement just-in-time inventory control systems, which together reduce drastically overall levels of work-in-process (WIP) inventory.[1]

Many of the savings in reduced working capital, factory storage, and materials handling from cutting WIP will eventually be reflected in lower total manufacturing costs. But many of the savings that arise from transactions *not* taken – less borrowing to finance inventory, for example, or less need to expand factory floor space – will not be reflected in these costs. Therefore, such direct measures as average batch sizes, WIP, and inventory of purchased items will provide much more accurate and timely information on a company's manufacturing performance than will the behavior of average manufacturing costs.

Productivity. Direct measures of productivity are a third important set of nonfinancial indicators. Even in companies publicly committed to productivity improvements, accurate measurement of productivity is often impossible because accounting systems are designed to capture dollar-based transactions only. Without precise data on units produced, labor hours used, materials processed, energy consumed, and capital employed, administrators must deflate dollar amounts by aggregate price indices to obtain approximate physical measures of productivity. But errors in approximation often arise that can easily mask any period-to-period changes in real productivity.

Alternatively, managers rely on partial productivity measures, such as value added per employee or output per direct labor hour, which attribute all productivity changes to labor. These mea-

sures tend to overlook gains from the more efficient use of capital, energy, and managerial effort and so encourage the substitution of capital, indirect labor, energy, and processed materials for direct labor. But as direct labor costs decline relative to total manufacturing costs, it becomes more – not less – important to focus on total factor productivity.

Nor can managers finesse these measurement problems by looking only at aggregate data on profitability. In the short run, product profitability may be caused more by relative price changes and holding gains not recognized by historical cost-based systems than by structural improvements in the production process. A temporary expansion of demand can, for example, enable a company to boost its prices faster than its growth in costs. In the long run, however, the higher wages paid by U.S. companies will lead to competitive difficulties – unless offset by higher productivity. During the 1960s and 1970s, many U.S. companies earned a comfortable profit and did not notice that their productivity had begun to stagnate or even decline. They are noticing now.

Innovation. Some companies choose to compete not by efficiently producing mature products that have general customer acceptance and stable designs but by introducing a constant stream of new products. Customers buy the products of these innovative companies because of the value of their unique characteristics, not because the products are cheaper than those of competitors. For innovating companies, the key to success is high performance products, timely delivery, and product customization. Attempts to impose cost minimization and efficiency criteria – especially early in the product development process – will be counterproductive.

Cost accounting systems, however, rarely distinguish between products that compete on the basis of cost and those that compete on the basis of unique characteristics valued by purchasers. Thus, it is difficult to manufacture new products in facilities that also manufacture mature products since plant managers, evaluated by an accounting system that stresses efficiency and productivity, find it disruptive to make products for which both the designs and the process technology are still evolving. Companies that cannot afford the luxury of a separate facility for new product manufacturing must learn to de-emphasize traditional cost measurements during the start-up phase of new products and to monitor directly their performance, quality, and timely delivery.

Work force. Another limitation of traditional cost accounting systems is their inability to measure the skills, training, and morale of the work force. As much recent experience attests, if employees do not share a company's goals, the company cannot survive as a first-rate competitor. Hence, the morale, attitudes, skill, and education of employees can be as valuable to a company as its tangible assets.

Some companies, noting the importance of their human resources, conduct periodic surveys of employee attitudes and morale. They also monitor educational and skill levels, promotion and training, and the absenteeism and turnover of people under each manager's supervision. These people-based measures are weighted heavily when managers' performance is evaluated. Meeting profit or cost budgets does not lead to a positive job rating if it is accompanied by any deterioration in these people-based measures.

In summary, the financial measures generated by traditional cost accounting systems provide an inadequate summary of a company's manufacturing operations. Today's global competition requires that nonfinancial measures – on quality, inventory levels, productivity, flexibility, deliverability, and employees – also be used in the evaluation of a company's manufacturing performance. Companies that achieve satisfactory financial performance but show stagnant or deteriorating performance on nonfinancial indicators are unlikely to become – or long remain – world-class competitors.

Improving control systems

Improving manufacturing performance requires more of accounting systems than the timely provision of relevant financial and nonfinancial data. Fundamental changes in management control systems are also needed. In particular, there is a need to rethink the way companies use summary financial measures like ROI to coordinate, motivate, and evaluate their decentralized operating units.

The ROI measure was developed earlier in this century to help in the management of the new multi-activity corporations that were then forming. ROI was used as an indicator of the efficiency of diverse operating departments, as a means for evaluating requests for new capital investment, and as an overall measure of the financial performance of the entire company.

Through the use of ROI control, early twentieth century corporations achieved a specialization of managerial talent. Managers of functional departments (manufacturing, sales, finance, and purchasing) could become specialists and pursue strategies for their departments that increased the ROI of the entire company. Senior managers, freed from day-

to-day operating responsibility, could focus on coordinating the company's diverse activities and developing its long-term strategies.

In practice, decentralization via ROI control permitted senior executives to be physically and organizationally separated from their manufacturing operations. For many years, this separation was a valuable and necessary feature that enabled corporations to expand into many diverse lines of business. Recently, however, problems with running corporations "by the numbers" – that is, with excessive reliance on ROI measures but without detailed knowledge of divisions' operations and technology – have become uncomfortably apparent.

Inflation & ROI

The financial executives who pioneered in the application of ROI measures were not concerned with the distortions introduced by inflation. After World War II, however, as ROI-based control systems came into widespread use, continuous price increases gave a steady upward bias to ROI.

When fixed assets and inventory are not restated for price level changes after acquisition, net income is overstated and investment is understated. Thus managers who retain older, mostly depreciated assets report much higher ROIs than managers who invest in new assets. Such apparent differences in profitability, of course, have nothing to do with actual differences in the rates of return of the two classes of assets.

Financial accounting mentality

Such distortions of economic performance are but one manifestation of a broader problem: the use in a company's internal reporting and evaluation systems of accounting practices and conventions developed for external reporting. This is, for the most part, a recent phenomenon and is more common in the United States than in other parts of the world.

In Europe, many companies have one department to collect and analyze data for internal operations and another to prepare external reports. Some companies, like Philips in the Netherlands, even report to stockholders on the basis used to evaluate internal operations. By contrast, contemporary practice in the United States is to use for internal purposes conventions either developed for external reporting or mandated by such external reporting authorities as the Financial Accounting Standards Board and the SEC.

Interest expense

For example, many American corporations regularly allocate corporate expenses – say, interest costs – to divisions and profit centers according to some arbitrary measure of a unit's assets or working capital. Now, it is sensible to charge divisions for capital employed. It is not sensible, however, to use a pro rata share of the interest expense reported on external financial statements as the appropriate internal cost of capital. Such a procedure implies that a company financed entirely through equity would allocate no capital charges to divisions since it has no recorded interest expense.

Consider, at the other extreme, an autonomous division engaged in real estate whose assets are mostly debt financed. The financial accounting mentality would have this division bear a higher interest charge against earnings than would divisions financed more heavily by equity. But does anyone believe that the cost of debt capital is higher than that of equity capital?

Divisions should be charged for their investment in net controllable assets through a divisional cost of capital, perhaps adjusted to allow lower charges for working capital than for higher-risk fixed assets. Few companies use such a method today, perhaps because the divisional capital charges will "overabsorb" a company's actual interest expense. But considering actual interest expense as a company's only cost of invested capital is more a limitation of contemporary financial reporting than it is a criticism of charging divisions for the full cost of their investments.

Pension costs

Financial accounting practices can also lead to bad cost accounting in the allocation of pension costs. For example, prior service costs are sunk costs. They represent an obligation of a company for the past service of its employees. No current or future action of the company can affect this obligation. The amortization of prior service costs must, however, be recognized as a current expense in the company's financial statements. Therefore, many companies allocate prior service costs to divisions.

One company allocated these costs in proportion to pension benefits accrued. A plant with an older work force received almost all of its division's prior service costs, but several newer plants with much younger workers bore almost none. This arbitrary allocation produced a $4 per labor hour cost penalty on the older plant. As a result, the company was shifting work from the older plant to the newer ones. In addition, the older plant was losing market share as it raised prices in an attempt to earn a satisfactory margin over its high labor costs. Put simply, the company became a victim of its financial accounting mentality: first it allocated a noncontrollable, sunk cost to its plants and then it relied on this arbitrarily allocated cost for pricing and product sourcing decisions.

Other distortions

Distortions created by the financial accounting mentality intrude on many other internal measurements. How often, for instance, do executives, when measuring a division's investment base for an ROI calculation, include leased assets only when, according to FASB regulations, they must be capitalized on the external financial statements? Are development and start-up expenses, including software development, considered part of the investment in a computer-integrated manufacturing process, or are these investments in intangibles expensed as incurred because, according to SFAS 2, this treatment is mandated for external reporting? Do companies translate operations in foreign countries according to their economic exposure overseas, or do they use whatever translation method the FASB happens to be mandating at the time?

The point, of course, is that companies seeking to compete effectively must devise cost accounting systems that reflect their investment decisions and cost structures. Internal accounting practices should be driven by corporate strategy, not by FASB and SEC requirements for external reporting. Surely, the cost of record keeping in an electronic age is sufficiently low that aggregating transaction data differently for external and internal purposes cannot be a burdensome task.

Financial entrepreneurship

Another set of difficulties with ROI-based measurements stems from the ability of executives to generate greater profits from financial activities than from managing their assets better. Sixty years ago managers knew that higher profits and ROI came from efficient production, aggressive marketing, and a continual flow of product and process improvements. During the past 20 years, however, as it has become more difficult to increase profits through selling, production, and R&D, some companies have looked to accounting and financial activities to generate earnings.

At first, these activities – switching from accelerated to straight-line depreciation, for example – did little harm. Occasionally they proved costly, as when companies opted to pay unnecessary taxes by delaying or refusing a switch to LIFO because of its adverse impact on reported profits. Today, however, the romance with these devices is in full swing: mergers and acquisitions, divestitures and spinoffs, debt swaps and discounted debt repurchases, debt defeasance, sale-leaseback arrangements, and leveraged buyouts.

Some of these activities may create value for shareholders (current research is still attempting to sort out the net effect of these financial activities). Still, it is hard to imagine that a focus on creating wealth through the rearrangement of ownership claims rather than on managing tangible and intangible assets more effectively will help companies survive as world-class competitors. Ultimately, wealth must be created by the imaginative and intelligent management of assets, not by devising novel financing and ownership arrangements for those assets.

Intangible assets

The final and most damaging problem with ROI-based measures is the incentive they give managers to reduce expenditures on discretionary and intangible investments. When sluggish sales or growing costs make profit targets hard to achieve, managers often try to prop up short-term earnings by cutting expenditures on R&D, promotion, distribution, quality improvement, applications engineering, human resources, and customer relations – all of which are, of course, vital to a company's long-term performance. The immediate effect of such reductions is to boost reported profitability – but at the risk of sacrificing the company's competitive position.

The opportunity for a company to increase reported income by forgoing intangible investments illustrates a fundamental flaw in the financial accounting model. This flaw compromises the role of short-term profits as a valid and reliable indicator of a company's economic health. A company's economic value is not merely the sum of the values of its tangible assets, whether measured at historic cost, replacement cost, or current market prices. It also includes the value of intangible assets: its stock of products and processes, employee talent and morale, customer loyalty, reliable suppliers, efficient distribution network, and the like.

Suppose this stock of intangible assets could be valued each period. Then, when the company decreased its expenditures on these assets, their subsequent decline in value would lower the company's reported income. We do not, however, have methods to value objectively intangible assets. Therefore, reported earnings cannot show a company's decline in value when it depletes its stock of intangible assets. It is this defect in the financial accounting model that makes the quarterly or annual income number an inadequate summary of the change in value of the company during the period.

The task at hand

Present cost accounting and management control systems rest on concepts developed almost a century ago when the nature of competition and the demands for internal information were very different from what they are today. When companies now make arbitrary allocations of corporate expenses to divisions and products, accounting systems may provide even less valid cost data than did the cost accumulation systems in use 50 years ago. In general, though, an accounting model derived for the efficient production of a few standardized products with high direct labor content will not be appropriate for an automated production environment where the factors critical to success are quality, flexibility, and the efficient use of expensive information workers and capital.

General managers must be alert to the inadequacies of their present measurement systems. It is doubtful whether any company can be successfully run by the numbers, but certainly the numbers being generated by today's systems provide little basis for managerial decisions and control. Managers require both improved financial numbers and nonfinancial indicators of manufacturing performance. Because no measurement system, however well designed, can capture all the relevant information, any operational system must be supplemented by direct observation in the field. The separation of senior management from operations that the ROI formula made possible 80 years ago will have to be partially repealed. Successful senior managers must be knowledgeable about the current organization and technology of their operations.

For their part, accounting and financial executives must redirect their energies – and their thinking – from external reporting to the more effective management of their companies' tangible and intangible assets. Internal management accounting systems need renovation. Yesterday's internal costing and control practices cannot be allowed to exist in isolation from a company's manufacturing environment – not, that is, if the company wishes to flourish as a world-class competitor. ⛉

Reproduced by permission of John Wiley and Sons Limited. "Strengthening Managerial Approaches to Improving Technological Capabilities," by Bela Gold, ***Strategic Management Journal****, Vol. 4, 1983.*

Strengthening Managerial Approaches to Improving Technological Capabilities

BELA GOLD
Director, Research Program in Industrial Economics, Case Western Reserve University, Cleveland, Ohio, U.S.A.

Summary

This paper draws on an extensive array of theoretical and empirical studies covering a variety of industries in the U.S. and elsewhere in order: (a) to review common shortcomings found in evaluating technological innovations both before and after adoption, as well as in generating proposals for new innovations; and (b) to suggest means of improving each of these efforts.

Regaining and maintaining technological competitiveness is essential to strengthening competitiveness in the market place. Even inspired marketing and creative financing cannot ensure long term survival for products which cost more to produce and which are less attractive to customers than those offered by rivals.

However, more than 25 years of analysing the problems of improving productivity as well as other sectors of industrial performance[1] have convinced me that implementing major advances in technology represents a far more difficult and far-reaching challenge to management than is generally appreciated. The key reason for this is the failure to recognize that basic technologies are built not only into the production machinery, but also into:

(1) the expertise of the technical personnel
(2) the structure and operation of the production system
(3) the economically feasible range of changes in product designs and product-mix
(4) the skills and organization of labour
(5) and even the very criteria used to evaluate new capital goods proposals.

Each of these represents powerful and mutually reinforcing commitments to preserving existing production and organizational arrangements, except for small, gradual and localized changes. The influence of such deep-rooted commitments was perceptively captured by the poetic lines:

> With their eyes firmly fixed upon the past,
> They backed reluctantly into the future.

[1] For a partial listing, see the References which include studies covering a variety of technological innovations and related productivity, cost and other effects in a variety of industries in the U.S. and abroad. For a perceptive earlier review of allocation processes, see Bower (1970).

It would, of course, be undiplomatic in modern industry to voice immediate opposition to proposals for major innovations, but such resistance can be exerted quite effectively through unencouraging evaluations of proposals by the very specialists on whom management depends for expert judgements, as well as by highlighting the uncertainties and difficulties likely to be encountered in applying and using new technologies.

Moreover, even when the need for improvement has come to be accepted, the targets set are often much too low to regain competitiveness. For example, recent efforts by the U.S. steel industry to begin catching up with the higher productivity levels of leading Japanese competitors (Gold, 1978) have generated a variety of company programmes seeking to increase output per man-hour by a long-unattained 5 per cent annually. However, the most recently completed integrated mill in Japan, Ohgishima, which had set a target for 1977 of 1000 tons of finished product per wage-earner (more than double the best U.S. level at that time), raised its actual output per man to 1800 in 1980, almost three times the best record in a U.S. integrated mill (*Wall Street Journal*, April 7, 1981).

The development of programmes to achieve major advances in technological competitiveness is likely to require basic changes in traditional approaches:

(1) to evaluating proposals for adopting technological innovations
(2) to generating promising proposals
(3) to evaluating the effectiveness with which resulting potentials have been harnessed.

Accordingly, the following discussion will review the limitations of widespread practices in each of these areas before offering suggestions for revising them on the basis of our field studies in a variety of industries in the U.S. and abroad.

APPROACHES TO EVALUATING PROSPECTIVE TECHNOLOGICAL INNOVATIONS

Revising common evaluation criteria

Prevailing approaches to evaluating proposals to adopt technological innovations commonly involve estimating their prospective contributions to cost savings and revenue increases as compared with current operations and then comparing the resulting gain in profits with the additional investment required (Gold, 1977), but such an essentially static perception leads to reliance on the wrong criteria.

The basic objective in adopting substantial technological innovations must be to safeguard or improve the firm's competitiveness over an extended future period. Instead of comparisons with current operating performance, therefore, relevant evaluation criteria should be developed by estimating needed improvement targets over at least the next 5–10 years. This would require serious efforts to appraise recent and prospective trends: in the availability and prices of needed inputs and possible substitutes; in required product capabilities and in product-mix; in the technologies likely to be employed by competitors; in new sources of competition; and in any relevant governmental policies—thus adding major additional analytical responsibilities to those which currently dominate many corporate planning groups. If significant changes from the present seem likely in any of these determinants of future competitiveness, decisions about prospective innovations might be quite different from those which would be counselled by comparisons only with current operating performance. Such evaluations of needed future improvements might also urge consideration of innovations other than those currently being proposed.

In addition, it is important to recognize the serious negative bias involved in evaluating major technological innovations on the basis of the purportedly sophisticated criterion of the 'net present value' of expected future returns. If such returns are discounted at 15–20 per cent annually, major projects which take even 3 or 4 years to build, 'debug' and bring to high levels of use would invite rejection in comparison with recent high, virtually riskless money market yields, but such a criterion is dangerously myopic because of its failure to consider the effects on long term competitiveness and profitability of resulting rejections of successive technological advances (Gold and Boylan, 1975). Hence, it would seem necessary to shift such capital budgeting evaluations from maximizing net present value to what I have called a 'continuing horizons' approach. This involves recognition that most major innovations achieve only gradually increasing use and benefits, experience an extended period of reasonable competitiveness and attractive contributions, only to succumb in time to yielding progressively declining returns. In order to maximize average returns on a continuing basis, therefore, it is necessary to plan for the periodic introduction of new sources of improved competitiveness, recognizing that they will not yield attractive short term returns, and despite the tendency to underestimate the net present value of their eventual contributions engendered by discounting them at high interest rates. Accordingly, one would seek to choose an array of capital projects such that, when the overlapping time paths of expected net returns from each are aggregated, and the 'net present value' is calculated as of *successive* 3 year intervals, the results promise to safeguard acceptable longer-term as well as short-term rates of profit (Gold, 1975: 24).

Common errors in estimating the prospective benefits

Our field research suggests that the substantial differences which are commonly found between pre-adoption estimates of the results of major innovations and actual outcomes can be attributed to two different sources: errors in estimating prospective gains in productive efficiency; and errors in estimating resulting cost and profit benefits.

Errors in estimating gains in productive efficiency

One of the important sources of such errors is the tendency to underestimate the time needed to achieve effective functioning of the innovation, often by a considerable margin. Common reasons include inadequate recognition of the delays and extended shortcomings likely to be caused by such factors as: needed modifications in the hardware and controls to adapt to the particular characteristics of the given plant; identification of the detailed operating requirements, capabilities and limitations of the innovation as the basis for developing guidelines for operating practices; inadequate engineering experience in ensuring rapid and effective adaptation of the innovation to changing production requirements; and development of appropriate maintenance procedures. For example, our study of a large flexible manufacturing system revealed that effective functioning was not achieved for more than two years, with substantial additional realization of its capabilities spread over several more years (Gold, 1981c). The utilization rates of continuous slab casters in a number of leading American steel mills have continued to fall far short of those attained in Japanese steel mills even five years after installation (Rosegger, 1980).

A second source of errors in estimating the prospective benefits of major innovations is the tendency to overestimate the average rate of use in calculating expected returns. In part, this often reflects a wilful ignoring of the realities of past fluctuations in total product demand levels and of the limitations on using particular operations due to the uneven effects of shifts in product-mix. Interesting examples of such effects are provided in Skeddle's study of early float glass installations in this country (Skeddle, 1980).

In addition, expected gains in productive efficiency are likely to be exaggerated because of reliance on an overly narrow evaluative framework which ignores the need to develop adaptive adjustments in preceding as well as subsequent operations in order to reintegrate the production process effectively. This may involve the reallocation of some operating tasks, possible changes in prior quality specifications (possibly reaching all the way back to procurement), alterations in production rates and work-in-process inventories, changes in inspection arrangements, and adjustments in machine capabilities and labour inputs. Extensive installations of robotics and of computer-aided manufacturing systems are among the more recent illustrations of the wider ranging impacts of major technological innovations (Gold, 1981c, 1982).

A fourth common source of erroneous estimates of net benefits derives from the associated tendency to concentrate on the prospective operating inputs and outputs of the innovative hardware, while giving inadequate consideration to the often substantial requirements for additional inputs by staff personnel. These may include production planning, quality control, systems engineering, maintenance specialists and even product designers—all of them seeking not only to facilitate integration of the innovation into the larger manufacturing and control system, but also to harness any further potential benefits to the system as a whole made possible by the emerging capabilities of the innovation. In tightly integrated plants, not only must each of these supporting specialties participate in adapting past arrangements to accommodate the new capabilities and requirements, but they must also reintegrate their respective modifications through reiterated system revisions. Our study of American experience with computerization in the steel industry illustrates the difficulties of trying to introduce such innovations on a piecemeal basis and from the bottom-up (Gold, 1978).

A fifth possible source of such erroneous estimates is rooted in the assumptions made concerning the problems of gaining labour acceptance of major innovations. The issues which seem to be encountered most frequently centre around: changes (usually reductions) in employment levels; changes (usually increases) in output quotas; shifts in job classifications (commonly affecting skill levels, seniority and associated relative pay rates); and adjustments in overall wage rates and fringe benefits (usually increases) to gain needed co-operation. Such problems tend to be much less serious than expected by managements during the early stages of innovation, when applications are few, limited in scope, and affect so few employees as to enable managements to ease any associated difficulties readily. It is only when early experiences have finally achieved technical success and substantial economic promise that the labour problems are likely to be generated by wider applications of innovative technologies. Hence, robotics and computer-aided manufacturing have not yet generated the labour difficulties that were associated with the introduction of many earlier advances in mechanization.

Errors in estimating the cost and profit benefits

The most common and most readily remediable among these involves ignoring the probable effects of resource-saving innovations on the price of the inputs affected. For example, increases in output per man-hour tend to engender largely or wholly offsetting increases in wage rates, whether through piece rates, incentives or trade union pressures. Similarly, reductions in unit material requirements often involve changes in qualitative specifications which entail higher prices. A second source of erroneous cost estimates involves overlooking the costs of underusing older facilities used for similar purposes in order to maximize use of the new.

Perhaps the most astonishing source of common errors is the assumption that resulting reductions in unit costs can be fully converted into profits. But this ignores the likelihood that efforts to gain such rewards through lowering prices, and thereby increasing sales, would be matched by competitors determined to retain market share, even if they have not achieved comparable cost-savings. Alternatively, seeking to achieve profit benefits by keeping prices unchanged is not likely to prove tenable for long because the diffusion of cost-saving innovations tends to trigger price reductions by competitors seeking to capitalize on their belated adoptions by regaining or increasing market shares (Gold, 1979: Chapter 14).

Another critical point relating to estimating the benefits of prospective innovations is that the more deviate they are from the expertise and past experience of the staff, the more vulnerable such pre-decision evaluations are likely to be.[2] This is true even of estimates of the effects of prospective major increases in the scale of facilities (Gold, 1981b: 14, 23).

Improving estimates of the prospective effects of technological innovations

Our field research has uncovered very little evidence of serious efforts to improve the accuracy of pre-decision estimates of prospective innovational effects on the basis of empirical evaluations of their shortcomings. The reason given most in answer to our enquiries is that each major innovation is regarded as unique along with the factors determining its prospective effects. A second common answer is that such estimates about the performance of hitherto unused innovations under uncertain future conditions are unavoidably subject to wide margins of error. In view of the large stakes involved, however—including future profitability and even survival—systematic efforts to identify the shortcomings of past pre-decision estimates as well as efforts to improve them would seem very much in order.

One of the important sources of such errors, and perhaps the most remediable, is to recognize and try to minimize the biases that have been built into the evaluations generated by different groups within the firm. Engineering development personnel who are under pressure to justify their efforts are understandably tempted to offer generous estimates of the advances which they have achieved. Manufacturing supervisors seeking to minimize disruptions to smoothly functioning production operations tend to underestimate the net benefits of proposed major innovations. Marketing executives tend to be more enthusiastic about advances offering improvements in product capabilities than those affecting production operations alone. And finance officials often have ample cause, on the basis of past experience, to regard estimates of the benefits accompanying such proposals with scepticism expressed in the form of reducing probable revenues and increasing probable investment requirements and operating costs. Such motivational cross-currents can obviously serve as a source of creative tension within a firm. However, results are likely to be far more valuable if efforts are made to modify traditional biases on the basis of more realistic perspectives.

One means of doing so would involve a thorough analysis of the actual results of several past innovations for comparison with *ex ante* expectations in order to identify the component estimates involving the largest errors and then probing for the causes of such misperceptions as a guide to remedial adjustments. The array of common sources of such errors which has been presented offers a starting point for such enquiries and also implies means of improving the various estimates. A second approach to improving pre-decision estimates would involve comprehensive efforts to probe the experiences of earlier adopters of the contemplated innovations. If domestic adopters are competitors who are unwilling to share such learning, it may be fruitful to explore the possible co-operativeness of foreign producers in sharing such experience. Additional means obviously include reviewing the expected advantages,

---

[2] For example, see the discussion of deviations between traditional expectations and those more likely to result from applications of computer-aided systems in manufacturing (Gold, 1981c: 10–16).

limitations and problems of using the prospective innovation with competing vendors and with specialized consultants.

The points being emphasized are that there are common as well as uncommon elements in appraising the prospective effects of even quite dissimilar innovations; that greater efforts are warranted to use costly past experiences as a means of improving future evaluational efforts; and that resulting insights may yield valuable improvements in such estimates even if they continue to encompass substantial sectors of uncertainty.

GENERATING PROPOSALS FOR TECHNOLOGICAL INNOVATIONS

Weaknesses of traditional approaches

In most firms, proposals for technological innovations are expected to emerge from, or at least to be approved by, the operating sectors most likely to be affected by them. This process derives from the assumption that the managerial and technical personnel in each sector are the most knowledgeable about its performance requirements and shortcomings, and are also best qualified to evaluate any prospective innovational improvements. Although there is considerable merit in these beliefs, they are less likely to be true in respect to major innovational possibilities. One reason is that the more unfamiliar the technologies to be considered, the less qualified the current operating staff would be to evaluate it. It should also be recognized that such personnel may feel that unfamiliar innovations threaten both the adequacy of their expertise and the security of their organizational status—thereby encouraging underestimates of the innovation's prospective benefits. In the cases of robotics and, to an even greater extent, computer-aided manufacturing, most operating staffs lack sufficient expertise to evaluate either prospective applications or the alternative systems which are available for adoption. Moreover, the readiness of current operating executives and engineers to add personnel with such specialized capabilities may be inhibited by the prospect that increasing reliance on such new technologies would tend to increase the newcomers' opportunities for advancement at the expense of those lacking their expertise.

Managements also frequently act as though the most promising innovational proposals are bound to keep flowing because of an unrestrainable urge of managers and technical specialists to maximize improvements. In fact, however, analysis suggests that both the volume and the magnitude of innovational proposals are closely responsive to the attitudes of top managements, as reflected not by their exhortations but by their decisions about the proposals which have been submitted. Thus, a high proportion of rejections of new kinds of innovational proposals is bound to discourage additional submissions; and evidences that management favours only modest scale or quick repayment proposals tend to evoke conforming adjustments in the characteristics of the proposals which are submitted. It may be of interest in this connection to contrast the steady and impressive progress of computerization in the integrated steel industry of Japan, with its top-down guiding strategy, as against the intermittent and more limited gains resulting from reliance on the earlier but essentially bottom-up efforts in the American integrated mills (Gold, 1978).

Means of intensifying technological improvement efforts

In order to intensify technological improvement efforts, the most important requirements are for senior management to recognize that ensuring technological competitiveness is absolutely necessary to profitable survival and to commit itself to achieve and maintain such capabilities.

Among the means for demonstrating such a commitment, consideration might be given to a variety of measures. To begin with, it would prove valuable to arrange for a thorough, objective and expert evaluation of the advantages and disadvantages of its current technologies, facilities and products in all major operations and product lines relative to its competitors. Although virtually all companies covered in our studies professed to be fully informed about such matters; few could offer any persuasive evidence in support of these claims. Yet it is obviously difficult to maximize the benefits of invariably limited investment resources without first identifying the sectors of most urgent needs and the magnitude of the lags to be overcome. Such a mapping of current advantages and disadvantages should then be supplemented by a comparably thorough exploration of prospective changes over the next 5 years or longer in the capabilities of foreign as well as domestic competitors, in input and output markets, and in any relevant governmental regulations—a second major source of guidance which is seldom subjected to comprehensive analysis. The results of these two surveys would then provide a reasonably solid base for defining a set of improvement targets for technological capabilities and productive efficiency.

In order to ensure progress towards such improvement goals, top management might well consider an array of interconnected measures. Instead of the all-too-common expedient of exhorting everyone to surpass past performance, it might be more useful to begin by establishing a special group, closely tied to top management, to undertake the current and longer term evaluations suggested above, to build on the findings as a basis for proposing preliminary improvement targets, and to draw on all kinds of internal and external expertise to evaluate alternative innovations capable of furthering progress toward such goals. Key questions might include: which advances should be licensed or bought from the outside and which should we seek to develop internally? How much of needed advances could be achieved through modifying or upgrading existing facilities and products as against the additional gains available only through shifts to new facilities and products in the next 2 or 3 years, and over the next 5 years? Issues like these require consideration of the proportion of current and prospective capital outlays which should be allocated to 'strategic investments', meaning those to be evaluated on the basis of expected contributions to longer term competitiveness and profitability rather than to 'net present value'.

Decisions about such matters are often rooted in basic philosophical commitments rather than in specified analytical procedures. For example, during the course of one of our field studies, we heard the director of manufacturing technology support a proposal for the acquisition of major new facilities and equipment with a comprehensive array of estimates of investment requirements, operating costs and revenues over the next six years, only to have them abruptly dismissed by the capital allocations official as 'nothing more than unreliable guesses which could not justify the large investments involved'. 'But', the latter then added, 'tell me why the company needs to make such an investment in order to safeguard its future competitiveness'. In response to our query about the basis for taking such an unusual position, he told us that, in order to maintain its outstanding position in the industry, this company consistently allocates 15–20 per cent of its capital expenditures to projects which are too new to permit persuasive analyses of prospective returns, but which seem of such great future promise as to warrant undertaking significant investments in pioneering their exploration and development.[3]

[3] A somewhat similar approach was reflected by the position of a senior research officer in a major European corporation in justifying his opposition to the detailed economic evaluation of research proposals and similar innovative activities. His management regards such efforts to reach beyond accumulated experiences as 'similar to an underground stream which wanders unseen for longer or for shorter distances to emerge eventually at some unexpected spot. "We do not know where it will emerge, and we do not know when it will emerge, but we want to be there when it does!" In extending his argument, he explained that the company would expect to be first every so often. But even when it was not, its research people would be so familiar with the surrounding technological terrain, and with the paths which had been found unfruitful as well as those which had not yet been explored, that it could in most cases duplicate the achievements of the successful pioneers, or develop a reasonable equivalent, in comparatively short order. And, in his view, it is this combination of insurance against catastrophic lags in whatever quarter the competition manages to leap ahead, plus the chance that it will itself achieve leadership with reasonable frequency, that constitutes the soundest justification of the research investments undertaken under such a cloud of uncertainties.' Quoted from Gold (1973: 136).

Progress towards such technological improvement goals also requires development of well-informed and responsive managers, technical personnel and workmen. Towards that end, it would be helpful to initiate educational efforts followed by specialized training programmes to achieve general understanding of the need for major advances, of their implications for future operations, and of the particular stresses likely to be confronted in the course of absorbing various kinds of innovations. At the same time, careful attention must be given to possible needs for changing existing organizational structures so as to facilitate the effective use of major innovations. For example, the introduction of computerized manufacturing systems requires more effective and continuous integration between computer-aided design and computer-aided manufacturing than is likely to be achieved when each represents a separate organizational unit, leaving the critical point of interaction between them at the periphery of each unit's responsibilities (Gold, 1981c: 31). Such organizational adjustments should also provide for detailed periodic monitoring of progress towards technological improvement goals, including evaluation of the causes of any shortcomings as well as consideration of possible shifts in targets in response to changing pressures and opportunities.

EVALUATING THE RESULTS OF INSTALLED TECHNOLOGICAL INNOVATIONS

In contrast to the voluminous literature on pre-decision evaluations of innovational effects, few publications offer serious evaluations of the actual effects of the innovations which have been adopted. The explanation seems to be that the relevant methods and interpretations are considered to be so obvious as to offer no analytical challenges. Our research suggests, however, that such efforts are confronted by significant difficulties concerning the appropriate criteria for evaluation; the timing of such appraisals; who should make them; and how the results should be used.

Criteria of evaluation

Most efforts to evaluate the results of major technological innovations tend to concentrate on financial measures, especially on comparisons of outlays and costs with original expectations. But such findings tend to be inadequate, and may even be misleading, unless account is also taken of all other changes engendered by the innovation. These may include one or more of the following: changes in the quality as well as in the quantities of inputs and outputs; shifts in operating tasks among production stages; alterations in product-mix and in the length of production runs; adjustments in reject rates; and increases or decreases in equipment down-time.

Moreover, effective appraisal also requires uncovering the causes of observed deviations from expected results. In particular, management needs to know the extent to which such changes were attributable to equipment characteristics; or to engineering modifications of production methods; or to maladjustments in input and work flows; or to changes in labour capabilities and efforts. In the absence of such analytical determinations, there seems to be a tendency to ascribe any observed deficiencies to external or to unpredictable factors—thereby ignoring internal remedial potentials. It should also be recognized that early experiences in introducing and using major innovations often alter original estimates of what tasks can be performed most effectively as well as prospective results. Hence, comparisons with later perceptions may be more meaningful than with the less informed original expectations.

In evaluating the cost effects of technological innovations, results may be heavily affected by decisions concerning whether the following are treated as increases in the investment charged to the innovation or as additions to operating costs: the cost of interruptions to production caused by the innovation; the cost of delays before achieving effective functioning of the innovation, including attendant costs of equipment modifications, 'debugging', operator training, and trial runs; additional outlays in order to improve the capabilities of the new facilities; and the costs and investments involved in readjusting preceding and subsequent operations in order to improve the capabilities and integration of the encompassing production process. In general, there seems to be a widespread tendency to charge such additional outlays to early operating costs, in respect to which overruns are likely to be regarded sympathetically as virtually inevitable, thereby holding down the investment base to be used in calculating future rates of return.

The timing of evaluations

Such firms as undertake 'post-audits' or 'make good' evaluations of installed innovations tend to rely on a single appraisal, usually within 6–12 months after project completion. These early appraisals often yield overly optimistic findings, however, because generous allowances are made to offset actual shortcomings on the assumption that these are attributable to temporary problems—such as excessive maintenance, inadequate labour experience, or underuse due to incomplete integration with adjacent operations.

If such appraisals are to be more effective, they would have to be made repeatedly over 3–6 month intervals for at least the two to three years which the complete absorption of major technological innovations seems to take (Gold, 1979: Chapter 10). Successive evaluations would reveal trends in various performance measures, demonstrate which shortcomings are transitory and which are not, and identify additional sources of improvement through progressively more thorough integration with the larger manufacturing and control systems.[4]

Responsibility for evaluations

One of the critical problems faced in evaluating major innovations is the pressure for biased findings. In the case of very large projects, favourable biases tend to be encouraged by fears that negative findings might reflect on the senior officials who made the basic decisions and be resented by them. Biases may also be generated by assigning evaluations to those who recommended the adoption decision, on the grounds that they alone have the requisite technological competence. Still another source of biases may be the desire of designated evaluators to maintain future co-operative relationships with the project officials involved. As a result of experiences with such problems, some firms have abandoned *ex post* evaluations entirely (Gold *et al.*, 1980: 313) whereas many others seem to have relegated them to relatively routine financial reporting exercises.

However, such carelessness, passivity or defeatism has already helped to undermine awareness of the shortcomings of past efforts to improve technological competitiveness. In addition, such attitudes have also helped to delay recognition of the need to organize more formal, comprehensive knowledgeable and regularized evaluations of the effects of the large investments in major technological innovations which may be major determinants of future

[4] For example, in one computer-aided manufacturing system, it was found that successive minor modifications in operating methods and control yielded a progressive increase in capacity of approximately 40 per cent over a 5 year period. But such increments were so gradual that management failed to alter production planning and output targets accordingly—leaving the additional capability unused until its belated discovery as the result of a strike.

profitability. Provision of the needed objectivity and range of relevant expertise may require that such evaluations be the responsibility of some senior official who is concerned not only with exacting appraisals of performance, but also with uncovering any possible means of improving results and with identifying the sources of errors in pre-decision appraisals and in the processes of introducing innovations so as to develop more effective guides to such efforts in the future. Such an executive could then secure the contributions of various relevant specialists, and surmount the common self-protective devices of lower-level supervisors and technicians, in seeking to uncover the sources and causes of shortcomings in performance as well as of remediable errors in pre-decision appraisals.

Formalizing responsibility for the comprehensive evaluation of major technological innovations might yield two additional benefits. First, by providing top management with systematic assessments from an influential official of the potentials, accomplishments and shortcomings of projects which have absorbed major capital allocations, such periodic reports may help to engender greater sensitivity to the importance of technological improvement efforts throughout the firm. Second, the very seriousness and scope of such efforts may help to overcome the present widespread failure to make any constructive use of even such post-adoption appraisals as are made, either in re-examining means of improving the current results of relatively recent innovations, or in strengthening the bases for evaluating future proposals.

It is also worth noting the seemingly universal avoidance in post-installation evaluations of estimates of resulting contributions to profitability. Estimates are often made of reductions in the direct operating costs of the specific operation affected by the innovation and of attendant savings in labour or materials inputs. But very seldom indeed are such appraisals extended to the point of estimating resulting changes in profits. This implies recognition of the complexities of tracing the effects of the innovation-induced adjustments within its sector of operations through subsequent interactions with other sectors, and through concurrent changes in the level, composition and prices of all other inputs as well as outputs, to determine its distinctive effects on net returns. However, if reasonably persuasive determinations of such contributions are so difficult to make, even on the basis of detailed actual performance data (Gold, 1955: 181), one cannot help wondering about the practical usefulness of continuing to base decisions to adopt or reject major innovations on the estimates of prospective profitability derived on the basis of the far more inadequate and vulnerable information likely to be available at that stage (Gold, 1977).

CONCLUDING OBSERVATIONS

1. There has been a dangerously prolonged underemphasis in a wide array of industries on supporting vigorous development of their technological capabilities. Resulting serious lags behind continuously advancing technological frontiers have been a major source of declining competitiveness.
2. Overcoming such lags is likely to prove far more difficult than is widely recognized. The major reason is that such recovery efforts require not only heavy investments in new facilities and equipment, but also substantial changes in managerial priorities and policies, in staff capabilities and in organizational arrangements. An additional reason is that during the extended period needed to overcome built-in resistances, to effectuate required adjustments and to build momentum towards catching-up, leading competitors may be expected to keep pushing further ahead.

3. Effective efforts to regain technological competitiveness require:
   (a) top management commitment to this objective
   (b) development of an array of improvement targets based on objective, competent and comprehensive evaluations of the current and prospective technological advantages and disadvantages of each sector of production as well as each major component of the firm's product-mix
   (c) establishing organizational arrangements:
      (i) to continue monitoring relative competitiveness and its determinants
      (ii) to evaluate alternative technological innovations proposed as means of progressing towards established targets
      (iii) to appraise the performance of adopted innovations periodically as the basis for uncovering developing shortcomings or additional potentials
      (iv) to ensure effective integration of various technological improvement efforts so as to maximize mutual reinforcement and to minimize maladjustments.
4. In order to encourage needed management commitment to, and support for, such recognizedly long term programmes, whose major benefits may not be realized for some years, it would seem necessary to alter the incentives and rewards which in recent years have motivated an emphasis on maximizing short-term profitability (Gold, 1979: Chapter 17).

REFERENCES

Bower, J. *Managing the Resource Allocation Process*, Graduate School of Business Administration, Harvard University Press, Cambridge, MA, 1970.

Eilon, S., B. Gold and J. Soesan. *Applied Productivity Analysis for Industry*, Pergamon Press, Oxford, 1976. Russian translation—Ekonomist, Moscow, 1980; Chinese translation—The Technical Economy and Modernization of Management Institute, Beijing, 1982.

Gold, B. *Foundations of Productivity Analysis*, University of Pittsburgh Press, Pittsburgh, PA, 1955.

Gold, B. *Explorations in Managerial Economics: Productivity, Costs, Technology and Growth*, Macmillan, London, 1971; Basic Books, New York, 1971. Japanese translation—Chikura Shobo, Tokyo, 1977.

Gold, B. 'What is the place of research and technological innovations in business planning?', *Research Policy* (London), No. 2, Summer 1973.

Gold, B. (ed. and co-author), *Technological Change: Economics, Management and Environment*, Pergamon Press, Oxford, 1975.

Gold, B. 'On the shaky foundations of capital budgeting', *California Management Review*, Winter (March) 1977, pp. 51–60.

Gold, B. 'Steel technologies and costs in the U.S. and Japan', *Iron and Steel Engineer*, April 1978, pp. 32–37. Japanese translation—*Jōhō Shūho* (Tokyo), July 1978.

Gold, B. 'Factors stimulating technological progress in Japanese industries: The Case of Computerization in Steel', *Quarterly Review of Economics and Business*, Winter (December) 1978, pp. 7–21. Spanish translation in *Cienca y Desarrolo* (Conejo National De Cienca Y Tecnologia, Mexico D.F.), November–December 1979.

Gold, B. *Productivity, Technology and Capital: Economic Analysis, Managerial Strategies and Governmental Policies*, D.C. Heath-Lexington Books, Lexington, MA, 1979, 1982.

Gold, B. (ed. and co-author) *Appraising and Stimulating Technological Advances in Industry*, Special Issue of *Omega: The International Journal of Management Science*, October 1980.

Gold, B. 'Approaches to strengthening the Swedish steel industry: A case study of industrial policy issues' in *The Future of the Steel Industry in an International Setting*, Royal Academy of Engineering Sciences, Stockholm, 1981a.

Gold, B. 'Changing perspectives on size, scale and returns: an interpretive survey', *Journal of Economic Literature*, March 1981b, pp. 5–33.

Gold, B. *Improving Managerial Evaluations of Computer-Aided Manufacturing*, National Academy of Science Press, Washington, D.C., December 1981c.

Gold, B. 'Robotics, programmable automation and increasing competitiveness', in *Exploratory Workshop on the Social Impacts of Robotics*, U.S. Congress Office of Technology Assessment, Washington, D.C., 1982.

Gold, B. and M. G. Boylan. 'Capital budgeting, industrial capacity and imports', *Quarterly Review of Economics and Business*, Fall 1975, pp. 17–32.

Gold, B., G. Rosegger and M. G. Boylan. *Evaluating Technological Innovations: Methods, Expectations and Findings*, D.C. Heath-Lexington Books, Lexington, MA, 1980.

Rosegger, G. 'Continuous casting: physical and economic performance' in Gold, B., G. Rosegger and M. G. Boylan, *Evaluating Technological Innovations: Methods, Expectations and Findings*. D. C. Heath-Lexington Books, Lexington, MA, 1980.

Skeddle, R. W. 'Expected and emerging actual results of a major technological innovation—float glass' in Special Issue of *Omega*, October 1980, pp. 553–567.

Reprinted by permission of Wickham Skinner, "Operations Technology: Blind Spot in Strategic Management," ***Interfaces,*** *Vol. 14, No. 1, January-February 1984,*

Operations Technology: Blind Spot in Strategic Management

WICKHAM SKINNER

Graduate School of Business Administration
Harvard University
Soldiers Field
Boston, Massachusetts 02163

Innovation in operations equipment and process technology can be used strategically as a powerful competitive weapon. It can bring to bear many other strategic factors besides achieving low costs — superior quality, shorter delivery cycles, lower inventories, lower investment in equipment, shorter new product development cycles, and new production economics.

The potential of aggressive innovation in operations equipment and process technology is frequently a blind spot in strategic management. This powerful competitive weapon is generally unused, neglected in both corporate operations and the professional literature. In fact, conventional practices in strategic management nearly inevitably doom technological development in equipment and processes to take place in a reactive and pedestrian mode.

In employing equipment and process technology (EPT) as a strategic tool often only one concept is considered: that of becoming "low-cost producer." But even that strategy, while listed as "generic" by Michael Porter [1980], is generally employed for protection and not used as an integrated, offensive strategy. Usually, it is perceived as a fortunate byproduct of having already achieved (somehow) a high market share, success, growth, and thereby, volume production. And it is the volume which delivers the mathematics of the learning curve and, hence, low production costs. This nice set of dominoing events is well known and it often seems to happen. But seldom do these events result from a proactive strategy, one which would say, "let's invest enough in EPT research and development to gain a competitive advantage from creative technological innovation" (and not just from sheer volume). But even this is a limited view of the strategic possibilities that await discovery in operations technology.

Planned, aggressive innovation can bring to bear many other strategic factors besides achieving low costs. Competitively unique EPTs can also produce superior quality, shorten delivery cycles, reduce inventories, minimize investment in plant and equipment, cut down the new product development cycle, make possible entirely new products, shift economies of scale so that short production runs are feasible, and create new production economics, allowing for a richer product mix, more product proliferation, and more customer specials.

One of the outstanding examples of the achievement of strategic advantage by investment in innovative equipment and process technology is in operation at John Deere's new tractor manufacturing complex in Moline, Illinois. New computer controlled machining, assembly and inventory processes are enabling the company to shorten lead times, economically offer customers unique products, and lift quality levels to new highs.

Each of these potential achievements can offer distinctive competitive advantages around which the corporation can plan its strategy [Lauenstein and Skinner 1980].

With this rich array of potential contributions it is surprising that production or operations technology is so little used in strategic management. Why is it given a back seat to new product development and marketing management as a competitive resource? Why is investment for research and development usually focused on *product* development rather than on *process* development?

The current situation in many US manufacturing industries offers some insights into these questions. During the past year I have been studying twelve American industrial firms, focusing on some of the new manufacturing technologies such as robots, computer-aided design (CAD) and computer-aided manufacturing (CAM), lasers and the like, in an effort to understand the "factory of the future" [Skinner 1981] and whether, when, and how it is to come into being. The twelve firms I selected are all large-scale leaders in advanced manufacturing in the forefront of their industries. I believe they offer good examples of the best, current practice. From this research some tentative explanations have begun to emerge as to why equipment and process technology continues to be a blind spot in strategic management, and why in spite of the enormous strategic potential of the new, largely microprocessor-oriented technologies, progress toward creating new competitive advantages through EPT development seems slow.

Industrial Technology — 1982

The US factory has come under increasingly severe pressure from foreign competition; it has frequently proven noncompetitive; it repels many capable young people; it seethes with a discontented labor force. Nevertheless, wholesale automation, mechanization, robotics, computerized, reduced-cost operations, and a radically improved quality of industrial working life (QWL), remain promises that have been just around the corner for 30 years.

Now that manufacturing performance is clearly critical to competitive survival in many industries, it is surprising and ironic that we still find ourselves with outdated factories and obsolete equipment. Worse, we are strangely slow to adopt the promising new manufacturing technology that could reverse or at least begin to improve the inferior strategic positions of many US industrial firms. In a few instances operations executives have developed new process technology that could become the foundation of a changed, competitive strategy. But these instances are few and far between.

Why is the American factory establishment so slow to adopt new manufacturing technology? Some predict that we will continue to slip in our ability to produce competitively and will end up as a non-industrial nation. Yet many technologists argue that the "factory of the future" can still save us. Both sides of the debate agree that the US factory is in peril.

The tantalizing promise of advanced mechanization and automation has been an elusive hope for a long time. Now, finally the much-heralded "robot" is making its appearance, but with more publicity than impact. Even numerically controlled machine tools and numerically controlled servo-mechanisms are being adopted by factories surprisingly slowly. Process control computers are widely used in some single-product industries, but even this remarkable tool has been adopted reluctantly and slowly in light of the number of possible applications. We hear a great deal about the potential of the computer in the factory, but the literature about computerized production controls is replete with disappointing results. Technological progress has been creeping instead of galloping as we might have expected in this era of exploding scientific and product technology.

It is abundantly clear that the equipment available and today's powerful technological ideas, concepts and processes have far outrun the ability of the factory system to absorb them. We have available advanced machine tools and processes, computers, controls, numerical control, direct numerical control, and computer-aided design in manufacturing far in excess of our ability to purchase and successfully use them. Somehow, management in the manufacturing world has not been able to fully exploit the new technologies that engineers and scientists have developed.

Over its history the factory institution has seldom been able to adopt new technology quickly, perhaps because absorbing new technology is as much an organizational problem as a technical problem.

Industrial History 1945 — 1982

What has been happening in our factories since 1945 helps us to understand why the factory has seldom made operations technology a strategic resource. The economic and societal factors which structured the factory affect its current ability to take advantage of new operations technology. One major change in this period is a vast increase in competition. The rise of the diversified corporation has resulted in more companies competing in more markets and more industries. International competition and international trade have been steadily rising in most industries. Manufacturing has increasingly spread all over the world, and Asian nations have gradually penetrated the traditional Western markets. As a result, no industry has escaped foreign competition. Instead of sharpening industrial competence, however, in many cases this competition has dulled it. A generation of dispirited "losers" has been created; investment funds have been withheld; and growth, new technology, and innovation have been dampened.

While losing its share of the worldwide market and running down its equipment, the factory has been forced to cope with a trend toward shorter product life cycles. Accelerated changes in technology and the proliferation of products in many industries have combined with increased competition to reduce product life cycles.

This combination of increased competition and shorter product life cycles has affected the factory in many ways. It must produce more customer specials, and shorter runs; order quantities are lower; set-up costs rise as a percentage of total costs; and problems of managing inventories of parts and materials have become more critical. The result has been more stress and complexity in operations management.

Adding another straw to the camel's back are sweeping changes at the corporate level. In the 1950s companies began to adopt modern, scientific marketing methods, placing more emphasis on new product design, development, product management, and advertising. In the 1960s, modern financial management methods came into use, featuring accurate and eagle-eyed financial controls, yearly operating plans, and strict monthly budgets, quarterly budgets and precise monthly controls against these budgets. Exacting financial controls and measurement systems with the loopholes closed one by one became a central force of corporate management affecting executive behavior and providing a dominant yardstick for factory performance, not only for inventory control, but for making investment decisions where paybacks on new machinery and equipment were set by corporate goals. The financial orientation at the top caused operations managers more administrative hassle and difficulty in justifying the purchase of new machinery and equipment. The hurdle rates in many industries climbed to 35-50 percent, meaning that new machinery and equipment were often simply not purchased.

The late 1960s and 1970s saw the appearance of strident employee demands and new worker expectations. The work force frequently demonstrated its discontent with low quality, low productivity, and lack of cooperation. Different age groups required different supervisory tactics. More pressure was placed on factory managers to keep workers "happy" and productive. As factory management was squeezed from above to lower costs and make better products, employees and unions pushed upward. Factory managers were caught in the middle.

But there was scant recognition of the forces behind these pressures on factory management, and few, if any, new management tools and concepts were developed to cope with these new waves of problems and demands. Most factories were managed in the 1970s very much as they had been in the 1940s and 1950s. Manufacturing management was dominated by engineers and a technical point of view. This may have been adequate when management issues largely centered on efficiency and productivity and the answers came from industrial and process engineers. But the problems of operations managers in the 70s had moved far beyond mere physical efficiency.

Operations managers' jobs had become much more complicated. Once simply charged with "efficiency," operations managers now had to manage more demanding workers, to handle more complexity in product mix, production, inventory, cost controls, and scheduling, to introduce more new products, to change products and volumes flexibly, and to successfully compete in a more intensive competitive environment. Manufacturing managers faced economic problems, control problems, work force problems and confounding dilemmas trying to satisfy top management on a multitude of conflicting performance criteria, mainly financial in orientation.

These issues required a new breed of multidimensional manufacturing managers but the factory system had produced too few such people. The age of mass production and industrial engineering is over. Today, fewer products are produced strictly on a mass production basis, and a single-minded focus on productivity is no longer adequate. Technical skills and insights are no longer sufficient to cope with strongminded, top management applying the new and demanding financial tools. Many previously successful factory managers have been outgunned by sophisticated professional top managers while being perplexed and hurt by the apparent alienation of a once-loyal cadre of workers.

Given 35 years of increasing pressures and declining competitive performance, it is little wonder that factory management has been in a reactive mode and hence unable to create and sell a corporate strategy to innovate equipment and process technology. If there is a blind spot in strategic management about operations technology, it is less because of myopia at the top than because of functional operations managers' inability to promote operations technology as their own competitive weapon. How is corporate top management to know about this promising but untried factor in competition if their own operations people do not burst out to proclaim and champion the new idea?

Obstacles to Change

Past history suggests that three roadblocks stand in the way of change for the future:

(1) If operations continue to be measured mainly by short-term, financially oriented standards, they will stay in the dark ages, held back by excessive requirements for justifying new equipment and processes; our production technology will grow steadily older. The short-term financial point of view also dominates decisions concerning nonphysical investments in operations, those designed to improve

the quality of life at work, and to improve systems for quality and production control.

(2) A second roadblock to change is top management's perception of operations as a kind of "productivity machine" rather than as a potential strategic resource. Their view is that "efficiency" is the overriding criterion of performance. Top management asks "how can operations improve our financial results?" instead of asking, "how can operations make us a stronger competitor?"

(3) Operations managers need to become better at long range planning. Operations managers are skillful at operations, but not at strategic, long-term planning. The conventional emphasis has been on productivity, meeting schedules, and maintaining efficiency on a month-by-month basis. Often we have been unable to look at total factory systems and make wholecloth changes. Instead, changes are made one by one, piecemeal. As each part of the system is changed, it throws another into disequilibrium. And as each change takes a year or two to make (for example, a change in equipment or inventory control or organization), the whole is never quite right. Small wonder, then, that too few operations managers are effective contributors to corporate strategic councils.

These three roadblocks have been industrial facts of life; they have left the keepers of operations technology so impotent as to be unable to use operations technology as a strategic weapon.

In addition to these managerial roadblocks, history suggests that the aggressive use of operations technology for corporate strategy has been delayed by the state of the art as much as by any myopia of strategic planners. The equipment and process technology (EPT) available over the past 30 years in many industries has frequently been inadequate to form a competitive weapon. Too often the EPTs available have simply not met the needs of business.

Progress has been limited by the technology available. New and modern EPTs have moved in on a grand scale in certain process industries such as chemicals, paper, steel, plywood, and petrochemicals, where they offered giant economies of scale. But in fragmented industries where enormous productive units have been a competitive impossibility (for example, metalworking, assembly, electronics, furniture, textiles, construction, shoes, apparel, plastic molding, printing, packaging, electrical machinery, transportation equipment, boat building, foundries, and toys), the new technologies offered have been merely evolutionary and creeping in their rate of progress. The new EPTs have generally failed to meet the strategic needs of the business because they have been:

— too costly, relative to their benefits,
— inflexible, in terms of product and material variety, changeovers, and setups,
— designed to be efficient only on long runs and high daily volumes, and
— full of "bugs" and the problems inherent in new processes.

Operations management has not been offered the right equipment, systems, technology, and tools. Superb individual technologies have been developed (for example, automated machine tools and direct numerical control). But when these magnificent new innovations have been applied, usually they have been slow in becoming effective and economic. They take years to improve and the payoffs are usually quite a bit lower than projected. Radical new equipment often does not mesh with existing equipment. Managements have grown prudently chary.

It seems strange that market forces have not induced "better mousetraps." But the available EPTs have generally been developed by specialized equipment producers to one part or step in a process. The individual EPTs may have performed satisfactorily but they were not coordinated, nor did they cover enough process steps to eliminate people and overhead and thereby produce major economies. Machine tool companies for example, concentrate on their specialties: milling, turning, and drilling. Some firms have offered machining centers but few have offered fully integrated and coordinated total operations sequences, including transfer mechanisms and computerized controls for low-volume manufacturing and for sufficient product and volume flexibility. Needed but not available is the kind of designer-contractor prominent in the chemical process industries offering the "turnkey" new facility.

For the most part, the industrial equipment industry has apparently been unable to interpret customers' strategic needs and develop product lines accordingly. In spite of the decline of mass production, and the trend toward shorter runs and more customer specials, too few EPTs satisfactorily solve the cost, quality, and technical problems found outside the high-volume process industries.

Finally, strategic investments in operations technology have been discouraged by conventional premises and practices:

(a) Corporate performance and reward systems are based on performance against annual budgets. Well written-off facilities look more profitable on paper, and reward systems in most companies reinforce this view.
(b) Capital budgeting is based primarily on return on investment, rather than on strategic analysis.
(c) The inexperience of top managers in manufacturing makes them unreceptive to major innovations in factory technology. To many executives, manufacturing and production are necessary nuisances — they soak up capital in facilities and inventories; they resist changes in products and schedules; their quality is never as good as it should be; and their people are unsophisticated, tedious, detail-oriented, or unexciting.
(d) Manufacturing executives are averse to the risk involved with major new innovative investments. They have seen too many new pieces of equipment fail to be fully utilized and not work as well as promised (requiring further investment in support facilities and overhead) to be enthusiastic about anything for their factory that is not well-tested and proven, preferably by somebody else. "Let my competitor take the risk," is a common,

not altogether irrational outlook. Why be number one in taking risks for new EPTs when the penalties for failure are so high?

(e) Manufacturing managers tend toward cautious conservatism. Typically, their careers have been insular and their attitudes and behavior are conditioned by years of vulnerability to criticism. They get the blame for falldowns and they bend every effort to be certain that things won't go wrong.

(f) Manufacturing managers must focus on the short term to meet the monthly schedule, maintain good quality and delivery, and solve personnel problems. They find it difficult to develop a strategic point of view in the midst of their struggle with existing EPTs and the basic in-place systems; they do not see alternatives, and they often lack the perspective to see what is possible or even what other companies are doing. Manufacturing strategy is a new art.

To look ahead, how will the future differ from the past? Can and will operations technology play an enlarged and more strategic role in the industrial firm? What is changing?

Several Pivotal Changes

During the past five or ten years the operations technology available has begun to change enormously in nearly every industry. If offers new capabilities and no longer clashes with such marketing realities as short product life cycles.

Radical, order-of-magnitude changes in EPTs are now appearing. The new technology, microelectronics, is offering a new wave of EPTs in virtually every industry. EPTs that can think, react, self-correct, adjust to product and volume changes, offer general-purpose flexibility and special-purpose economies, coordinate with other process links, reduce scheduling, paperwork, indirect labor, and overhead costs and offer a whole new range of versatility and even cost reduction are becoming increasingly available. In addition to microelectronics are new materials, advanced software capabilities, new optical sensors, and the increased effectiveness of employees armed with more information and offered more opportunities to participate; we begin to see potential answers to many of the drawbacks and problems in the EPTs available in the past.

Key to new strategic advantages are the new technologies that allow operations to become much more versatile and flexible through the use of microprocessor computers and controls. They make job shop, short-run, product-specialized factories operate more like process-type, long-run, high-volume factories, requiring fewer setups and changeovers, less indirect labor, and easier scheduling and sequencing. Complete printed circuit board facilities such as those at Hewlett-Packard and Lockheed, for example, furnish evidence of the practicality and competitive leverage of self-contained, microprocessor based new manufacturing technologies.

The microprocessor, coupled with computer-aided design and flexible machining and processing equipment, can make changes in products and materials in minutes rather than hours. Setup and changeover times are drastically reduced. They can make product proliferation and customer specials an affordable marketing strategy. The factory will have a whole new set of economics.

The new technologies appear to supply the answer to the most fundamental manufacturing dilemma of the 1960s and 1970s, namely, that while manufacturing technology was making greater economies of scale possible with mass production, markets, driven by economic and competitive pressures, were moving away from mass production towards more specialized products.

Greater knowledge and sophistication are required to select, acquire, and implement the new technologies. They test the abilities of engineers and managers and rapidly outdate a generation of technical people. They require a systems approach; if a system is poorly designed or based on only partial information, the new integrated technology systems can be disastrous. Organizational power will shift to managers who understand these systems.

With the new technology at hand, are the associated management problems going to disappear? Probably not, but these problems are at least being recognized. Management analysts are stressing the damaging effect of emphasizing the short term at the expense of innovation and long term development. In an effort to broaden their perception of the strategic potentials of operations functions, many large firms such as General Electric, TRW, American Standard, and Hewlett-Packard are now training their top executives in manufacturing strategy.

This is not just "management development," but an effort to shift the corporation's basic emphasis in manufacturing from operations to policy, and from tactics to strategy. By focusing on structural decisions — EPT choices, the number and size of plants, capacity policy, and make or buy decisions — instead of on costs and short-term productivity, manufacturing can become a competitive weapon. With poor structural decisions, no amount of able production management can achieve competitive results.

Another change in some companies is promising. At the Copeland Corporation, one of the world's largest refrigeration compression manufacturers, instead of projecting only their investments' quantitative effects, they are beginning to assess investments in nonquantitative terms such as:

- flexibility for product change,
- flexibility for volume change,
- effects on quality and reliability,
- impact on morale and attitude,
- impact on the ability to compete, and
- impact on the ability to survive long term.

Many corporations, such as General Electric, are beginning to require such nonfinancial goals in order to build superior competitive resources for the long term. Their yardsticks are not limited to dollars, but measure performance in terms of developing people, EPTs, QWL, morale and competitive assets and structure.

A final hopeful factor is the possibility that the new factory will be such a different place to work that its people will become a competitive resource as well as its technology. The movement for better quality of work life is gathering momentum and has brought about hundreds of experiments in new working environments. Better and more interesting jobs are being created and human resource management is being improved. Companies that accomplish a revolution in the work place will create a resource in their operations that strategic planners cannot ignore.

The ultimate question is, are these changes profound enough to make companies use operations technology as an element in strategy?

The New Industrial Competition

An extraordinary amount of change is taking place in industrial operations. Factors that have had a largely negative impact for 35 years are changing. Changes in these factors may produce an order-of-magnitude change in the factory. It is no exaggeration to say that the factory is being reinvented in the 80s.

The burst of new operations technologies that has occurred simultaneously with the new global industrial competition [Abernathy, Clark, and Kantrow 1981] seems an extraordinary coincidence. But this competition is essentially technological. Operations technology and product technology are propelling vigorous new competitors to a larger share of the market. Industrial management today depends on managing technological change.

The flood of new products and operations technologies demands a new level of organizational performance for handling substantial change. Organizations must learn rapidly, adapt quickly, and solve a stream of problems without crisis. For example, the auto industry needs to learn how to bring out a new model in two or three years rather than six or seven. Further, it needs to replace the massive, inflexible capital equipment that has kept its product strategy captive to its operations technology.

The auto industry is typical of many industries where economics and inflexible operations technology collide with the need for new products to compete with those produced by on-rushing technology. That a burst of global competition has coincided with the development of new operations technology is not a coincidence but two sides of a single phenomenon; the new competition is based on that technology. The question for corporate strategic planners is not whether operations technology can become a strategic resource, but how to make it one, just as rapidly as possible.

References

Abernathy, W.; Clark, K.; and Kantrow, A. 1981, "The new industrial competition," *Harvard Business Review*, Vol. 59, No. 5 (September/October), pp. 68-81.

Lauenstein, Milton and Skinner, Wickham 1980, "Creating a strategy of superior resources," *Journal of Business Strategy*, Vol. 1, No. 1 (Summer), pp. 4-10.

Porter, Michael E. 1980, *Competitive Strategy*, The Free Press, New York.

Skinner, Wickham 1981, "Factory of the future — Always in the future," *Toward the Factory of the Future*, American Society of Mechanical Engineers, New York.

Cost Justification: The Overhead Dilemma

Stephen Hunter
Northern Telecom Limited
Mississauga, Ontario, CANADA

Abstract

Manufacturing engineering groups in many industries experience difficulty in justifying flexible manufacturing systems. Common themes regarding this issue are extracted from a survey of industry sources.

INTRODUCTION

Justification of standalone robotic installations is becoming increasingly straightforward. The costs and benefits to be considered have been widely written about. Displaced labour was seen as the major benefit. Savings in indirect costs also resulted, but were secondary effects.

As companies gain experience with the new technologies, larger and more ambitious automation projects are being designed. The justification problem resurfaces with the larger systems (Flexible Manufacturing Systems utilizing robots). As the level of integration of these systems increases, reduction in indirect costs (overhead) becomes the dominant saving, making the justification problem more acute.

This paper seeks to understand the problems inherent in justifying such systems and is based on a survey of the current industry literature.

The problems involved in justifying systems or technologies whose major benefit is overhead reduction are similar. For this reason, insight may be gained from articles addressing any of the emerging technologies (CIM/CAD/CAM/FMS/Robotics) as the bibliography will indicate. The understanding gained is directly applicable to large, highly integrated robotic systems; systems which will be referred to in this paper as Flexible Manufacturing Systems (FMS).

INDUSTRY PRACTICE

A survey of the literature on problems associated with justifying CIM/CAD/CAM/robotics reveals that many companies are struggling with the same issues. Numerous articles have been published (see bibliography) trying to identify methodologies to be used to ease the task of justification.

The authors of these articles were from three sources primarily: vendors of robots/automation; consultants; and academics. No one has identified a panacea to the problem. No one is advocating generating general rules of thumb or ratios identifying indirect savings: direct savings. It is interesting to note the limited number of articles published by end users of the automation systems. In fact, a spokesman for IBM (37) indicated that internally at IBM, justification isn't the issue it was three to four years ago.

Early articles addressing the subject were of two types - either very theoretical in their approach (authored by academics) or very simple educational articles geared to single robot applications. The latter type of article highlighted checklists of costs and savings. Direct labour was usually identified as the primary source of benefits.

More recent articles are addressing the more subtle difficulties associated with identifying indirect savings. The widely recommended approach boils down to indepth financial analysis.

In the articles surveyed, there was widespread agreement that the criteria used for evaluating automation investment decisions should be based on a discounted cash flow method (NPV or IRR). Two articles, (25) (37, General Dynamics), specifically highlighted that the evaluation should conform to the company's normal framework for deciding on equipment proposals.

Throughout these articles, several recurring themes/problems have been identified. These themes summarize the feeling of the "industry" today and the remainder of this section will review them (Reference numbers correspond to the numbers in the bibliography). Personal discussions with engineers from different industries and companies have reiterated the themes presented here.

The first prerequisite of successfully justifying a flexible automation project is the development of a top-down long-range automation strategy. (2)(5)(8)(14)(21)(24)(25). The critical step is understanding the key need of the business and structuring the project to serve this need (2). It is imperative that projects are developed which tie into corporate goals (17). Companies which are successfully implementing new technologies and moving from a labour-based to a technology-based environment have been found to use a structured master plan (9). Include technology modernization planning within the strategic planning process (7). Develop specific and staged implementation plans after analyzing the business from the top-down (5). Managers allow - and even encourage - CAM related proposals to rise from the ranks of line engineers and supervisors.... This bottom up approach will not work for CAM.... Buying into CAM represents a strategic effort to boost the effectiveness of a total manufacturing system (25).

Financial evaluation should be geared to evaluation of the total 'system'. When dealing with the justification of major assembly systems, it is crucial to view the ROIs from the "system approach".... It would be counter-productive and even damaging to the justification of a system by approaching it piece-meal fashion and try to justify each component of the system individually (24). CAM is different because it is a "contagious" technology - that is, it offers progressively greater benefits as it integrates more sectors of a plant's operations (25). One of the easiest comparisons is to gut the entire plant, start over, you get a real comparison.... It is extremely difficult to do it in little bitty chunks.... it's hard to identify the savings (37, G.E.).

A detailed understanding of your current manufacturing costs, particularly overhead, should be a prerequisite for a flexible automation proposal. Defining the problems is at least as important as defining the solution (2). The kind of discipline needed to implement a robotic system forces you to understand your entire manufacturing operation from incoming parts through assembly and test and out the door as finished products... Then some real cost-saving measures can become apparent (21). Cost management needs to be an integrated and dynamic part of the technology management planning, control and measurement process (7). Economic justification first requires an in-depth analysis of obvious, and not so obvious, factors that influence both product and plant overhead costs (28). It is the company's responsibility to take the initiative and quantify those seemingly intangible variables that represent real differences in cost (11).

Current accounting methods do not provide the necessary info to readily understand the true manufacturing cost. Many cost accounting systems no longer reflect the manufacturing process... cost control efforts need to be redirected with added focus placed on indirect costs (8). We don't do a good job tracking indirect costs... This is the area Flexible Manufacturing is attacking (37, General Dynamics). With the decreasing importance of variable labour costs, companies that allocate the fixed, sunk cost of equipment and information systems according to anticipated production volumes will distort the underlying economics of the new manufacturing environment.... It is unlikely, however, that any cost accounting system can adequately summarize a company's manufacturing operations (1). Many of the savings areas are omitted because they require more effort to quantify... Indirect savings that are more difficult to quantify are often not included in the justification analysis (13). Try to establish a solid product cost basis for comparison... This is a tough assignment (17).

Developing and selling the proposal for a flexible system is a non-trivial assignment and requires an individual with a broad understanding of manufacturing, particularly engineering and accounting. The ability to prove convincingly a robot or automation system is a good investment seems to require the talents of a "super engineer" (30). Knowledge of the financial factors will make the engineer a more effective salesman of the CAD approach (32). The more difficult the justification, the more need there is for a champion to help sell the project (17). With increasing demands for improved utilization of expensive machinery, the shift to more capital intensive production will require a combination of technical and financial expertise which does not appear to be widely available... The present UK situation appears to be one with a more or less complete division of labour between the two aspects and overtones of confrontation between the separate professional groups of engineers and accountants (20). Today's managers must have a working knowledge of economics, accounting, and finance if they are to successfully evaluate robotic application alternatives, determine the value of their implementation efforts, and have significant influence on reducing plant manufacturing costs and improving the profits of manufacturing organizations (19). There is a weakness on the part of manufacturing people to express their case in a concise way... need engineers better equipped to deal with the financial community (37, Ford).

Multi-disciplinary teams are needed to develop and evaluate flexible manufacturing systems. The best person to suggest and evaluate an FMS investment may no longer be the manager directly concerned with the application... Because FMS's operation and benefits cut across so many company functions, a broader team is needed to evaluate CIM projects in general (14). This task force with members whose experience and responsibility include engineering, manufacturing, financial and data processing functions, will be responsible for seeing that the project meshes with the overall corporate goals and strategies (22). A certain amount of investigation and assistance from various company departments is required to put dollar figures on indirect savings (13). The use of multi-disciplinary teams will ensure a proper and complete assessment of improvement opportunities, a thorough understanding of critical issues and potential barriers, and will contribute to user commitment (9). The evaluation, justification, and implementation of a FMS is a major undertaking that has considerable impact on every department in a company... To be successful, the project will require support, ideas, and commitment from every part of the organization, so many representatives should be involved as soon as possible in the early planning phases of the project (17).

Economic evaluation of a flexible manufacturing system should include the best estimates available for all benefits, tangible and intangible. Failure to do so implicitly assigns a value of zero to that benefit. In lieu of confirmed figures, educated estimates are the best sources of quantifiable data... Despite the resistance that may be encountered from financial people within your organization, these estimates should not be thrown out due to the lack of accurate records. If they are not used in the justification of a project and the savings truly do exist, the analysis is more incorrect than it would have been if the educated estimate had been used (13)(30). Management tends to irrationally deal with intangibles by often ignoring them and by so doing places a zero value on them (31).

ECONOMIC EVALUATION OF FLEXIBLE MANUFACTURING SYSTEMS

The problems associated with justifying an FMS have been widely discussed in the industry literature. Insufficient detail on overhead spending and lack of models to explain overhead spending are industry-wide problems. Large Flexible Manufacturing Systems should be aimed at reducing overhead throughout the plant. Although overhead costs are budgeted and tracked, in most areas this is done at a macro level. The micro detail of what each material handler, production control clerk, test engineer does is not available. It is not surprising that direct labour is a ready target - each operation is recorded and tracked in great detail. The practical result of this situation is that most projects initiated are aimed at direct (readily identifiable) labour and then attempts are made, as best as possible, to estimate impacts to overhead areas.

Adherence to the key recommendations of the literature would benefit companies and result in more effective projects being developed and smoother project justification.

The common themes presented in the literature are presented again in abbrieviated form.

1 - top down long range manufacturing strategy
2 - understand, in detail, your current costs
3 - current accounting methods do not provide necessary info
4 - financially analyze total system
5 - project manager must have broad understanding
6 - multi-disciplinary teams needed
7 - include best estimates available for all benefits

These themes have been culled from numerous sources and when looked at as a group identify the basic problem existing for companies today and present a pragmatic solution to the problem.

The micro-detail of overhead expenses is very difficult to comprehend. Companies who rely on projects which percolate up from the bottom of the Manufacturing Engineering ranks will possibly miss larger opportunities. Engineers will attack those costs which are most readily visible, understandable and which they feel they have control over - direct labour and material. Companies who are serious at introducing systems into manufacturing aimed at reducing the overhead component of manufacturing cost must drive these projects from the top.

Overlooked in the literature is the fact that the traditional organization structure of most companies compounds the justification problem. Historically, Manufacturing Engineering have been seen as having control over the specification of manufacturing processes and tools aimed at reducing direct labour hours. These structures evolved when labour was the dominant manufacturing cost. Today, with overhead costs overshadowing direct labour, cost reduction activities need to be redirected at functions over which Manufacturing Engineering has historically had no authority material handlers, production controllers, material expediters, inspectors, inventory levels, and all other staff groups. This organization issue necessitates projects intended to impact overhead to be driven top-down.

A starting point for any co-ordinated cost reduction is understanding, in detail, current manufacturing costs. This is the major stumbling block preventing more projects from being developed which are aimed at overhead. It is difficult to design a project aimed at a cost component which is not well understood. The engineer finds himself attempting to justify 'soft' benefits which are based on 'soft' costs. Once there is a clear understanding of the 'true' cost there should be little dispute over a proposed project's impact on it.

Due to the fact that the current accounting system, and most likely any cost-effective alternative, cannot provide the detailed information necessary, the remaining themes identified in the literature present pragmatic solutions to today's problem.

The selection of a project manager with a broad understanding and a reliance on multi-disciplined teams is probably the key. Projects under consideration should be known to all functions impacted and their 'expert' inputs used. The detailed current understanding may only come from the people in each function.

The predominant practice within the industry is for Manufacturing Engineering to initiate a project, develop a financial analysis, and submit this for review with their local Finance/Accounting organization. Usually, Finance has had little prior involvement. Attempts to quantify overhead benefits may be questioned both for the methodology used as well as the specific financial claims. The latter questions may result from the fact that Finance is exposed to different information than that upon which Manufacturing Engineering developed their estimates. This isolation between Manufacturing Engineering and Finance can be compounded when the grass roots engineer and financial analyst sit down to review a proposed project. The manufacturing engineer may not be strong in presenting his ideas in financial terms and the financial analyst may have difficulty understanding the technical thrust of the project. Finance should be involved in a pro-active role from the beginning - identifying what reports may be helpful, suggesting how to better quantify estimates and identifying opportunities.

Involvement from all of the 'expert' areas of the company should minimize the 'softness' to the benefits. Longer planning horizons and more comprehensive projects will necessitate developing financial returns for the total project.

Estimation of overhead impacts should be developed on a function by function basis. The company's budget represents the best source of information on overhead spending ('real' spending) and as such should be a focal point for understanding overhead spending and determining project impacts.

OBSERVATIONS AND CONCLUSIONS

- The same issues are being addressed at many companies.
- No one has identified a panacea to the problem of justifying systems intended to have a major impact on overhead costs - the widely recommended approach boils down to indepth financial analysis.
- Widespread agreement in industry that financial indicator used for evaluating flexible automation investment should be based on discounted cash flow (NPV or IRR).
- Common themes extracted from industry:

  1) The first prerequisite of successfully justifying a flexible automation project is the development of a top-down long-range automation strategy.

  2) Financial evaluation should be geared to evaluation of the total 'system'.

  3) A detailed understanding of your current manufacturing costs, particularly overhead, should be a prerequisite for a flexible automation proposal.

  4) Current accounting methods do not provide the necessary information to readily understand the true manufacturing cost.

5) Developing and selling the proposal for a flexible system is a non-trivial assignment and requires an individual with a broad understanding of manufacturing, particularly engineering and accounting.

6) Multi-disciplinary teams are needed to develop and evaluate flexible manufacturing systems.

7) Economic evaluation of a flexible manufacturing system should include the best estimates available for all benefits, tangible and intangible. Failure to do so implicitly assigns a value of zero to that benefit.

- Projects which are initiated at the lower manufacturing engineering levels will continue to concentrate on direct labour reduction due to availability of 'as-is' detailed information.
- Organizational structure makes it difficult for any one function to initiate a project intended to impact overhead in numerous functions.
- Top-down initiative needed to develop effective overhead reducing manufacturing strategy.
- Rely on multidisciplined team to design the new 'system'.
- Finance should be involved in a pro-active role from the beginning of the system design.
- Financial evaluation of overhead should concentrate on expected impact to real spending (department budgets) as opposed to standard cost.

SUMMARY

Manufacturing strategies targeted at reducing the overhead element of manufacturing cost must be initiated top-down. Reliance on project initiation percolating from the lowest ranks of Manufacturing Engineering will, of necessity, be aimed at reducing direct labour. This preoccupation with direct labour stems from the availability of 'hard' 'as-is' labour standards and the shortage of manufacturing engineers with engineering/business backgrounds. This problem is compounded by the lack of authority Manufacturing Engineering has over other functions overhead spendings.

Justification of appropriate Flexible Manufacturing Systems will be made easier by:

1. Top-down initiation.
2. An understanding of current overhead costs <u>before</u> the system is designed.
3. Project Manager with broad manufacturing background.
4. System designed by a multi-disciplined team.
5. Early pro-active involvement by Finance.

BIBLIOGRAPHY

1984

1. Kaplan, R.S., "Yesterday's Accounting Undermines Production", Harvard Business Review, pp. 95-101, July/Aug. 1984.

2. Klipstein, D.H., "Practical Considerations in Quantifying the Purchase of Automated Vision Systems", Proceedings of Robots 8, Vol 1, pp. 2.63-2.73, Detroit, Michigan, June 4-7, 1984.

3. Estes, V., "Robot Justification - A Lot More than Dollars and Cents", Proceedings of Robots 8, Vol 1, pp. 2.1-2.11, Detroit, Michigan, June 4-7, 1984.

4. Rogers, P.F., "The Economics of Robotic Arc Welding Workcells", Robotics Today, pp. 46-48, June 1984.

5. Close, D.E. "Productivity & Technology - Top Management's Vital Role", Manufacturing Engineering, pp. 101-103, April 1984.

6. Sepehri, M., "Cost Justification Before Factory Automation", P&IM Review and APICS News, pp. 40-48, April 1984.

7. Michaels, L.T., Muir, W.T., and Eiler, R.G., "Technology Cost-Benefit Tracking", Material Handling Engineering, pp. 95-102, March 1984.

8. Michaels, L.T., Muir, W.T., and Eiler, R.G., "Improving Technology Cost-Benefit Analysis", Material Handling Engineering, pp. 49-54, February 1984.

9. Michaels, L.T., Muir, W.T., and Eiler, R.G., "The Relationship Between Technology and Cost Management", Material Handling Engineering, pp. 58-64, January 1984.

10. Adlard, E.J., "Computer-Integrated Manufacturing - Its Application and Justification", Engineering Digest, pp. 16-20, January 1984.

11. Barlow, S.E., "Cash Flow Analysis in Automated Storage and Retrieval System Justification", Automated Material Handling and Storage, Auerbach Publishers Inc., 1984.

1983

12. Muir, W.T., "Cost-Benefit Analysis and Cost-Benefit Tracking", Proceedings of AUTOFACT 5, pp. 8.34-8.47, Detroit, Michigan, November 14-17, 1983.

13. Potter, R.D., "Analyze Indirect Savings in Justifying Robots", Industrial Engineering, pp. 28, 29, 94, November 1983.

14. Airey, J., "Economic Justification - Counting the Strategic Benefits", Proceedings of 2nd International Conference on FMS, London, U.K., October 26-28, 1983.

15. Staples, G.R., "FMS - Convincing the Board", Proceedings of 2nd International Conference on FMS, London, U.K., October 26-28, 1983.

16. Dempsey, P.A., "New Corporate Perspectives in FMS", Proceedings of 2nd International Conference on FMS, London, U.K., October 26-28, 1983.

17. Tepsic, R.M., "How to Justify your FMS", Manufacturing Engineering, pp. 50-52, September 1983.

18. Miller, R.K., "The Bottom Line - Justifying a Robot Installation", Robotics World, pp. 32-35, April 1983.

19. VanBlois, J.P., "Economic Models: The Future of Robotic Justification", Proceedings of Robots 7, pp. 4.24-4.31, Chicago, Illinois, April 17-21, 1983.

20. Fleck, J. "The Adoption of Robots", Proceedings of Robots 7, pp. 1.41-1.51, Chicago, Illinois, April 17-21, 1983.

21. VanBlois, J.P., and Andrews, P.P., "Robotic Justification: The Domino Effect", Production Engineering, pp. 52-54, April 1983.

22. McDonald, J.L., and Hastings, W.F., "Selecting and Justifying CAD/CAM", Assembly Engineering, pp. 24-27, April 1983.

23. Rodgers, R.C., "How to Join the Robot Revolution-Part I", Foundry M&T, pp. 24-26, March 1983.

24. Andrews, P.P., "Justification Considerations for Robotic Systems", IBM Robotic Assembly Institute Presentation, February 1983.

1982

25. Gold, B., "CAM Sets New Rules for Production", Harvard Business Review, pp. 88-94, Nov/Dec 1982.

26. Carlsson, J., and Selg, H., "Swedish Industries' Experience with Robots", The Industrial Robot, pp. 88-91, June 1982.

27. Naidish, N.L., "Justifying Assembly Automation, Part 2", Assembly Engineering, pp. 48-49, May 1982.

28. Naidish, N.L., "Justifying Assembly Automation, Part 1", Assembly Engineering, pp. 46-49, April 1982.

29. VanBlois, J.P., "Robotic Justification Considerations", Proceedings of Robots 6, Detroit, Michigan, March 2-4, 1982.

30. Meyer, R.J., "A Cookbook Approach to Robotics and Automation Justification", Proceedings of Robots 6, Detroit, Michigan, March 2-4, 1982.

1981

31. Litecky, C.R., "Intangibles in Cost-Benefit Analysis", Journal of Systems Management, pp. 15-17, February 1981.

1980

32. Powell, R.E., "Justification and Financial Analysis for CAD", Proceedings of the 17th Design Automation Conference, pp. 564-571, Minneapolis, Minnesota, June 23-25, 1980.

33. Engleberger, J.F., Robotics in Practice, pp. 101-110, Amacom, 1980.

1979

34. Behuniak, J.A., "Economic Analysis of Robot Applications", SME Technical Paper MS79-777, 1979.

1978

35. Ciborra, C., and Romano, P., "Economic Evaluation of Industrial Robots - A Proposal", Proceedings of 8th International Symposium on Industrial Robots, pp. 15-23, Stuttgart, West Germany, May 30-June 1, 1978.

36. Govsievich, R.E., "Determining the Economic Effectiveness of Industrial Robots", Machines & Tooling, Volume 49 No. 8, 1978, pp. 11-13.

Videotape

37. "Justification, Application and Predictions", SME Video Magazine, 1983.

    - a Forum including spokesmen for Ford, IBM, G.E., General Dynamics, Ex-Cell-O and Eaton.

JUSTIFICATION PROCESSES

This section looks at the justification process itself, notes some dangers, and describes ways to conduct it. The major point is that the task must not be approached as a stand-alone problem. Rather, it is a lengthy process and involves a number of complex steps, considerable time, and continuing interaction with a variety of individuals.

The first paper, by McDonald and Hastings, identifies its focus as CAD/CAM, but the justification process they suggest is generic in its application and well-considered. Lardner deals with the first two steps in the overall process recommended by McDonald and Hastings. The paper describes how Deere & Company initiated its automation process and focuses on the "as is" study they conducted.

The paper by Bernard addresses the project form of justification analysis and explains how this form operates. Again, the justification issue itself is only lightly touched upon; the focus is on the value of a project team in addressing the need for new technology and the tasks the team must complete.

The last paper, by Meredith and Suresh, provides a structure for the justification methodologies presented in the following three sections. It also rationalizes the great variety of justification techniques seen in the literature and provides guidance on when to use each type.

Reprinted from ASSEMBLY ENGINEERING, April, 1983. By Permission of the Publisher

Selecting And Justifying CAD/CAM

Better planning and a larger up-front investment can help avoid costly mistakes later on in implementing CAD/CAM.

James L. McDonald and
Dr. Warren F. Hastings
Productivity International, Inc.

The benefits to be derived by implementing computer aided design/computer aided manufacturing (CAD/CAM) technologies for improving productivity and product quality are real. They have been proven by numerous companies throughout the last ten years. However, except for general assertions about the benefits, and generally unsupported claims of productivity increases ranging from 2:1 to 20:1, very little reliable data has been published to guide companies contemplating an investment in CAD/CAM. Why is this? Two reasons: the first being that once installed, very few companies go back to confirm the benefits and verify the estimates used to justify the purchase. Either that, or they may not have accurate records that can be used for baseline comparison. The second reason is those companies that have adequate records and do make comparisons may consider the financial data and other information confidential, and hence not release it.

The seven step procedure for the justification, specification and selection of CAD/CAM technologies—known as The CAD/CAM Audit—has been refined and verified over several years. Although the method described here assumes that a suitable product is available from the turnkey CAD/CAM system vendors, the procedure is just as valid for situations where unique hardware and/or software is required. In the latter case, however, much more effort is required in the needs analysis and systems specification stages of the process. The overall objective of the procedure is to allow a company to define its total system needs, determine which parts can be bought from vendors and which parts have to be customized, determine the costs and benefits of each step, and prepare an implementation plan for the system. This methodology can be used just as effectively to evaluate and plan for a CAD/CAM system, a shop floor control system, advanced manufacturing technologies such as DNC/CNC and robotics, word processing, process planning, or an entire system encompassing all of these areas.

The advantage of The CAD/CAM Audit approach is seen in Fig. 1. By expending more resources (funds) in the initial stages of the project, a better system is obtained at lower total cost. In some cases, the extra up-front investment avoids costly changes in direction during later stages of the project, and may even mean the difference between success and failure.

The seven steps of the process are:

- Feasibility study
- Needs analysis
- Cost/benefit analysis
- Systems architecture design
- System specifications
- Benchmarking and selection
- Implemention and start-up

Each of these steps requires a different mix of personnel skills; from entrepreneurial in the early stages when the idea of change is first introduced, to practical problem-solving skills in the implementation phase. However, throughout the entire process there is the need for a multi-disciplinary task force to maintain the direction of, and dedication to, the project. This task force must be supported by senior management and

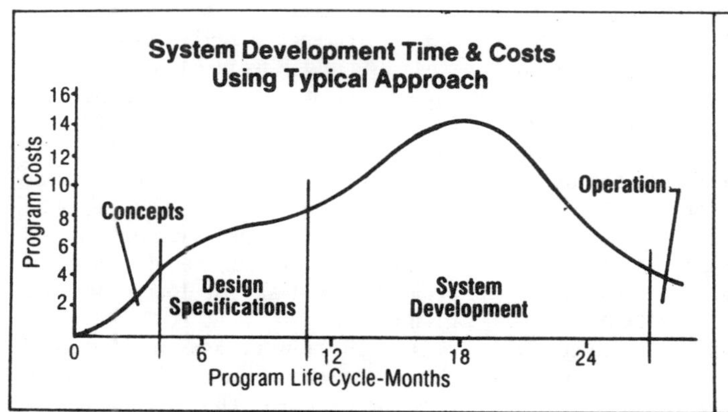

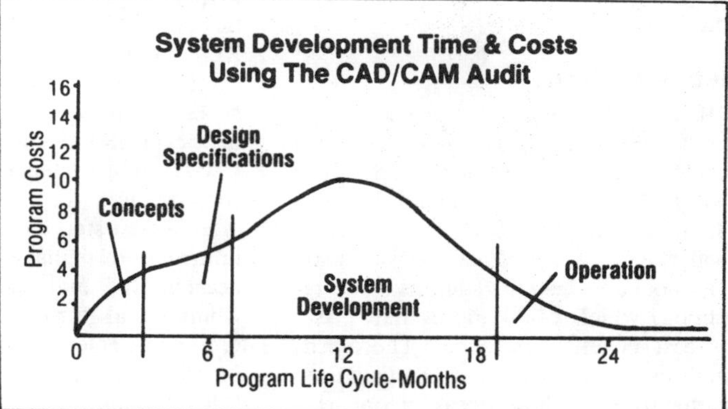

Fig. 1

have the help of some person or group with extensive knowledge of the technology under study. This help could come from a specialist corporate group, or external sources such as consultants.

Feasibility Analysis

To see how the methodology is used, let us apply it to the case of a company which is wondering if it should purchase a turnkey CAD/CAM system, and, if so, which one. The first step will be to identify the opportunities (the Feasibility Analysis), to see if further effort and expenditures are justified. The Feasibility Analysis entails a review of current methods, equipment and practices to see if new or different ways of doing things would be beneficial. This step should identify those areas and activities which are candidates for the new technologies. Fig. 2 shows how CAD/CAM impacts almost all phases of the product cycle, from design through manufacturing and inspection. Thus, when considering the acquisition of a CAD/CAM system, consideration should include product design and drafting, facilities layout and design, product and process R & D, N/C programming, process planning, tool engineering, quality assurance, customer and vendor interfaces, and also Bill of Materials maintenance.

At this stage it should be possible to obtain broad brush figures for costs and benefits. Don't get into lengthy vendor discussions or negotiations, as you have not yet established the viability of the application or your real needs. Get approximate hardware and software costs and offset these against expected savings over a five year timeframe. (A shorter term return on investment is possible, but five years is more realistic.) Helpful information on various systems and their capabilities can be obtained from trade publications and periodicals which print product reviews. "A Survey of CAD/CAM Systems," 3rd Edition, available from Leading Edge Publishing, Inc., Dallas, TX is also a useful source of comparative data on CAD/CAM systems.

Needs Analysis

If estimated costs don't exceed benefits by more than a factor of two, you should proceed to the next stage—The Needs Analysis. You should proceed with the evaluation because there is a good chance you have overlooked some not-so-obvious benefits such as scrap reduction, which could more than offset the cost/benefit differences. The Needs Analysis should comprise a structured examination of those areas of opportunity identified in the first phase. The data to be gathered in this step will be used for cost/benefit analyses, system design and specification, management planning, and training and implementation steps.

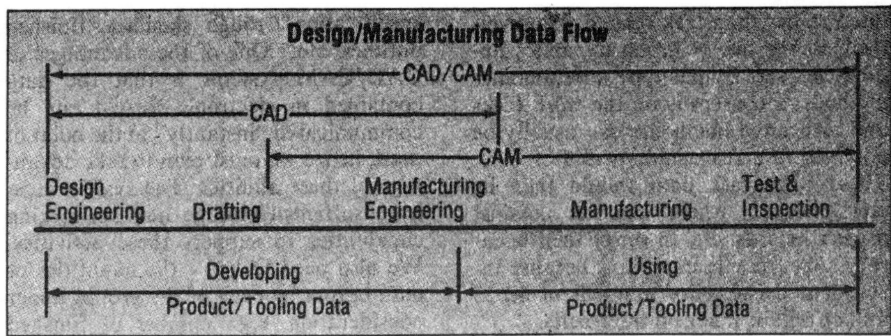

Fig. 2

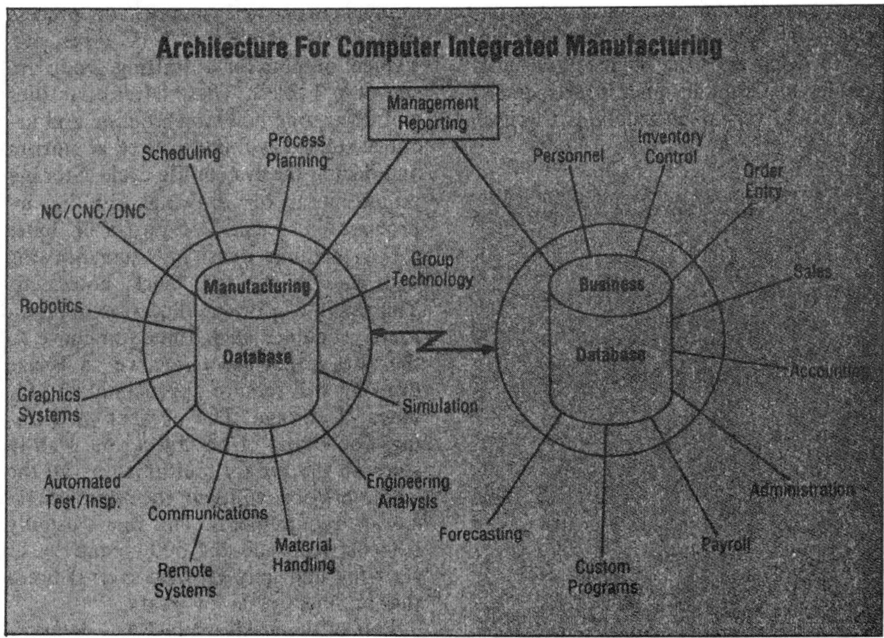

Fig. 3

© Productivity International, Inc. Dallas, Texas

The multi-disciplinary task force should be formed now. This task force, with members whose experience and responsibility include engineering, manufacturing, financial and data processing functions, will be responsible for seeing that the project meshes with the overall corporate goals and strategies, and satisfies the established requirements for ROI, etc. This task force should comprise four or five people with the necessary knowledge and time to devote to the project. Remember, a move into CAD/CAM systems usually has major consequences for a company, as it is often a first step toward computer-based information management in the engineering and manufacturing disciplines. Commercial turnkey CAD/CAM systems are elements of a total architecture for computer integrated manufacturing (CIM) as outlined in Fig. 3.

The Needs Analysis begins with an in-depth technical review of the identified opportunity areas. Each of these must be carefully analyzed to identify both the work tasks and data flows involved in each job function which could be beneficially impacted by the proposed system. For example, the design and drafting function is examined in detail to isolate generic functions—not product- or project-specific tasks. These could, for instance, be subdivided into meetings, liaison activities, calculations, design sketches, quotations and specifications, use of reference materials, creation of parts lists and Bills of Materials, detail drawing, print and copy making, checking and ECO (Engineering Change Order) control, lost time, and so on. Each of these categories could be further subdivided depending on the particular activities within the organization or group being studied. Table 1 shows a typical work task list for a drafting group.

Once all the work tasks are identified, the manhours associated with each task must be recorded. This is not easy, as most department records are project related rather than task related. However, through the use of group dynamics approaches and Delphi type questionnaire methods, a consensus on the work tasks and associated manhours can usually be obtained in a reasonable period of time. These work task data should then be cross-checked where possible against project records, etc. to verify their accuracy. Accuracy is important because investment decisions of upwards of $0.25 million will be made on the results.

Along with the work task and manhours data, it is necessary to determine what elements of data pass from one person or section to another. For instance, what does the engineer/designer pass to the detail designer/draftsperson? Are they verbal instructions, written specifications, rough sketches, finished outlines, etc.? One of the advantages of CAD/CAM systems is that the data contained in electronic format can be communicated "instantly" to the point of need, hence we must plan to take advantage of these abilities. The system must have sufficient storage and networking capabilities to support these activities. We also need to know the quantities of data output by each of the groups under consideration, e.g., number of finished drawings (categorized by sheet size and complexity for storage calculations), number of copies, number of process plan sheets, number of N/C tapes, etc. Typical outputs for a drafting group are shown in Table 1. These latter quantities will determine how much online and archival storage will be required at startup and during the system life cycle. Storage requirements for drawings can be approximated by using a figure of 8K bytes for a medium density A-size drawing with dimensions, title block, notes, etc. The required storage allocation approximately doubles each time you move to the next larger sheet size, i.e., a B-size drawing will require approximately 16K bytes of storage. The number of drawings (or views) of an object you wish to store on the system multiplied by all the active projects will give the approximate storage requirements. This figure usually runs between 100 and 500 megabytes at start-up, and may grow to several times that number over a few years.

Table 1
Categories of Work Tasks

Drafting	
A. Drafting and Layout	**Hours**
1 *Liaison*	
Coordinating Drafting and Layout	620
2 *Reference Material*	
Catalogs	15
Engineering Standards Manual	44
	59
3 *Group Standard Materials*	
Layout Bill of Material	147
Reference Old Layout	294
Reference Old Bill of Material	59
Old Job Files	59
Assembly List Files	59
Drawing Files	59
Stock Books	59
	736
4 *Drafting Time*	
Draw New Drawing	
Electrical Wiring Diag., Schematics Harness	2690
Electrical Mechanical Components	2690
Mechanical Detailing	2690
ECN Incorporation	1360
Other-Civil/Arch	2690
	12120
Drafting Layout	830
5 *Print and Copy Running*	
Check Prints	140
6 *Lost Paid Time*	
Administrative	1835
7 *Checking*	300
Total Hours	16640
Group Output	
(Number of Drawings and Write-Ups)	
A. Original "A" (8.5 x 11)	55
Original "B" (11 x 17)	450
Original "S" (28 x 40)	432
B. Pages of Original Bill of Material	450
C. Pages of Copy of Bill of Material	1 copy

Part of the needs analysis is a determination of the manhours which could be saved by utilizing the proposed CAD/CAM based technology. This involves the identification of each work task which could be impacted by the proposed system. For the impacted tasks, appropriate productivity ratios are assigned and the equivalent manhours using a CAD/CAM system are calculated. The sum of all the appropriate CAD/CAM manhours determines the number of terminal hours/year required to handle the current activities. The difference between the manual hours for those tasks, and the CAD/CAM terminal hours to complete the same effort is the direct labor savings attributable to the new technology. A typical summary is shown in Table 2.

An examination of the work task also identifies which application packages will be required. Some of the applications which are available from the vendors include mechanical design/drafting (including solids modeling), electrical, electronic design for PWB and IC applications, numerical control, process planning, classification and coding for design retrieval, etc. Usually the vendor provides a basic design/drafting package and the customer has to purchase all other application packages at extra cost. These costs range from about $5,000 to $50,000 per application, depending on such factors as complexity and development costs.

So far, we have identified the needs, in terms of hardware and software, of the basic unit. Next, it is necessary to consider the need for communications to remote sites, or to other computer systems. Generally, any extensive calculation requirements, such as simulation and finite element analysis, will require communications with a mainframe processor to avoid unnecessary delays. Attempts to do such processing on the graphics central processing unit (CPU) will severely degrade its performance. Parts lists and Bills of Materials generation may require communication with existing mainframe computer facilities.

As a result of the Needs Analysis, we are now able to generically identify all required systems hardware and software. Also, the required distribution of hardware (terminals, plotters, hard copy units, etc.) can be determined. The actual numbers will depend on the way in which the system will be operated (dedicated operators or open shop approach), the number of shifts to be worked and the type of work for which the system will be used.

Cost/Benefit Analysis

The next step in the process is the Cost/Benefit Analysis. This step draws heavily on the information collected in the preceding phase. Costs for the identified hardware and software are obtained from vendors, together with estimates of the costs for installation and facilities preparation. Maintenance costs also must be obtained, and these usually run at about 12% of the base system cost per annum. This cost includes both hardware and software maintenance and software updates. Other costs to be identified include those for training personnel, new personnel required to operate and manage the system, and any startup costs involved with loading initial information into the system.

These costs are offset by the estimated savings derived from the direct labor reductions mentioned earlier, additional income or savings derived from shorter design-to-market cycle time, scrap and material savings, etc. In estimating savings, you should look at expected business growth, increased labor requirements which may be avoided or substantially reduced through the use of CAD/CAM systems, and other less tangible benefits. Obviously, trends in wage growth rates should be factored into the savings to be derived in future years.

These costs and savings are entered into standard discounted cash flow (DCF) and return on investment (ROI) calculations using the guidelines and figures established by your company. If the investment satisfies the company criteria, then you should proceed to the next phase, which is the system architecture and specification.

System Architecture & Specification

Should the Cost/Benefit Analysis fail to satisfy company investment criteria, management should not discard the notion of investing in CAD/CAM without further examining the subject. The failure to satisfy the investment criteria may have resulted from an insufficient knowledge of the potential savings to be derived from CAD/CAM systems or the selection of inappropriate technologies for the particular operations under study. Given the wide variety of systems available, and the successes of many companies with CAD/CAM, it is likely that an alternative approach may prove to be right for your organization. However, even if an appropriate technology cannot be found, the effort to date has not been wasted. Experience has shown that a careful study of the work task data collected in the Needs Analysis phase will reveal many opportunities for improving and streamlining present methods. In some cases, it may be found that process control documentation is inadequately formatted and/or contains the wrong information for certain stages of the manufacturing process. Inefficient work procedures may cause unnecessary iterations in the design drafting cycle, and in other cases it may be discovered that much time is lost in trying to find the appropriate revision level of a specific document. By taking advantage of the information obtained in the work task data you may be able to make considerable savings in present methods with virtually no investment.

The specification of a systems architecture may be relatively simple if a standard vendor product is suitable for the application. Or it can be complex if the installation will involve multiple CPU's located at different sites and networked together. The need to communicate to other computing systems, such as the company's MIS and financial applications, will further complicate the systems architecture with requirements for gateways to various networks, communications protocols, etc. Remember that the time spent developing the systems architecture will not be wasted. Future manufacturing technology will replace today's series of loosely coupled discrete processes with fewer integrated processes controlled by automated information flow. The system you are designing should form part of that future manufacturing architecture.

Table 2
Impact of CAD/CAM on Manhours in Various Departments

Group	Total Hours	Potential Hours Impacted	Actual Hours Impacted	Equivalent Terminal Hours	New Total Hours	Man Hours Saved	% Of Total Hours Saved
Design	43,299	29,165	19,017	9,510	33,792	9,507	22%
Analysis	28,000	13,028	11,188	2,366	19,178	8,822	32%
Tool Design	20,747	13,777	12,567	4,509	12,689	8,058	39%
N/C	7,036	4,076	3,886	1,749	4,899	2,137	30%
Process Planning	27,696	18,763	18,763	6,254	15,187	12,509	45%
Quality Control	72,589	25,316	17,721	5,900	60,768	11,821	16%
Totals	199,367	104,305	83,142	30,288	146,513	52,854	27%

Benchmarking, Selection and Startup

Now that you have identified all of your hardware, software, communications and other requirements, you are almost ready to begin your serious discussion with the vendors. However, before doing this, crystallize your ideas and concepts by putting together a specification for the hardware and software you wish to purchase. The specification should detail your requirements in terms of software and hardware capabilities and performance, equipment conformance to appropriate industry standards, criteria for acceptance of the system, system uptime and performance requirements, and vendor response to maintenance calls. This detailed document, which may run from 50 to 100 pages in length, should also include a brief summary of the selection procedures and benchmark requirements. By this time, you know what you want, and are in a position to negotiate effectively with the vendors. Your specification should be sent to those vendors with products which you consider suitable for your requirements. The vendors will need 30 days to respond to this Request for Quotation.

Upon receipt of the vendors' quotations, they should be examined in detail and all responses weighted according to the degree to which they satisfy your requirements, and the relative importance of each requirement for your function. As a result of this process, you should end up with a short list of two to three vendors who are potentially able to satisfy most of your requirements.

The next step in selecting your CAD/CAM system is the benchmark. If you have done your needs analysis and system specification work thoroughly, this important step should present few difficulties. The benchmark should not only identify the best candidate system for your organization, but also identify those areas of requirement which will not be satisfied by the vendor and, therefore, require additional development effort inhouse. To be successful, the benchmark must accurately reflect your requirements in terms of basic and specialized capabilities. For instance, if you must run mechanical and electrical applications simultaneously out of the same design database, then your benchmark should include tests which reflect this requirement. Accuracy, resolution, the need for color, problems you have to solve through the use of CAD/CAM and any other characteristics which are essential for a successful installation should be tested during the benchmark.

By looking at the work task data and the associated direct labor savings through the use of CAD/CAM, it is also possible to consider cost/benefit tradeoffs. Savings to be derived through the availability of certain applications packages or the advantage of one vendor's products over those of another vendor can be evaluated from this data. The result of this step should be the identification of that system which best fits your present and future needs.

Finally, you should set up procedures for monitoring the progress of your new investment. Specific performance criteria should be set up and reviewed periodically to ensure that the new installation is performing to expectation. The estimated labor savings used in the justification procedures should be included as part of this checklist. —*RW*

Reprinted courtesy of the Society of Manufacturing Engineers. Copyright 1983, from ***Manufacturing Engineering,*** *August 1983.*

The Corporate Transition to Superior Manufacturing Performance

An industry leader takes a look at manufacturing today and urges managers to simplify their operations to manageable levels. The goal: economic survival

J. F. LARDNER
Vice President
Manufacturing Development
Deere & Company

FACED with increasingly successful offshore competition and sharp declines in shares of world markets, U.S. managers are beginning to take a hard look at the need to achieve superior manufacturing performance in order to guarantee survival.

This management concern represents an important change from the time when American companies dominated world markets for manufactured goods. From the early 1950s until the latter half of the 1970s, there was little reason to believe that U.S. manufacturing did not represent a standard of manufacturing performance unequalled in the world. However, events of the last several years have shaken this belief and today manufacturing is struggling to establish standards by which superior manufacturing performance can be measured in a world of global competition.

Based on what we are learning from the new group of competitors, I am convinced many of the measurements used to judge manufacturing performance in the past 30 years need to be altered or perhaps abandoned. This will not be an easy task.

For several years, Deere & Company has been attempting to achieve superior manufacturing performance. Progress has been encouraging and we are optimistic about the future. While we still have a long way to go to reach our goals, we have learned much from our experience. We have also learned a great deal from companies that have seen what we are doing and from other companies engaged in similar programs. We have now begun to identify some of the important elements that must be present if a program of superior manufacturing performance is to succeed.

Environment: An Important Factor

Perhaps the most important element is the environment in which the effort takes place. Two conditions must exist within this environment. First, there must be a visible, top-down commitment to the program by senior management, who should understand the strategic implications of the undertaking. They must insist on a broadly distributed, well-coordinated introduction of the technology and of the organizational changes required to achieve success. They must also ensure that adequate resources are available to get the job done. Achieving superior manufacturing performance is not a short-term project.

The other critical factor is how management views manufacturing. It may base its views on the nature of manufacturing principles which hold that manufacturing can be divided into a series of small, easily definable elements according to established specialties within the manufacturing spectrum. Advocates believe that if each of the separate elements is optimized, the whole manufacturing operation will automatically be optimized as well. This is the approach we have been following for the last 30 or 40 years and is the one that has produced the kinds of manufacturing facilities and organizations we have today.

James F. Lardner

There is, however, a new (or perhaps revisionist) school that believes in a single, indivisible whole, one that includes much more than the material-transformation activity commonly thought of as manufacturing. This school holds that manufacturing begins with product and design concepts and ends with servicing of the product in the field. Furthermore, though manufacturing is infinitely complex in all of its fine details, no single part can be treated in isolation.

This view of manufacturing is driving efforts to integrate CAD and CAM and is behind programs to achieve computer-integrated manufacturing, which is expected to be the foundation for the factory of the future. It seems the view closest to that of the Japanese, who make use of just-in-time manufacturing, the integration of material handling into the manufacturing operations, the successful pursuit of quality, and the integration of design and manufacturing.

The idea that the task of optimizing manufacturing must be approached in an overall manner is not appealing to many operating managers and members of existing staffs. In fact, two of the most important departments, industrial engineering and cost accounting, have been instrumental in developing many of the standards for manufacturing performance that we use today. They will not easily be persuaded to change. Nevertheless, if there isn't a fairly broad, well-distributed recognition by management of the indivisible nature of manufacturing, it will be impossible to establish programs for superior manufacturing performance. Our own experience and the experience of others argues that a fractionalized approach just will not work!

People: Necessary for Success

Superior manufacturing performance demands a special mix of people in management and staff groups. They are not always found in manufacturing organizations. The ideal group of people would be the following:

▶ An adequate number of competent and experienced manufacturing managers and senior technical specialists. They are people who have carefully thought about the way we look at, and manage, manufacturing and they are constructively dissatisfied with the results.

▶ A group of engineers and technicians, trained in the use of computer-based technology, who can apply these skills to solving practical problems. This group is usually made up of younger people who lack depth of experience in operations.

▶ Outside consultants in a few highly specialized areas. It is advisable to search them out and evaluate their capabilities well before they are needed.

▶ Finally, and perhaps most important, people with broad experience in high-level manufacturing management and in problems of technology transfer. They provide the intellectual leadership in new and uncharted waters.

There is no formula for the exact mix of experience, skill, and talent needed. It will depend to a degree on the specific situation, but it is safe to say, however, that most early estimates of the resources required are optimistic — so be prepared!

The cost trail became lost in the wilderness of overhead accounts

From our experience at Deere & Company, I would say that the most difficult resources to find are those at the senior level, that is, broad-gage people who can provide intellectual leadership. Only slightly less difficult to locate are the experienced, mid-level managers and senior technicians who question the assumptions about manufacturing management and control, convinced that there is a better way.

Program Definition

It is essential to define the program itself. At the beginning, most likely, program definition will not be very detailed because no one knows what needs to be done to get "from here to there." The plan, then, must depend heavily on basic concepts for setting goals and directions. A good starting point is to decide what results are expected and then to work back from there.

When we started our program, we had an idea that this was the kind of factory we wanted. The project's two senior groups focused on a strategy, which we later found to be very successful. A review of the program's objectives by the group led to one important conclusion: if the goals could be achieved, manufacturing costs would be reduced substantially. Therefore, if we could identify the real manufacturing cost drivers, we could also identify the principal problems in achieving superior manufacturing performance.

The Experience at Deere & Co.

A number of the most experienced people in the program suspected that the rapid growth in manufacturing overhead accounts and manufacturing costs was not simply the result of increasing complexity in manufacturing operations. Rather it was caused, in large part, by the failure to respond effectively to a changing environment. The difficulty was that the objectives of the program had been defined by people who took a holistic view of manufacturing, while the available financial, operational control, and reporting systems were designed to deal only with the many small pieces that result from using a conventional, fractionalized approach to manufacturing.

When we tried to analyze the possible effects of the changes we were contemplating, and tried to establish baseline cost data from which to work, we found that the cost trail frequently became lost in the wilderness of the overhead accounts. This was particularly true for fixed overhead categories, where the link between overhead costs and the actual causes of these costs was tenuous or nonexistent. We needed a means of evaluating the proposals developed in the project so we could get a realistic estimate of the likely results. It was clear that we needed a way of looking at the *whole* manufacturing activity. We turned to a simplified version of value-added analysis, which took us from minute details to the overall view in one jump, and gave us the beginnings of the insight we wanted.

As a way of looking at the whole manufacturing activity, we turned to value-added analysis

We began with the total Factory Selling Price and divided it into three parts: Direct Material, which is all material in any form, purchased to become a part of the finished product; Transformation Costs, or all costs related to transforming the Direct Material into the end product; and Margin, or the difference between Factory Selling Price and the sum of Direct Material plus Transformation Costs.

Though we recognized that the cost of the Direct Material was influenced by product design, and considered product design an important factor in achieving superior manufacturing performance, we decided to treat it as a longer-term problem to be addressed after examining Transformation Costs. Because Material Transformation is commonly thought of as *the* manufacturing operation, it seemed a good place to begin.

Examining this breakdown in more detail, we learned a great deal about where the problems were and why we had them. Though the conclusions derived from this exercise have not been universally accepted, they are being accepted by our key manufacturing executives and operational managers.

The Problems

It was the analysis of the Material Transformation Costs that gave us our first real clue. Though it is possible to break down Transformation Costs into a large number of classifications, we chose to

SUPERIOR MANUFACTURING PERFORMANCE

Here are the marks of superior manufacturing performance:

▶ For a given output, performance will permit a significant reduction in plant size. Savings of one-third to one-half are possible.

▶ For a given output, performance will require far fewer people. The principal areas of reduction will be in the indirect and white collar workers, who are viewed as beneficiaries of the transition to the factory of the future.

▶ Factories will operate with minimal inventories and will be notable for high throughput velocity.

▶ Manufacturing quality will be high and defective material and work will approach zero. Postproduction costs, primarily warranties due to manufacturing defects, will be low.

▶ Factories will demonstrate great flexibility in responding to changes in schedules and product mix.

▶ These factories will be capable of rapid startup at much lower costs when introducing new or revised products.

▶ Products will be manufactured at a substantially lower unit cost offering greater value for the money — an important advantage in a world of global competition. ■

keep it simple and still identify the origins of the costs involved in making the product. Where additional information or analysis was required, special studies were inaugurated. We started with the following breakdown:

▶ Total employment costs for salaried employees

▶ Total employment costs for the indirect labor blue collar group

▶ Total employment costs for the direct labor blue collar group

▶ Material Transformation Costs other than employment costs or nonemployment costs.

We then divided nonemployment costs into the following categories:

▶ Interest and taxes

▶ Depreciation

▶ Energy costs

▶ Indirect materials and services

▶ Returns and allowances and all other charges and credits, including sales incentives.

Then we analyzed trends for the past 15-18 years in each of these categories.

The Findings

Costs that we could relate directly to the physical transformation of material made up only a small portion of the total costs charged to the transformation activity. In addition, there had been a relative reduction in these costs as a percent of all costs associated with the transformation process. During this period, however, remarkable growth occurred in the costs of salaried employees per unit of output and in nonemployment costs per unit of output, and to a lesser extent in indirect labor costs per unit of output. This was not what most managers had expected to find. The next step was to find out why.

Savings have reached into all phases of our operations: material handling, utilization, throughput, etc.

The Reasons

We began by assuming that everyone on the payroll had a job to be done. We also assumed that most people were working with diligence and effort. The question to be answered then became, What caused all the extra work?

When we analyzed what people were doing and how they were allocating the money in the category of nonemployment costs, we found something interesting. The increase in workloads appeared at first to be the result of a tremendous increase in the complexity of manufacturing over a 15-year period. However, there were some people who found this answer too simple, and they asked, How much of the increase in complexity is unavoidable and how much is a result of the way we are dealing with the manufacturing operation? At first, the possibility that we might be needlessly compounding our problems seemed unlikely.

We learned a great deal about where the problems were

Confirming the Worst

I will not dwell on the details of how we tracked down the causes of unnecessary manufacturing complexity. Suffice it to say that we confirmed the worst suspicions of those who raised the question originally. In pursuing the "divide and optimize" theory, we had lost sight of what we were really trying to do. For example, in separating design and manufacturing, we obscured the fact that we not only needed a product which could perform satisfactorily and have features customers wanted, but we had to make it at a profit and service it in the field.

Typical introductions of new products involved a sequential process: products were designed; manufacturing plans were prepared; tooling and equipment were procured; and then the products were produced, sold, and serviced. The result of this approach was that tooling costs exceeded estimates; product introductions were late; or, if the target dates were met, extra costs and efforts were incurred. Product launches were followed by volumes of engineering changes by the shop. Startup scrap and rework was customarily excessive.

We also pursued specialization on the shop floor. The majority of our departments were organized by machine tool function, not by the parts to be produced. The result divided the material transformation process on most production parts among so many people that no one felt responsible for the end result. Furthermore, by doing a small part of the total transformation in each of several departments, we greatly compounded the problem of production control and scheduling. An additional penalty incurred with this kind of factory operation was that our material-handling costs became a large part of our overall expense in manufacturing.

Conclusions

We can see now what we should have recognized before. If you have 10,000 parts and you make them in 10,000 different ways, you will have a far greater management and control problem than if you identify 20 part families in the 10,000 parts and are able to make them in only 500 different ways. In addition to reducing the complexity of the manufacturing problem, the probability of errors and omissions is also reduced.

As we unraveled the story, it became clear that complexity was directly reflected in the increased number of overhead personnel. We found that the majority were engaged in data and information processing. It also became clear that if the complexity of the operation was reduced, the data generated would be reduced, and fewer people would be needed. Simplification of the problem would also allow greater use of computers.

The road to superior manufacturing performance does not start with large capital investment programs. Nor does the road begin with the wholesale adoption of new and exotic technology. It starts, instead, with an examination of the total manufacturing operation and with a program that reduces operations to human dimensions by reintegrating them into a comprehensible, manageable whole.

We are now beginning to measure the results of our programs to reintegrate manufacturing; little by little we are picking up the pieces and putting them back together. We finally have design and manufacturing engineers who produce manufacturable designs the first time. We have seen tooling costs halved and reductions in manufacturing costs of 10-20%. We are rearranging large portions of our factories to establish flow-through manufacturing. Our savings have reached into all phases of the transformation operation: material handling has been reduced 40%; utilization is up 15%; throughput up 15%; and defective material and work are down 50%. And, finally we are welcoming back the worker, who was ejected as a decision-maker by scientific management years ago. ■

Reprinted from **Industrial Engineering**, March 1986.

Structured Project Methodology Provides Support For Informed Business Decisions

By Paul Bernard
Litton Integrated Systems Technology

"Strategic decision-making in today's increasingly more complex business environment must be supported by a justification analysis which focuses a company's perspective outward toward competitive issues rather than inward toward the short-term payback goals of the past."

The above quotation could have been made by practically any manager in industry today. It reflects the growing realization that decisions to allocate limited resources to those areas with the highest expected return to the firm can no longer be based solely on mathematical approaches which ignore strategic business considerations.

The accounting profession has developed highly structured methods for allocating these resources in cases which lend themselves to a traditional financial analysis. There are a number of payback analyses that can be used to calculate a precise economic justification as long as costs and savings can be quantified. However, these methods do not work well for strategic expenditures when returns may be low or negative in the early years and when intangibles such as competitive advantage or flexibility are primary reasons for investment.

As a consequence, convincing top management of the need to invest in new technology and then justifying it is often more of an art than a science. Rather than relying on some type of "creative" accounting procedure to reduce all of the intangibles to dollars and cents in order to calculate a precise economic answer using imprecise or highly subjective source data, one approach is to use some type of structured project methodology to analyze the relative advantages and disadvantages to the firm. Such an approach is intended to provide management with all of the information necessary to make an informed business decision when decisions are dependent on managerial judgment rather than on hard and fast mathematical equations.

Project approach

A project approach is simply a structured methodology for analyzing a situation which has a variety of inputs, variables and possible outcomes. The thrust of the approach is to segregate the analysis of a subject that is complex when taken as a whole into well defined and manageable sub-parts which can then be quantified and compared to predetermined business standards, norms, goals and objectives. The result is typically one or more projections of the environment (business, technology, competition, etc.) as well as recommended actions to be taken based on each.

The decision to invest in new technology (defined here as new to the company) is not one to be made lightly. Total cost can run in the six to seven figure range, and project duration can be years from start to finish.

New technology may even affect the way a company is managed (just-in-time or manufacturing resource planning), how costs are determined (allocate overhead to sales rather than direct labor when labor is drastically reduced or eliminated), design and reliability of the product, and even how the company is perceived by customers, investors and competitors (is the company progressive or stodgy?). Justification must take *all* pertinent factors into account, not just those associated with direct and measurable costs and savings.

No two justification analyses are ever exactly the same, and a project methodology such as that shown in Figure 1 allows a standardized and proven approach to be customized to fit the need. The following steps are typical of those used to evaluate a firm's requirements and the impact that new technology may have on the competitive posture of the business:

- *Project team*—Assemble a team of people from the company who understand how the business operates and can provide the expertise necessary to perform the justification analysis. Assign a team leader who has the time and experience to manage the project. If the technology is indeed new to the company, it may be worthwhile and even necessary to complement the in-house team with contract personnel who are familiar with the technology or are experi-

Figure 1: Project Approach for Justifying and Implementing New Technology

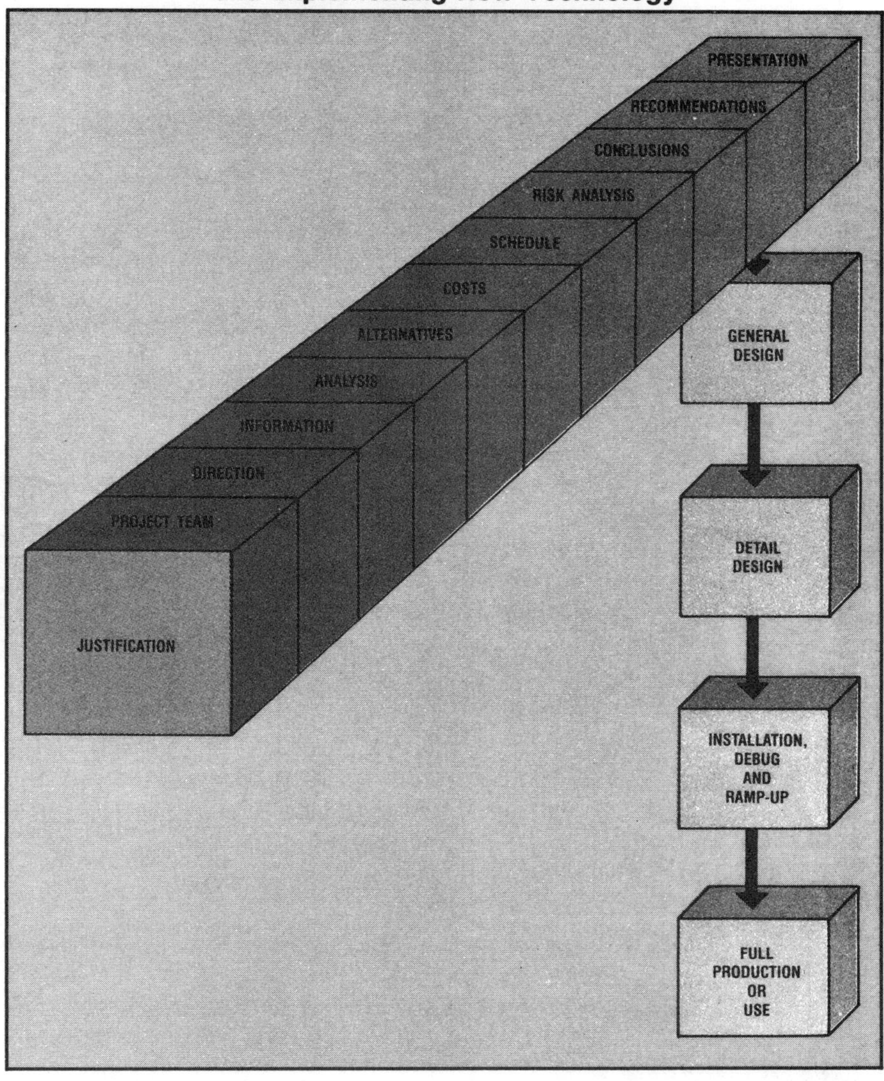

enced in managing projects of the magnitude and complexity being considered. Outside personnel may also be able to bring time and analysis tools such as simulation and modeling to the project that are unavailable in-house.

● *Establish project parameters*—Once the team is formed, the next step is to discuss the project with company management. The outcome of this should be an understanding of management's expectations and a fairly good idea of how the team should proceed. It is very important at this point to identify any pre-established expectations and to resolve any conflicting goals or objectives.

● *Direction*—Information obtained from these discussions will allow the team to formulate the project goals, objectives, constraints and qualifiers and to make individual team assignments. The scope of the project will determine which areas to review and in what detail, and this will provide the basis for establishing an overall schedule and individual milestones for tracking progress. An operating budget should also be established at this time to cover travel, R&D, prototypes, contract personnel and administrative expenses.

● *Information gathering*—In order to justify any kind of technology, it is first necessary to learn as much about it as possible. This involves understanding its areas of strength and application as well as its weaknesses or limitations. Sources of information include vendors, literature, users, colleges and universities, and R&D facilities such as the National Bureau of Standards and technology assessment centers.

Information must also be accumulated on other areas related to the project. This might include performing product and process analyses, evaluating existing equipment and personnel capabilities, identifying present operating costs and support requirements, and doing a partial or comprehensive business analysis and marketing study.

● *Analysis*—Once the team has gathered, documented and verified all of the supporting data and each member has gained a good working knowledge and understanding of the technology and all of the related business and technical issues, an in-depth analysis to determine the effect on the company can be initiated. As a start, this may require producing layouts and process plans and re-concepting the product to enhance its manufacturability.

The impact of the new technology on each business area can then be assessed in order of effect, from those that are most directly affected (product cost and quality, for example) to those which may be only indirectly affected (ability to introduce new products to the market faster). This is the hard part, because many of the areas are subjective in nature and it is difficult to properly identify and prioritize all of the areas affected.

Failure to identify *all* areas of impact can seriously jeopardize not only the credibility of the team, but also the results of the analysis and any conclusions. Brainstorming is often an effective way to identify and prioritize these areas, and the analysis may call for a system of weights to account for relative importance across the range and variety of items being analyzed.

● *Alternatives*—One of the major problems with justifying new technology lies in the simple fact that it is new. Another problem is that no projection can be made with certainty. As a consequence, there is no guarantee that any one scenario

Figure 2: Justification Phase

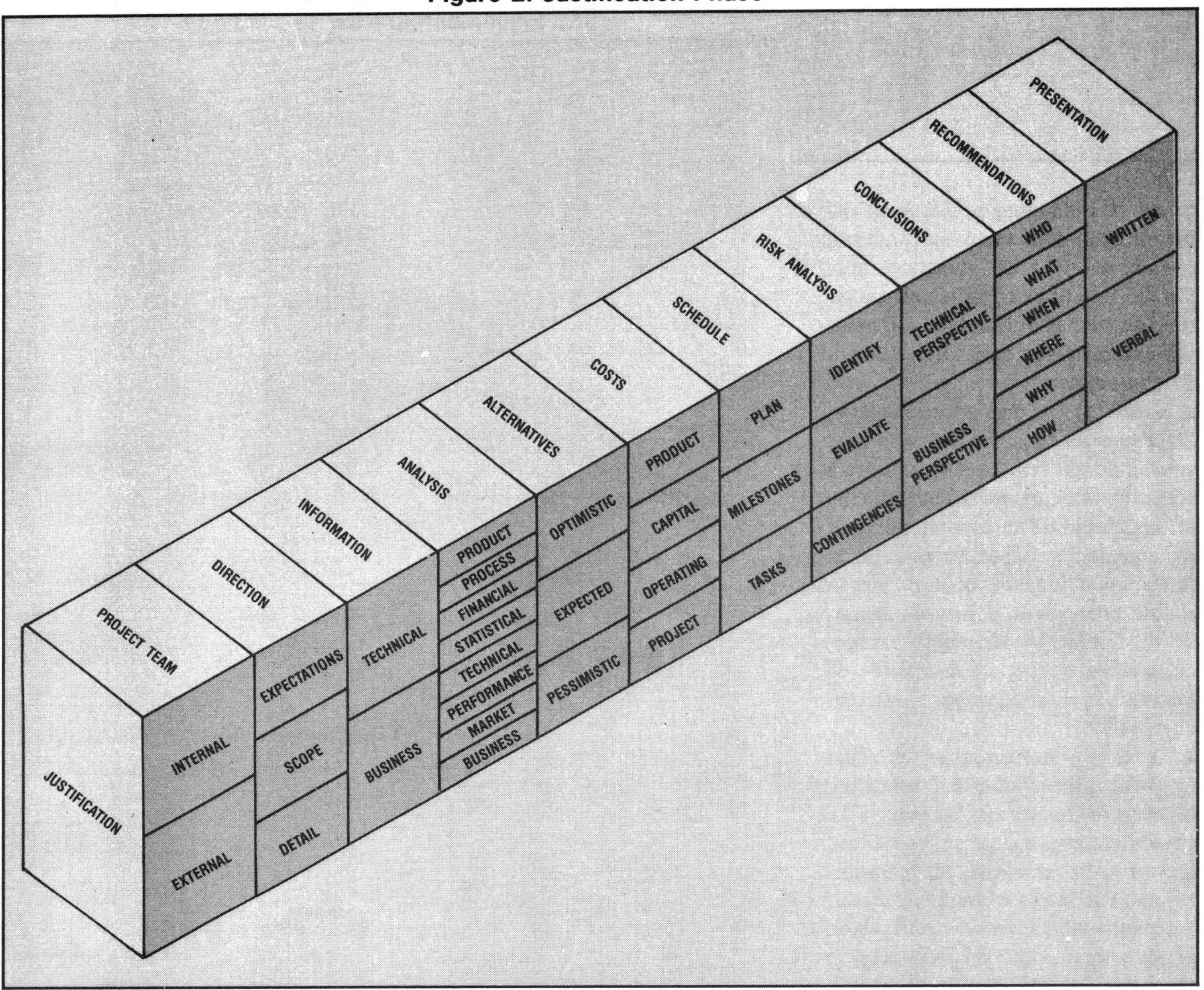

which must account for the interactive effects of cost, quality, reliability, flexibility, demand, competition, governmental regulations, obsolescence, personnel, product redesign, process change, other capital and strategic projects, lead time, etc., will be correct. Therefore, it is usually necessary to construct two or more scenarios based on either high probability events (product cost will decrease by some percentage, for example) or high impact situations (no reaction from competitors versus quick and decisive reaction).

Developing scenarios is important from an analysis perspective because optimizing each individual factor may not optimize the whole. Scenarios also allow the gray areas in an analysis to be simulated, given the expected probability of occurrence of certain events or reasonable ranges for cases in which absolute values cannot be identified with any certainty. Ranges can be a reflection of statistically calculated confidence intervals or merely an admission that some leeway is required in the analysis.

To the extent that results do not change substantially throughout a range of values, the results can be accepted as fairly stable. If, however, there is any kind of significant change, either the need for a more in-depth analysis is indicated or the final decision becomes that much more subjective.

- *Costs*—The scope of the technology analysis will determine the detail with which costs are developed. It will also identify which areas of cost are important for decision-making and when in the project these costs will be incurred. These costs will typically fall into the categories of product cost (material, labor and overhead), capital cost (equipment and related engineering), operating cost (space, utilities, perishables, etc.) and project cost.

Internal company costs can usually be calculated to a fair degree of accuracy for material and labor once the impact of the new technology has been determined. Calculating overhead is not as straightforward, and companies should resist the temptation to simply use some percentage of direct labor or sales. Don't burden a project with overhead from other areas of the company unless a corresponding reduction is made in these other areas (and the savings are documented as part of the justification for this project). Once *real* overhead costs have been determined, any method can be used to allocate them to individual units.

Vendor prices for capital equipment will generally be quoted as an order of magnitude (about $1 million), at a certain level of budgetary accuracy ($1 million plus or minus 15%), or as firm priced ($1,112,386 . . . no one ever quotes round amounts) for a given time period (commonly 30-60 days). The accuracy will be a function of the amount of detail provided and the level of detail requested.

Prices should include hardware, software, vendor test and run-off, insurance, shipping, installation, spares, test and debug, training and education, documentation, taxes, warranty, tooling and fixtures. Some of these won't be included in the vendor's purchase price, but must be accounted for nevertheless.

One of the inputs required to calculate overhead is the operating cost. The complexity of this calculation is directly related to the size and extent of the technology project. Cost elements which may have to be taken into consideration can include utilities, perishable tooling or materials, additional technical or support personnel, maintenance supplies, floor space, ongoing training and education, and maintenance contracts.

Any costs associated with evaluating and justifying the technology, with identifying equipment and vendors, or with managing the project fall into a category of general project costs. These may include site costs, installation, costs associated with production slowdowns or stoppages during installation, engineering, integration, simulation, project management, and intangibles such as contingency and risk.

Costs don't have to be exact to justify new technology. By the same token, management should expect and get costs which are reasonably accurate, fairly complete and verifiable. It is incumbent on the project team to identify all of the sources, and the various levels of accuracy, even if a cost is estimated. Don't give management a reason for killing the project due to questionable development of the costs. Backup is no guarantee that the cost is correct, but at least it provides a basis for discussion and further analysis.

- *Schedule*—Once the various tasks associated with costing the technology project have been completed, a schedule and action plan can be established. Information should include the major phases involved with the project (general design/detail design/installation, debug and ramp-up/full production or use) as well as breakouts of important or critical path tasks within each phase.

Different versions of the schedule might show optimistic, expected or pessimistic outcomes. The graphical layout of the schedule should be accompanied by a written implementation plan which describes in the level of detail required how the technology will be specified, purchased, installed, integrated and used.

- *Risk analysis*—Simply obtaining the new technology is no guarantee that it will perform as expected, and all the planning in the world will not guarantee that the project will proceed on schedule and within cost. A complete justification must include an evaluation of all areas of risk associated with the project according to the probability of occurrence and the magnitude of effect on the project. It should also identify the various factors that may either increase or decrease the risk, and establish a strategy for monitoring and controlling these factors.

Areas to evaluate when performing a risk analysis may include cost and vendor performance, product redesign, manufacturing, financial performance, reliability, personnel, amount of integration required, obsolescence, environmental considerations, productivity and throughput, customer satisfaction and acceptance, and outright failure of the technology to achieve expected results.

A comprehensive risk analysis will prove to management that the project team understands the ramifications to the company of proceeding with the technology procurement. It is also a necessary prerequisite for evaluating and developing contingencies to account for deviations from plan.

• *Conclusions*—When the project team reaches this point, some conclusions must be reached about the technology and project viability. This could include such factors as whether the technology is cost justifiable from a financial accounting perspective (ROI, cash flow, present value, payback, etc.), whether it will meet performance and quality requirements, whether it is the right technical solution and, most importantly, whether it is the right *business* solution.

• *Recommendations*—The next to last task is to enumerate all of the various recommendations identified by the project team as integral to the successful evaluation and implementation of the project. Each recommendation should be stated in enough detail that some specific action can be taken. Avoid recommendations which cannot be quantified, such as "become more competitive."

Structure the list by department, by product or product line, by process, by project phase, by person, or in any other manner which provides top management with a clear and concise plan of action. Then define how to implement the recommendation, when to do it in relation to everything else, who to assign responsibility for it, what it will cost, and what the benefits will be.

• *Sell management on the results*—The last step of the justification analysis is to package the work that was performed and results achieved in a professional, comprehensive and interesting format. Use plenty of figures, tables, charts and graphs (color is a plus). Provide a description of the actual project methodology which was followed and include any source data, calculations, simulations or analyses.

Then, don't just hand a study document to management and expect it to be self-explanatory . . . it won't be. Put together a presentation, sit down in a conference room, discuss the results and *sell* them on the technology.

Stress the fact that using the technology makes good, sound business sense for the company, and get the decision-makers to agree with the validity of the results, even if this means adjourning the meeting to allow time to modify portions of the analysis. Once the study results have been accepted and agreed to, it becomes a bad business decision not to go ahead with the project.

Figure 2 provides a summary of the project approach just described. The key is to make sure the justification analysis is thorough and believable and then to sell the project to management only when they are ready to buy it. A certain comfort level with the project must be achieved first, a level which is directly correlated to the amount of resources required and risk to the firm. It doesn't help to try to force quick acceptance, because management knows it is easier to change a no to a yes than vice versa.

There is no equation which can be used to calculate one magic number that will tell management whether a new technology is viable or not. If there were, management involvement wouldn't even be required in the decision.

What *is* required is an analysis that provides management with all of the elements that are needed to make an informed business decision, knowing all of the expected benefits and related risks. The greater the impact of new technology on a company, or the greater the effect of hard-to-quantify intangibles, as illustrated in Figure 3, the greater is the need for an approach which combines traditional financial criteria with strategic, marketing, process, product and technology analyses.

Be cautious of justification methods which are based on converting all of the decision elements to equation

Figure 3: Justification Curve

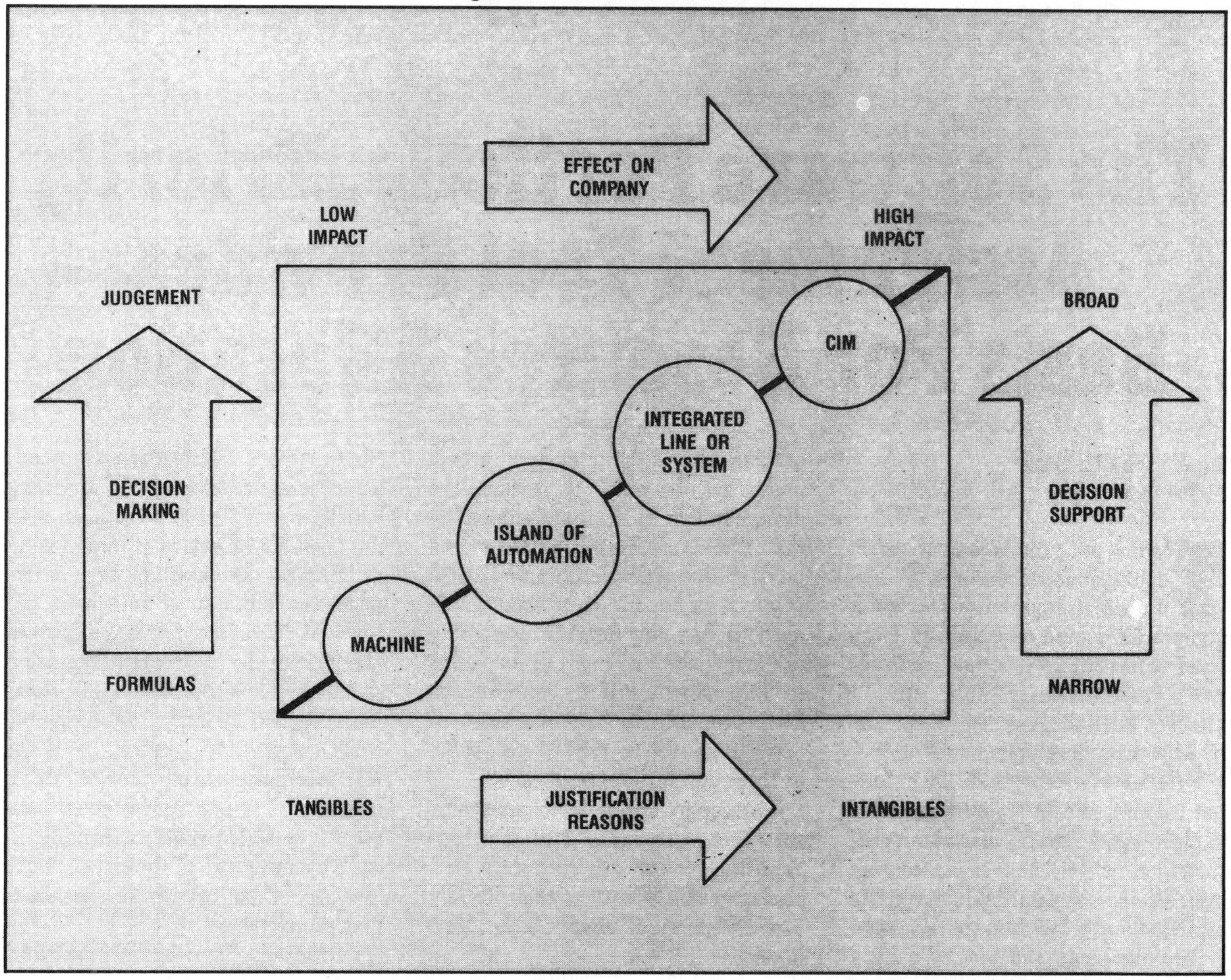

EFFECT ON COMPANY

LOW IMPACT

HIGH IMPACT

JUDGEMENT

DECISION MAKING

FORMULAS

MACHINE

ISLAND OF AUTOMATION

INTEGRATED LINE OR SYSTEM

CIM

BROAD

DECISION SUPPORT

NARROW

TANGIBLES

JUSTIFICATION REASONS

INTANGIBLES

form. After all, wasn't it an equation which proved scientifically that bumblebees can't fly?

For further reading:

Clark, J.T., "Selling Top Management—Understanding the Financial Impact of Manufacturing Systems," APICS Conference Proceedings, 1982, pp. 265-272.

"Economic Justification—How to Cut the Risks," *Modern Materials Handling,* Cahners Publishing Company, July, 1985, pp. 57-59.

Gould, L., "Why Implement CIM," *CIM Strategies,* Cahners Publishing Company, August, 1985, pp. 9-10.

Kerr, J., "The CIM Sell: A Boardroom Decision," *Electronic Business,* November 15, 1985, pg. 113.

"Making The Most of Your Automation Dollar," *Datamation,* Technical Publishing, October 15, 1985, pp. 80C-D.

"Performance Measurement and Strategic Investment," *Outlook,* Harbor Research, September, 1985.

Ramalingam, P., "Selecting The Best Capital Investment Proposal—A Practical Approach," *Production and Inventory Management,* Second Quarter 1983, pp. 101-116.

Paul Bernard is a senior systems engineer with Litton IST (Integrated Systems Technology), a division of Litton Industries Inc., where he is responsible for identifying, evaluating and solving manufacturing-related problems which require a total business perspective. He obtained a BSME from Clarkson College and an MBA from Canisius College and is certified at the fellow level by the American Production and Inventory Control Society.

Justification techniques for advanced manufacturing technologies

JACK R. MEREDITH* and NALLAN C. SURESH**

A major problem in the adoption of advanced manufacturing systems for the automated factories of tomorrow is the prerequisite justification process. Many worthwhile projects have been turned down because the qualitative benefits could not be included in the justification procedure while the direct cost savings were insufficient to meet the financial hurdles set by the firm. This paper identifies and demonstrates the range of techniques that have been used by firms to justify automation investments and describes the conditions under which it is most appropriate to employ them.

Introduction

It is now well recognized (see, for example, Kaplan 1984 or Rosenthal 1984) that the major roadblocks to automating our factories are not engineering shortcomings in the equipment or manufacturing processes but rather managerial attitudes and policies. Foremost among such roadblocks is "the justification problem," as identified by Curtin (1984), Michael and Millen (1984), and others.

The basic problem is that many of the advantages of these new manufacturing technologies lie not in the area of cost reduction but rather in more nebulous, "strategic" areas such as shorter lead times, simpler scheduling, and more consistent quality. Yet, since these manufacturing systems are largely equipment based, and manufacturing equipment has historically been justified on the basis of cost reduction or capacity expansion (for example, see Grud 1984, McDonald and Hastings 1983, Meyer 1982, Muir 1984), these systems are typically expected to be justified on these same measures. In some cases that we identify later, this expectation is reasonable but in others it is not.

Our aim here is to identify those situations where economic justification policies are suitable and those where other justification procedures are more appropriate, according to a conceptual scheme developed (Meredith and Hill 1985) by matching the range of justification procedures observed with the intended use of the technology. In the process, we describe some new justification approaches that have been used by firms and give examples to illustrate their utility and methodology. First, however, we describe the automation technologies to which we are referring.

Revision received October 1985.
* Department of QA & IS, University of Cincinnati, Cincinnati, Ohio 45221-0130, U.S.A.
**Department of Management Science and Systems, State University of New York at Buffalo, NY 14260, U.S.A.

Advanced manufacturing technologies

As described in Meredith and Hill (1985), new manufacturing technologies can be considered to span a continuum (see Fig. 1) in terms of level of integration from stand-alone equipment to full computer-integrated manufacturing (CIM). Robots and numerically controlled (NC) machine tools are often in the category, although they can obviously be computer integrated into other systems and equipment also, such material handling systems of manufacturing cells. The purpose of such equipment acquisitions is often to replace worn out or obsolete existing equipment.

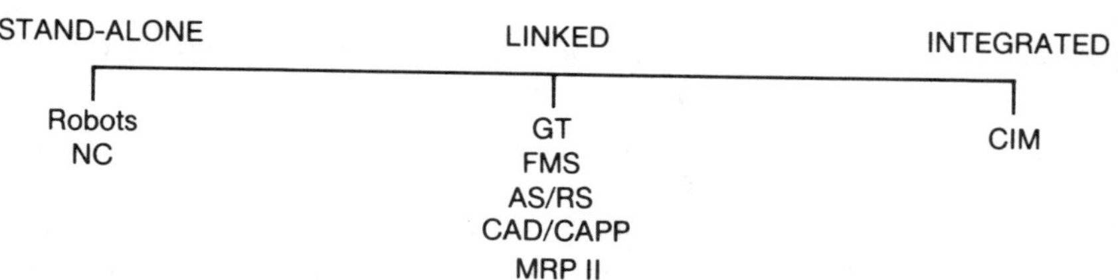

Figure 1. Advanced manufacturing technology continum.

When the stand-alone systems are linked together into cells, such as in group technology (GT) lines or flexible manufacturing systems (FMS), or more loosely, such as computer aided design (CAD) with computer aided process planning (CAPP), then an intermediate level of integration is achieved that exhibits a synergy between the independent systems. Other examples of such linked islands are automated storage/retrieval systems (AS/RS) with automated guided vehicle systems (AGVS) and manufacturing resource planning (MRP II) where the individual computer information systems are linked together.

When the design, planning, materials handling, manufacturing, and support systems (e.g. order entry, cost accounting, purchasing) are all linked together through computer control the factory is considered to be fully integrated, commonly known as CIM.

Two characteristics of all these advanced manufacturing technologies make their justification process more complex than such equipment has required in the past. First, these technologies are much more flexible, in most cases reprogrammable, than equipment has ever been before. As Gold (1982) points out, this flexibility maintains the value of the equipment over the long run, rather than letting its value depreciate. Fotsch (1983) reinforces this argument with the observation that companies are buying such equipment now because they believe they won't be spending for more equipment later. But the advantages of this flexibility are not easily captured in simple economic justification procedures.

The second characteristic of these new technologies that requires special consideration in the justification process was referred to earlier, their synergy when linked together. Users consistently report qualitative benefits from such linked systems, such as faster response to customer requests, that are deemed far more important than the normal cost savings. As Meredith (1985) has shown, when such synergy is properly accounted for in the economic justification formulae, a significant increase in the calculated return on investment can be demonstrated.

Such outstanding benefits are not attained without risk, however, and the risk involved in the acquisition of these enormously expensive systems is substantial. The risk is not only financial, but organizational as well since the entire company infrastructure (see Meredith 1986) must often be changed to obtain the benefits these systems offer. Consistent quality of input materials, new costing and payroll systems, and altered managerial structures are only a few of the many changes in the core fabric of the firm that are commonly required. The result is that the risks, as well as the benefits, are also inadequately considered in the economic justification procedures.

Three justification categories

Corresponding to the three categories of new manufacturing technologies in Fig. 1, three separate approaches to the justification issue seem to exist. For stand-alone systems where the purpose is the straightforward replacement of old equipment, even if some economic benefits not usually considered (such as inventory or space reductions) are obtained, the standard economic justification approaches can be used with an allowance for the additional economic benefits or costs.

When synergy, flexibility, risk, and non-economic benefits are expected, as with the linked systems, more analytical procedures are needed. In some cases, subjective estimates of probability distributions, or at least point estimates, are obtainable and can be included in the analysis.

Last, with systems approaching full integration, clear competitive advantages and major increments toward the firm's business objectives are usually being obtained. Here, strategic approaches are needed that take these benefits into consideration, although tactical and economic benefits may be accruing as well.

Each of these three justification categories spans a number of approaches, as illustrated in Fig. 2. In the remainder of this paper we will describe these approaches and discuss their pros and cons.

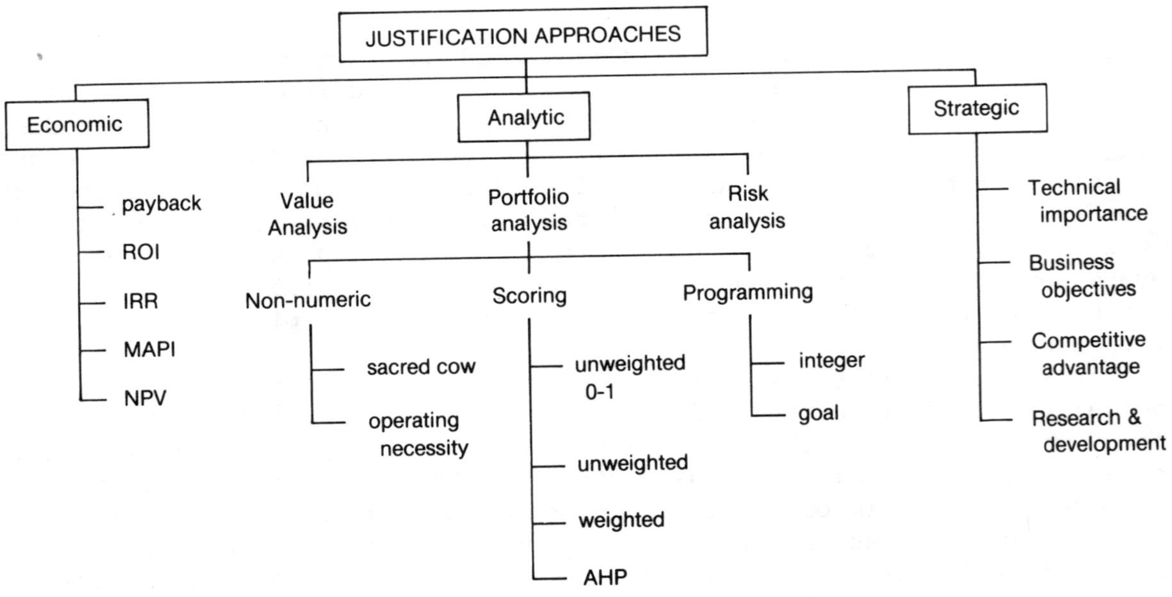

Figure 2. Classification of justification methodologies.

Economic justification approaches

There exist a number of formulae and approaches that firms use for the economic justification of equipment. Examples include breakeven analysis, MAPI (Machinery and Allied Products Institute) method, incremental rate of return, accounting rate of return, net present value, ROI (return on investment) and payback (or payout). All of these methods are well documented in the literature (e.g. Tombari 1978).

Fotsch (1983) has determined, however, that the payback and ROI methods are used by the overwhelming majority (91%) of firms so only these two approaches will be briefly described here and may be considered typical of the economic justification approach.

The data to illustrate the use of the payback and ROI methods are as follows. We consider a potential capital investment with an initial cost of $80,000. The annual operating cost is $5000 per year throughout its estimated life of five years. The annual expected benefits, costs, and resulting cash flows are given in Table 1.

Year	Benefits($)	Cost($)	Net cash flow($)
0	0	80000	(80000)
1	40000	5000	35000
2	40000	5000	35000
3	30000	5000	25000
4	15000	5000	10000
5	5000	5000	0

Table 1. Economic analysis example data.

Payback period

Adding the net cash flows until we reach $80,000, we obtain

$$35000+35000+x(25000)=80000$$

$$\text{or, } x=10000/25000=0.4$$

so the payback period includes the first year, the second year, and the proportion x=0.4 of the third year, for a total of 2.4 years.

Return on investment

ROI calculations are sometimes made with depreciation being netted out and sometimes not, we will show both methods. Based on five years, the annual depreciation will be 80000/5=16000. The average annual benefit of the investment is

$$(35000+35000+25000+10000+0)/5=21000$$

Using depreciation, the ROI is thus (21000-16000)/80000=6.25%. Without depreciation, the ROI is 21000/80000=26.25%.

The advantages of the economic approaches are their simplicity, clarity, apparent "bottom-line" impact, and ease of data collection. But their disadvantages include their inability to capture non-economic and strategic benefits. The primary disadvantage, however, is their use of a single value for decision making--a complex decision simply cannot be reduced to a single number and still contain the essential information needed for the decision. To do so gives the impression that the problem itself is simple, but gaining the complex benefits of advanced manufacturing technologies is reported by Gold (1982), Rosenthal (1984) and others, to involve serious repercussions throughout the organization.

Analytic justification approaches

The analytic techniques described in this section are again largely quantitative but more complex than the economic approaches. Also, they tend to capture more information and frequently consider uncertainty and multiple measures and effects. Their advantage is that they are more realistic, taking more factors and subjective judgments into account, and hence better reflect reality as understood by knowledgeable managers. Their disadvantages are that more data are required and the analysis is considerably more complex and time-consuming, though the use of a computer can minimize this difficulty. (Comments about the data requirements for all the justification methods are made in a later section.) Nevertheless, the complexity of these methods is a deterrent to their use, both because of execution as well as understanding.

Three major approaches are typical of this category and we describe each in turn.

<u>Value analysis</u>

This justification approach, described by Keen (1981), is used by some managers when assessing technical innovations. It consists of a two-stage process as depicted in Fig. 3. In the pilot stage the project is treated as an investment in research and development (R & D) rather than as a capital investment (described in the strategic justification section). Here, <u>value</u> to the firm is considered first and then expected cost is determined to see if it is acceptable. In the second "build" stage, assuming the pilot stage was successful, the expected <u>cost</u> is considered first and then the expected benefits are evaluated for acceptability. These two stages are further described and illustrated below.

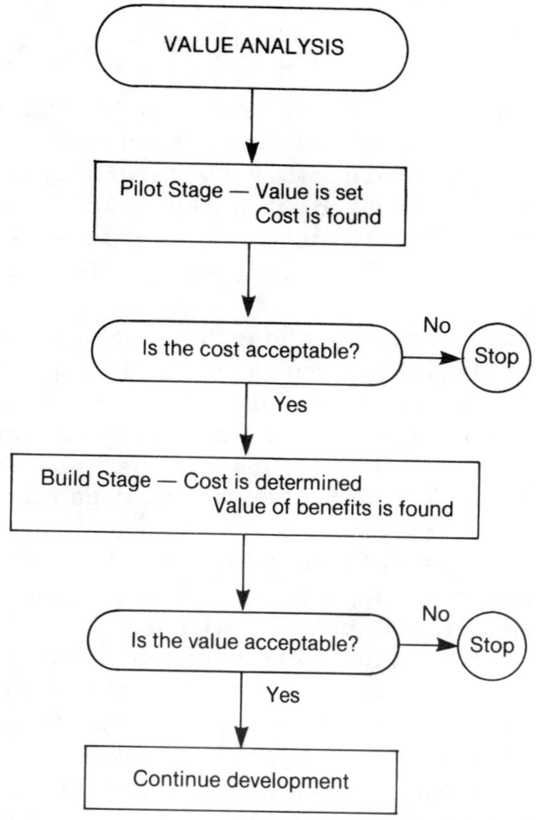

Figure 3. The value analysis process.

(1) Pilot stage

The decision to proceed with the pilot is based on an assessment of the expected, but not necessarily quantified, benefits. The pilot involves a small scale system, complete in itself but limited in its functional capability, to assess only a few of the expected benefits. The cost of the pilot is kept very low, for example, less than $20,000, to keep it in the R&D realm. When the pilot stage is finished, the benefits are evaluated to verify their utility to the firm.

One example of this would be the development of a part classification and coding system with only a few digits (and thus limited ability) to determine the reduction in design, drafting, and process planning time. The reduction in lead time to delivery of new products, or decreased response time to customer inquiries might prove highly significant. Or perhaps the reduced hassle in design engineering, or smoother flows between design and process engineering, might be more valuable than expected.

Another example would be the use of a simple robot for limited production tasks such as spraying or material handling. The improvement in morale, or decrease in accidents, or reduction of pollution in the plant could be more important than had been surmised. Or perhaps less important.

And another example would be the installation of a simple manufacturing resource planning (MRPII) system, say on a microcomputer, that could coordinate production operations but had limited capabilities in terms of generating purchase orders and so on. Reduced scrap, better planning, smoother product flows, and so on might prove to be extremely valuable to the firm.

(2) Build stage

If the pilot stage was successful, that is, if the expected benefits were realized, then the full development is considered next. At this stage the costs for the full system are evaluated very carefully and the expected value of the benefits obtained are compared with the cost to determine if they are justified. Clearly, various levels of costs can be evaluated to determine value thresholds of the extended benefits, or "bells and whistles" as they are sometimes called.

To continue the examples above, the coding system would consider additional digits to further detail the parts and allow additional benefits in design, engineering, drafting, process planning, and possibly manufacturing also. More capable and expensive robots would be considered for more complex production tasks. And a larger MRP II system, perhaps for the mainframe, would be considered with significantly expanded capabilities for supporting other functions and departments.

Note that the pilot stage defined above was "broad" rather than "deep" and not the typical incremental approach used in engineering. That is, its functional capability was limited but not its breadth of application. The other way of handling a pilot stage, which can also be useful, is to reverse the process. For example, conduct the classification and coding for only a limited number of the firm's products, say, all turned parts. But the full set of digits would be used for these parts. And in the third example situation above, the firm would implement only one portion of the MRP II system, for example, the routings package. But it would be used on the mainframe and for all products made in the shop.

Either of these methods can help the firm approach factory automation with more understanding and less risk. The essence of value analysis is to separate the cost and the value derived to let managers intuitively ascertain if the value of the benefits obtained is worth their cost. It also provides an incremental approach to the automation process that lets managers control the cost and thus not let the risk to the firm get out of hand.

Portfolio analysis

The scenario for portfolio analysis consists of a number of projects competing for limited capital funding. The task is to choose the best set of projects for implementation. The selection is made by creating a portfolio of projects that either maximizes value to the firm subject to capital investment or risk limitations, or minimizes risk or capital subject to attainment of a certain level of value. The value can be based on any number of factors such as return on investment, reduction in scrap, improvement in quality, and so on.

Three general types of portfolio models exist: non-numeric models, scoring models, and programming models. All three are discussed and illustrated in Meredith and Mantel (1985). We will describe each of the types in turn.

(1) Non-numeric models

These models, as their title implies, are justified on bases other than the usual trade-offs between costs and benefits. They typically exhibit an overwhelming characteristic that deems them exceptions to the general justification process.

The "sacred cow"--this project is one that has been suggested by a very senior and powerful official in the organization. It is "sacred" in the sense that the project will be maintained until successfully concluded or until the official personally recognizes the idea as a failure and terminates it. Virtually any of the automation technologies may become a sacred cow in some executive's eyes: MRP II, group technology, robotics.

The operating necessity--in this case, the project is required in order to keep the production system operating, or even for the very survival of the firm. The only question is whether the system is worth saving at the cost of the project. If so, the costs will be controlled to be kept as low as possible but the project will be authorized. Machine replacement and maintenance projects often fall in this category, particularly if they have been overloaded for extended periods or run with minimal maintenance.

(2) Scoring models

As with non-numeric models, there are a number of versions of scoring models, many of which are included in Dean (1968). We will illustrate one and describe the remaining major types in terms of it.

The simplest scoring model is called the "unweighted 0-1 factor model." A set of relevant factors is selected and one or more rates score each project on each factor depending on whether or not it qualifies for that criterion. The total number of qualifying factors are summed for each project and this then serves as the ranks for the projects. The advantage of such a process is its simplicity and consideration of a number of relevant criteria. The disadvantages are that it assumes all criteria are of equal importance and it allows for no gradation of the degree to which a specific project meets the various criteria. An example is shown in Table 2.

If a simple linear measure is used for the degree to which a project meets a criterion, say on a five-point scale, then an "unweighted factor scoring model" is obtained with, again, each factor being weighted equally. The x's in the table would thus be replaced by numbers ranging from 1 to 5, where 5 is perhaps "very good" and 1 is "very bad." Summing the values would again give a ranking of the projects.

	Project Number			
Factor	1	2	3	4
ROI 20%	X		X	X
Compatibility		X	X	
Volume flexibility	X	X		X
Product flexibility			X	X
Quality improvement		X	X	
Total score	2	3	4	3

Table 2. Unweighted 0-1 factor model.

Extending the idea one more step by putting weights on each of the factors to indicate their importance to the firm results in the "weighted factor scoring model," perhaps the most common of all scoring models. Usually the weights are normalized so they represent the proportion of total weight accorded to each of the factors. The score for each project is found by multiplying the weight for each factor times the project's score on that factor and adding all the results together:

$$\text{Total}=w_1 s_1+w_2 s_2+ \ldots$$

where w_i is the weight for factor i and s_i is the score on factor i.

Saaty (1980) has developed a weighted factor scoring model called the "analytical hierarchy process" (AHP) that corrects for the often-found inconsistency in human judgment when evaluating projects on various factors. The AHP procedure allows the rater to compare factors against each other to ascertain their weights, and similarly, to compare projects against each other on individual factors to ascertain their scores. It is believed that more accurate estimates of the rankings can be obtained by this pairwise comparison method.

(3) Programming models

A number of programming models can be framed, based on the equation given above for the weighted factor scoring model. For example, using integer programming each project can be represented by a 0-1 variable and those projects selected that maximize the set of project total weighted scores, subject to constraints on various resources such as capital facilities. For the formulation of this problem refer to Meredith and Mantel (1985).

Similarly goal programming formulations that consider the various factors as goals to be attained, subject to resource constraints, are possible. Again, weights are used on the goal deviations to give importance to each of the factors. Refer to Ignizio (1976) for examples.

Risk analysis

The approach of risk analysis (see, for example, Hertz 1964 or Turban and Meredith 1985) is to simulate the projects under consideration to determine the variables of interest--benefits, costs, yields, capacity, and so on--and describe the outcomes statistically or graphically. Cumulative distribution functions are determined for each variable of interest showing the likelihood of achieving a certain profit, capacity, return on investment, and so on. Various automation projects can thus be simulated beforehand and the results compared. Using the concepts of stochastic dominance as described by Whitmore and Findlay (1978), inferior policies or projects can be eliminated and only the most promising implemented (e.g., see Suresh and Meredith 1985). We illustrate the approach with a simple machine tool investment example below.

There are two potential investment alternatives to replace an existing manufacturing system that is old and overworked. The capital costs, hourly operating costs, and annual maintenance costs are given in Table 3. There are four products (A, B, C, and D) and several minor parts (P) currently being made on this equipment. The typical operation times required for each of the products and parts are given in Table 4 and the anticipated sales volume and price over the next eight years are listed in Tables 5 and 6, respectively. The sales volume estimates have an accuracy of plus or minus 250 units and the price accuracies are as shown.

	Present equipment	Alternatives 1	2
Capital cost	(salvage value, 5000)	40000	32500
Hourly operating cost	1.5	0.2	0.75
Annual maintenance cost	1000	1400	1250

Table 3. Equipment cost elements.

Product	Present equipment	Alternative 1	Alternative 2
A	0.055	0.010	0.020
B	0.055	0.008	0.015
C	0.055	0.009	0.025
D	0.175	0.050	0.070
P	0.050	0.005	0.009

Table 4. Operation times.

	Year							
Product	1	2	3	4	5	6	7	8
A	1500	1600	1700	1750	1750	1750	1600	1500
B	1600	1000	1200	1200	1200	1300	1200	1200
C	1100	1000	1000	1000	1000	1000	900	800
D	800	800	800	800	700	700	600	600
P	1500	1400	1400	1300	1200	1200	1100	1000

Table 5. Anticipated annual sales volumes.

Product	Year 1	2	3	4	5	6	7	8
A	191(2)	191(3)	190(3)	190(3)	188(3)	188(3)	186(3)	185(4)
B	195(2)	194(2)	194(2)	190(4)	198(4)	198(4)	198(4)	196(5)
C	194(2)	194(2)	194(2)	194(2)	194(2)	194(2)	194(2)	194(2)
D	191(2)	191(3)	190(3)	190(5)	188(3)	188(3)	186(3)	185(4)
P	192(1)	192(1)	192(1)	190(3)	189(2)	189(2)	189(2)	188(2)

Table 6. Anticipated annual sales price and accuracies.

Given the uncertainties in sales volumes and prices, a simulation risk analysis to determine factors such as net present value (NPV) and machine utilization of each of the alternatives is warranted. In the simulation model (written in SIMSCRIPT II.5), the sales volume is taken as a uniform random variate between the 250 unit accuracy limits. Prices are determined in the same manner. Following this, machine operating hours, equipment utilization, and then the cash flows over the eight year horizon are computed and the NPV found.

A number of factors may be analyzed in the equipment selection decision process such as sales volume, revenue, idle time, and so on. Here we present, as examples, only two. In Fig. 4 the cumulative probability distribution functions of the NPVs of the two alternatives are graphed and in Fig. 5 the same is done for the equipment utilization.

As can be seen in Fig. 4, the NPV distribution for alternative 1 falls to the left of that for alternative 2. Also, there is a considerable area of overlap between the two distributions, from 10200 to 12400. It is clear from the figure that alternative 2 is preferred. That is, the cumulative probability of achieving less than any particular NPV is always greater for alternative 1 than alternative 2. Thus, alternative 2 dominates alternative 1 by "first order stochastic dominance" (Whitmore and Findly, 1978).

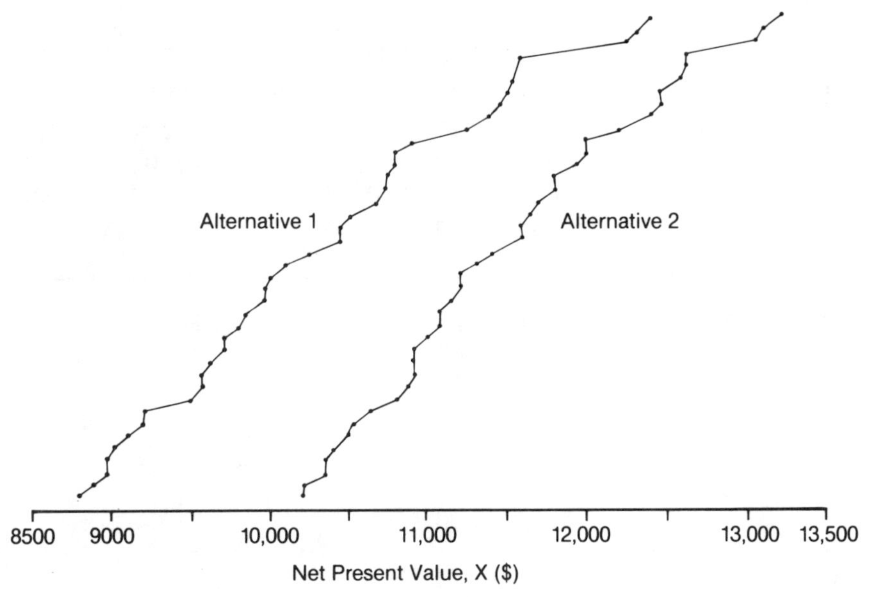

Figure 4. NPV cumulative distribution functions.

However, equipment selection should not be viewed as a single criterion decision and other factors should also be considered. For example, Fig. 5 shows that the distribution of machine utilization is also less for alternative 1 than alternative 2, thereby offering extra capacity for possible future market growth. Clearly, other such factors may also be considered critical in any particular technology decision and their distributions should also be factored into the analysis.

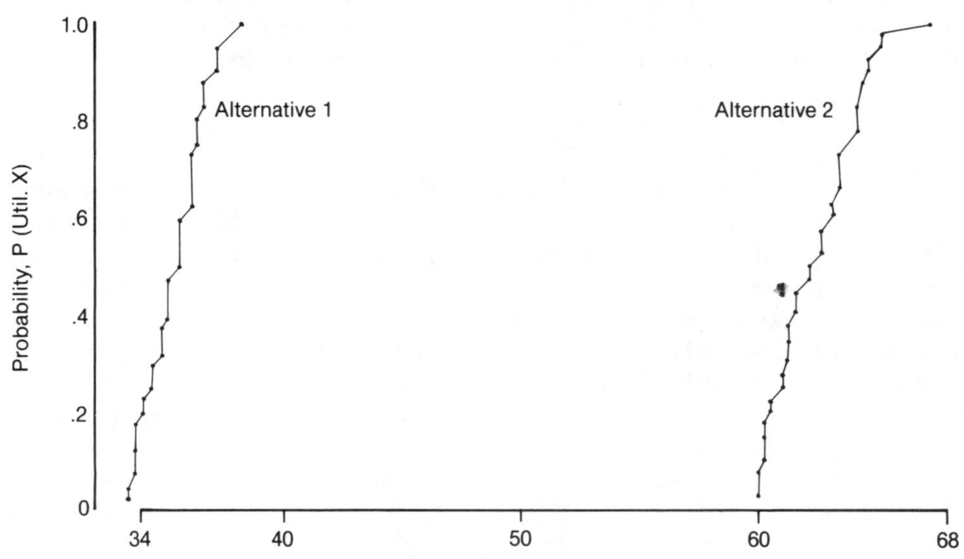

Figure 5. Utilization cumulative distribution functions.

Strategic justification approaches

The strategic approaches tend to be less technical than the two previous categories, though they are frequently used in combination with them. The advantage of the strategic approaches is their direct tie to the goals of the firm. A disadvantage is the possibility of overlooking the economic and tactical impacts of the projects, myopically focusing entirely on the strategic impacts. Economic justification calculations will commonly be made in combination with strategic considerations, but analytic evaluations are rarely included (usually due to their time and trouble). However, if a strategic approach is used, the economic and analytic implications should also be checked, simply for a clear understanding of all the impacts of the project.

Four main approaches are commonly used at this level.

Technical importance

From a strategic viewpoint, a desired end cannot be attained unless this project is undertaken first. That is, justification under the concept of technical importance implies that the project is a prerequisite for an important follow-on activity. Its return may be negligible, or even disadvantageous, but later, more desirable work cannot be attempted without implementing this activity first. It is common for activities such as these to be grouped with the desired follow-on project in a "package" that is approved en masse by the approval board.

Many examples of such activities exist among advanced manufacturing technologies. Firms planning to use cellular manufacturing usually find it necessary to conduct a part-family classification and coding analysis first, though the analysis itself may appear to have no value to them. And it is commonly stated that inventory and bill of material records must be 95% accurate before implementing material requirements planning (MRP) systems. And finally, it is often stated as a truism, though it is not necessarily true at all, that a firm must quickly start somewhere in the factory automation process in order to get onto the automation learning curve before the competition gets so far ahead of the firm it can never catch up.

Business objectives

Justification of a project because it directly achieves the firm's business objectives is a clearly strategic approach. "Key indicators" or surrogate measures of this achievement are often used to verify the attainment of these objectives, and to measure when the firm is losing control in these areas and needs to intensify its efforts.

Examples in automation abound. For instance, just by employing the most sophisticated computerized production processes, firms give the impression to customers, vendors, visitors, and the media that they are progressive, advanced, and an up and coming organization. (Of course, this may not be true at all.) As another example, automation may allow the consistent attainment of uniform product quality, a top business objective in many firms these days. Or again, automation may drastically reduce lead times and allow significantly better customer service, perhaps a strategic business objective of a company that is fighting foreign price competition.

Competitive advantage

In the competitive advantage justification approach an opportunity may exist for the firm to gain a significant advantage over its competitors by implementing this project. The advantage may not have been one of the strategic business objectives of the firm but it is too important for the company to pass up. The opportunity may have arisen from a unique set of circumstances or may be an outgrowth of a slight competitive advantage the firm already holds.

This situation occurs frequently in all areas of technology. A firm may hold a crucial patent that allows it to build on an existing base for a significant advantage over its competition. Also, many cases of automation in today's factories are raising opportunities for competitive advantages due to totally unexpected benefits such as reduced space requirements, better processing quality, higher performance capability, shorter design times, and so on.

A subcategory of this approach is "the competitive necessity." Here, the project is mandatory if the firm wishes to remain competitive in a particular market. This is the message so frequently spoken regarding all forms of automation these days and captured in the cliche "automate, emigrate, or evaporate."

Research and development

Treating a project as a R & D investment admits that it may fail but it holds sufficient strategic promise to justify the investment. The point is that one of many such projects will eventually come through and provide returns to the firm to reimburse all the failures. Without risk, nothing is gained.

A clear example of the R & D approach is the first stage of the value analysis approach described earlier, the pilot project. Another example is setting up one group technology line, or one manufacturing cell, to see how well it works, its costs, its problems, and its benefits. Firms often try to use this approach for promising automation ideas, but because of the risk, minimize the resources provided to the pilot project, whereupon it fails and the second stage of full implementation is abandoned. Companies must be careful not to pre-ordain failure by withholding needed resources at the pilot stage.

Comment on Data Requirements

With all these justification techniques, data acquisition can be a major problem since the technologies are so new. One solution is the pilot stage approach to data collection, as described in the section on value analysis. Another solution is to use the data reported in the literature (e.g., Rosenthal 1984; Meyer 1982), though caution must be advised. For example, some articles are relatively unspecific about not only the conditions under which benefits have been attained but even the technology employed that gave the benefits, calling every technology "CIM." In many of the justification approaches, such as risk analysis, goal programming, portfolio analysis, and economic analysis, management's own subjective estimates are required concerning importance ratings, limitations on resources, goals, and probable effects of the automation on existing operations.

Although it may be true that the justification of these new technologies is difficult only because of the lack of data, as was originally also the case with the office automation, numerical control, and even computers, we believe that there is a larger issue here because of the extensive interrelationships required for these technologies. When successfully used for strategic purposes, they may well alter the entire infrastructure of the organization, and even the organizational structure itself. Such extensive impacts have rarely been experienced with past technologies, thus the difficulty in assessing their benefits and justifying their use.

Conclusion

Economic justification techniques have historically been used to gain approval for capital equipment expenditures. However, this approach cannot cope with the nature of the benefits offered, such as flexibility and synergy, and the risks inherent in today's advanced manufacturing technologies. Faced with this inadequacy, some firms have developed new justification approaches, the range of which has largely been identified here. The approaches were categorized as

(1) economic--appropriate primarily for stand-alone replacement equipment with strictly economic benefits.
(2) analytic--appropriate for systems with both economic and non-economic benefits and risks, particularly if the probability distributions can be subjectively estimated, and
(3) strategic--appropriate for systems that contribute directly to the firm's business objectives.

When higher level (more strategic) approaches are employed, the lower level methods should also be used to reveal the full impacts of the decision. For example, a large machine tool manufacturer justified an FMS investment with a strategic level approach (business objectives) but nevertheless also conducted a risk and economic analysis of the investment to better understand its expected results. Obviously, stand-alone investments for economic reasons that have limited local impact need not be augmented with higher level justification analyses, since there are not expected to be any higher level impacts.

By recognizing the nature and purpose of the manufacturing systems they are contemplating and then using appropriate justification techniques, we hope that firms may be better able to justify the new manufacturing systems available to them today. By doing so they may then be able to avoid the pitfall of failing to "economically justify" new manufacturing systems that might well determine whether they will become a competitive force in the market or disappear from it.

Un des problèmes majeurs dans l'adoption des systèmes avancés de fabrication pour les usines automatisées de demain est le procédé de justification préalablement nécessaire. Beaucoup de projects méritant d'être retenus one été abandonnés parce que les avantages qualitatifs ne pouvaient pas être inclus dans la procédure de justification tandis que les économies effectuées sur le coût direct étaient insuffisantes pour répondre aux obstacles financiers établis par l'entreprise. Cet article identifie et démontre la gamme des techniques qui ont été utilisées par les entreprises pour justifier les investissements nécessaires à l'automatisation et décrit les conditions dans lesquelles leur utilisation s'avère la plus appropriée.

Ein Hauptproblem bei der Einführung fortgeschrittener Fertigungsverfahren für die automatisierten Fabriken von morgen ist der dazu erforderliche Wirtschaftlichkeitsnachweis. Viele lohnende Vorhaben wurden abgelehnt, weil ihre qualitativen Vorteile nicht in den Wirtschaftlichkeitsnachweis einbezogen werden konnten, während die Höhe der direkten Kosteneinsparungen nicht genügte, um die von der Firma errichteten finanziellen Hürden zu nehmen. In dieser Abhandlung wird eine Reihe von Techniken identifiziert und vorgeführt, die von Firmen zur Rechtfertigung von Investitionen in Automatisierungsvorhaben angewandt wurden, und werden die Bedingungen beschrieben, unter denen ihr Einsatz am zwechmäßigsten ist.

References

Curtin, F.T., 1984, The executive dilemma: how to justify investment in new industrial automation systems. CIMCOM Conference Proceedings, Society of Manufacturing Engineers, Dearborn, Michigan, U.S.A.

Dean, B.V., 1968, Evaluating, Selecting, and Controlling R & D Projects (American Management Association, New York, NY, U.S.A.)

Fotsch, R.J., 1983, Machine tool justification policies: their effect on productivity and profitability. Journal of Manufacturing Systems, 3, No. 2.

Gold, B., 1982, CAM sets new rules for production. Harvard Business Review, November-December.

Grud, J.M., 1984, Manufacturing Systems Change be Justified?, Production and Inventory Management, 2nd Quarter.

Hertz, D.B. 1964, Risk analysis in capital investment. Harvard Business Review, July-August.

Ignizio, J.P., 1976, Goal Programming and Extensions, (Lexington Books)

Kaplan, R.S., 1984, Yesterday's accounting undermines production. Harvard Business Review, July-August.

Keen, P.G.W., 1981, Value analysis: justifying decision support systems. MIS Quarterly, March.

McDonald, J., and Hastings, W.F., 1983, Selecting and justifying CAD/CAM. Assembly Engineering, April.

Meredith, J.R., 1985, The economics of computer integrated manufacturing. In Nazemetz, J.W. et al. Computer Integrated Manufacturing Systems: Selected Readings, Institute of Industrial Engineers. Norcross, Georgia, U.S.A.

Meredith, J.R., and Hill, M.M., 1985, Justifying advanced manufacturing systems. Working Paper, University of Cincinnati, U.S.A.

Meredith, J.R., and Mantel, S.J., 1985, Project Management: A Managerial Approach (New York: John Wiley & Sons).

Meyer, R.J., 1982, A cookbook approach to robotic and automation justification, Proceedings, Robots IV Conference, Society of Manufacturing Engineers, Dearborn, Michigan, U.S.A.

Michael, G.J. and Millen, R.A., 1984, Economic justification of modern computer-based factory automation equipment: a status report, Proceedings of the ORSA/TIMS First Special Interest Conference on FMS, August.

Muir, W.T., 1984, An Alternative for Evaluating CIM Investments--A Case Study, CIMCOM Conference Proceedings Society of Manufacturing Engineers, Dearborn, Michigan, U.S.A.

Rosenthal, S., 1984, A survey of factory automation in the U.S. Operations Management Review, Winter.

Saaty, T.L., 1980, The Analytical Hierarchy Process (New York: McGraw-Hill).

Suresh, N.C., and Meredith, J.R., 1985, Justifying multi-machine systems: An integrated strategic approach, Journal of Manufacturing Systems, November.

Tombari, H.A., 1978, To buy or not to buy? Weighing capital investments, Production Engineering, March.

Turban, E., and Meredith, J.R., 1985, Fundamentals of Management Science, Business Publications, Inc., Plano, Texas, U.S.A.

ACCOUNTING METHODOLOGIES

Basic economic/financial approaches of justification, called "accounting" approaches because of their primary attention to accounting data, are dealt with here. Brief descriptions of the basic methodologies and some pitfalls in the definitions and uses are also noted. Extensive examples are given for three major types of automation, and the quantification of more abstract benefits and costs are described also.

The opening article, by Mayer, illustrates the variety of definitions for the accounting methodologies and the confusing variations that can result. The articles by Tombari and Hackamack and Hackamack illustrate how to apply these methods and the results that are likely to occur with different investments.

The next three articles illustrate the application of the methods to automation projects. Rygh illustrates the justification of an automated storage and retrieval system and comments on some qualitative issues. In a now-classic article, Mayer discusses a comprehensive justification analysis for a complex robotic or automation project. Klahorst describes the Kearney & Trecker justification system as applied to an FMS project and illustrates the quantification of tooling, inspection, and other such costs.

The last two articles address the more abstract benefits and costs of automation. Phillips gives an overview of FMSs and identifies their many advantages, and some hidden costs and problems as well. In the final article, Primrose and Leonard use the concept of a "zero machine" to quantify these types of benefits, and then describe the computer program they constructed that embodies this concept.

Reprinted from ***Industrial Management,*** *March-April 1983.*

Pitfalls in Equipment Replacement Analyses

Raymond R. Mayer

National and international competitive forces are compelling organizations to increase productivity by, among other things, replacing deteriorated and obsolete production facilities. But given the high cost of money, as represented by current interest rates, it is imperative that a correct determination be made of whether the rate of return on the extra investment called for by new equipment is adequate. The purpose of this presentation is to describe the most common errors made in equipment replacement analyses and to point out how such errors can be avoided.

It is necessary that errors in equipment replacement studies be avoided because of the nature of their consequences. In some cases, the result is a decision to retain the currently owned asset, that is, the old equipment when it would be better to replace it; in others, the result is a decision to procure the proposed replacement, that is, the new equipment when it would be better not to do so. With either outcome, the firm will experience excessive costs and lower profits.

Probably the most common mistake in equipment replacement analyses is to use a deficient method of analysis. The other mistakes are attributable to an incorrect application of a sound analytical technique. A prerequisite for an understanding of why a given method is deficient or why a given application is incorrect is a knowledge of what characterizes a sound approach. So let us begin with what, of necessity, must be a brief and general description of such an approach.

A Sound Approach

There is more than one way of determining whether a proposed investment in new equipment is justified. Each of these calls for stating the future cash flows that the firm expects to experience with the alternative of retaining the old equipment and with the alternative of procuring the new equipment. These cash flow paterns can be developed only by describing every alternative in certain terms.

RAYMOND R. MAYER, Ph.D., is Walter F. Mullady Professor of Business, Loyola University of Chicago.

First, the captial investment represented by each of the units must be set forth. For the currently owned asset, this will be its *market value* because a decision to retain the asset is a decision not to convert it into an amount of money equal to its market value. For the proposed replacement, the capital investment will be its first cost, which might comprise a purchase price, delivery charges, and installation expenses.

Second, a service life for each alternative must be selected. Insofar as the old equipment is concerned, this will be the number of years the equipment will be retained if it is not replaced now. The life of the new equipment is the number of years it is likely to be operated if it is obtained. In either case, an attempt should be made to select the *economic* life, which is the life that yields the lowest average annual cost for the asset.

Third, the terminal salvage values must be estimated. These are the market values of the respective facilities at the end of their service lives.

Fourth, the *annual* recurring expenses must be forecast; these will consist of operating costs, administrative expenses, revenue disadvantages, and income taxes if an after-tax analysis is to be performed. Incidentally, the usual case is one in which it can be assumed that the activity involved will be carried out with either the currently owned asset or with its proposed replacement. That is, consideration is not given to the alternative of discontinuing the activity by retiring the old equipment and not replacing it. Under these circumstances, there is a need to consider only those recurring expenses that will be affected by the choice of equipment. The others are irrelevant because they offset each other.

Fifth, it will probably also be necessary to describe *future* replacements in the foregoing terms. This need arises from the fact that equal time periods must be considered when evaluating investment alternatives, and it is unlikely that the remaining service life of the old equipment will be as long as the service life of the new euqipment. Therefore, the cash flows that will be experienced with future replacements must be either estimated or assumed until the same point in time is reached for each alternative. An intuitive explanation of this is that the nature of the new equipment that will be available in the future could be such that, in the longer term, it would be more economical to retain the old equipment for another year or two, even though its costs during that period exceed those of the currently available new equipment.

Finally, the cost of financing the capital investment in each of the alternatives must be ascertained. This cost is referred to as the "cost of money" and will be expressed in terms of a percent per year. It will also represent the company's minimum rate-of-return requirement. To determine its value, management will consider what proportion of the investment in each alternative consists of borrowed money, the interest rate at which money is borrowed, what proportion of the investment in each alternative consists of equity funds, and the opportunity cost which is the rate at which equity funds can be invested elsewhere. Consequently, with ¼ debt financing, ¾ equity financing, a 16 percent interest rate, and a 24 percent opportunity cost, the average annual cost of money would be

$$1/4(16\%) + 3/4(24\%) = 22\%$$

After each of the alternatives has been described in the aforementioned terms, the resultant cash flow patterns must be processed in a way which will serve to identify the most economical course of action.

Processing Cash Flows

As mentioned earlier, there are a number of acceptable methods by means of which the processing of cash flows can be accomplished. We shall now consider the three basic ones.

In the first method, a year-by-year comparison is made of the two cash flow patterns. As a result of this comparison the company will know what extra, or incremental, investment is called for by the new equipment and will learn what the consequences of this extra investment will be. Then, by means which are beyond the scope of this presentation, these consequences can be converted into an equivalent rate of return on the required extra investment. If this return is less than the organization's cost of money, the extra investment in new equipment is not justified, and the old equipment should be retained. If the prospective return is equal to or greater than the cost of money, the old equipment should be replaced. Of course, irreducible, or intangible, factors must also be taken into consideration. Be that as it may, this approach is known as the *discounted cash flow method*.

In the second basic method, known as the *uniform annual cost method*, the two cash flow patterns are treated in a manner which yields the average annual cost of each alternative. If there are no relevant irreducible factors, the equipment which has the lower average annual cost is said to be the more economical alternative. It might be noted that one of the elements of the calculated average annual costs is the firm's cost of money. Therefore, if the new equipment happens to have the higher average annual cost, it can be concluded that the return on the additional capital investment the equipment requires is unsatisfactory. Conversely, if the new equipment proves to have the lower average annual cost, it can be concluded that the return on the extra investment the equipment requires is satisfactory.

The third, and last, basic method is the *present worth method*. That approach calls for calculating what total cost, or single expenditure, at the present time is equivalent to all the cash flows that will be experienced with a given alternative. The resultant total costs, which are referred to as present worths, are then compared, and, in the absence of intangible factors, the equipment which generates the lower total cost is said to be the more economical. As was true of the calculated average annual costs, the calculated total costs include the firm's cost of money. From this, it follows that, if the new equipment has the higher total cost, the return on the extra investment in that equipment is inadequate and that, if the new equipment has the lower total cost, the return on the extra investment in that equipment is adequate.

Having reviewed the characteristics of a correct approach to the evaluation of equipment replacement proposals, we shall find it easier to recognize the deficiencies of what is actually done in many cases.

The Payoff-Period Criterion

It was mentioned that the most common mistake in equipment replacement studies is the use of a faulty method of analysis. Given that there is an almost unlimited number of ways of doing something improperly, any attempt to enumerate the deficiencies of each of those ways usually proves to be a formidable task. Fortunately, that is not the case in this instance because, with a negligible number of exceptions, the faulty method employed is the *payoff-period approach*.

The approach can be easily described. In general, a determination is made of the number of years needed for the new equipment to "pay for itself." This number is obtained by dividing the "required investment" in the new equipment by the "annual saving" that will be experienced with the equipment. The result is then compared with an established payoff-period requirement. In the absence of irreducible factors, the investment in the new asset is said to be justified if the requirement is met; if it is not, the replacement proposal is rejected.

The Required Investment

The ease with which the method can be described suggests that its application is simple and straightforward. Yet, if equipment replacement analysts from six different companies that employ the method were confronted by the same set of circumstances, it would not be surprising to find that no two of them obtained the same result. One reason for this is that there is no general agreement with regard to what amount represents the required investment in the proposed replacement. Among the values used by various organizations are the following: (1) the first cost of the new equipment which, as we defined it earlier, would include the purchase price, delivery charges, and installation expenses, (2) the first cost of the new equipment minus its terminal salvage value, (3) the first cost of the new equipment minus the present salvage value of the old equipment, (4) the first cost of the new equipment minus the *sum* of its terminal salvage value and the present salvage value of the old equipment, (5) the average investment in the new equipment as obtained by calculating the average of its first cost and its terminal salvage value. Suffice it to say that there are still other definitions of what amount represents the required capital investment in the new equipment.

The Annual Saving

A variety of definitions also exists for the "annual saving" that will be generated by the new equipment. This saving is understood to be the difference between the annual recurring expenses that will be incurred with the old equipment and the annual recurring expenses that will be incurred with the new equipment. But some representative specific definitions that one will encounter are as follows: (1) the annual reduction in only the direct labor cost, (2) the annual reduction in the direct labor cost and in the maintenance cost, (3) the annual reduction in all direct manufacturing costs, (4) the annual reduction in all direct manufacturing costs and in overhead expenses, (5) the annual reduction in certain operating costs adjusted to reflect the tax consequences of the investment, (6) the annual reduction in certain operating costs adjusted to reflect the cost of financing the investment.

To this, we might add that some organizations, when determining the length of the payoff period, will estimate and use the saving that will occur during the first year of the new equipment's life. Other companies will estimate and use the *average* annual saving that will be experienced during the entire life of the new equipment. Furthermore, some of the latter companies, in the course of estimating the average annual saving, will assume that, if retained, the ofd equipment will continue to be operated for a number of years equal to the expected life of the new equipment; others will take into account that the remaining life of the old equipment is likely to be significantly shorter than the service life of the new equipment.

One should also keep in mind that, in a given replacement study, any one of the many definitions of the "annual saving" can be paired with any one of the many definitions of the "required investment" to obtain many different values for the number of years in which the proposed replacement will pay for itself.

The Required Payoff Period

We have seen that the calculated value of the payoff period depends on the definitions adopted for the relevant factors and that there is no general agreement with respect to what the correct definitions are. The problem this creates is compounded by the fact that there is no logical way of determining what the payoff-period requirement should be. As a result, one firm may require that new equipment pay for itself in 3 years, while a comparable firm has a 5-year requirement.

It should be recognized that the purpose of a payoff-period requirement is to provide an organization with a means for allocating capital in the most efficient manner. In this respect, it is similar to the minimum rate-of-return requirement established by firms that apply one of the sound analytical techniques considered earlier. But as we saw, there is a systematic and logical way of ascertaining a firm's cost of money, which then becomes its required rate of return. Unfortunately, this cannot be said of the payoff-period requirement. Instead, management is compelled to rely on intuition and hunch in the course of establishing that requirement. Very often, the results leave something to be desired. To illustrate, the firm may decide to adopt a single value for the requirement. If that value is 4 years, an investment can be justified in new equipment that pays for itself in 3 years and has a 6-year life, but not in equipment that pays for itself in 5 years and has a 20-year life.

Of course, some firms will have more than one payoff-period requirement to reflect the fact that different types of assets have different lives. But how does one adjust the requirement to compensate for variations in the expected service life? The answer is that no one knows. This is because there is no way in which, say, a 2-year payoff for an asset with a 10-year life can be compared with a 5-year payoff for an asset with a 20-year life for the purpose of deciding which of those two investment opportunities will yield the higher return.

We might also note that the selected payoff-period requirement is sometimes deficient in the sense that it is not adjusted or is adjusted inadequately to reflect differences in the degree of risk to be associated with various investments and differences in the tax treatment to which various investments are subjected.

Finally, it is not unusual to find that many managements tend to confuse a required payoff period with a required rate of return and to confuse a calculated payoff period with a calculated rate of return. In their minds, a 5-year payoff requirement is synonymous with a 20 percent rate-of-return requirement, and a calculated 4-year payoff for a new asset is synonymous with a 25 percent return on the investment in that asset. Novertheless, this kind of relationship does not exist between a payoff period and a rate of return. As an example, an asset which will pay for itself in 4 years but has only a 3-year life is probably not capable of yielding a return of 25 percent.

A tendency to confuse payoff periods with rates of return can have unfortunate consequences. Specifically, it encourages the adoption of an unduly short payoff-period requirement as a means for increasing the return on the investment in new equipment. What is not recognized is that there is an inverse relationship between the length of the required payoff period and the length of time that inefficient currently owned equipment will be kept in operation. In other words, a reduction in the length of the required payoff period serves to increase the number of years during which the firm might be incurring excessive operating costs. Admittedly, this is a way to increase the return on the *extra* investment in new equipment, but it is not a way to increase the return on the *total* investment. If it were, a responsible management would begin a new business by purchasing the most deteriorated and obsolete equipment available and then would replace that equipment immediately by efficient new assets. This, of course, is not done because to do so would be to maintain that it is desirable to create losses in order to be able to eliminate the cause of those losses.

More could be said about the deficiencies of the payoff-period approach, but those considered suffice to demonstrate that the use of the method in equipment replacement studies is not advisable.

Mistakes In Application

The adoption of an improper method of analysis is likely to result in incorrect replacement decisions, but there is no certainity that the adoption of a sound method eliminates the risk of error. This is true for two reasons. First, mistakes can be made when forecasting the cash flows that will be experienced with the alternatives; it will be recalled that it is these cash flows that are processed to obtain uniform annual costs, present worths, or the return on the extra investment in new equipment. The chances of making such mistakes can never be eliminated, but they can be reduced by utilizing qualified estimators.

A sound analytical technique can also lead to an incorrect decision because the technique was not applied properly. Unlike mistakes in forecasts, mistakes in application can be eliminated. We shall now consider the mistakes of this type that are commonly made.

Capital Invested

Some organizations that make a replacement study with a sound method of analysis use the book value of a currently owned asset as the amount which represents the capital investment in the asset. This is not objectionable when the book value is equal to the market value, but otherwise it is. Given that the market value represents the amount of money to which the old equipment is equivalent, the market value is the relevant figure.

An asset's book value at a certain point in its life depends on the depreciation method to which the asset was subjected and on the service life and salvage value used when applying that depreciation method. If the application of the method yields depreciation charges which reflect actual changes in the asset's worth, the resultant book value will be equal to the market value. When the two are not equal, the explanation is that the depreciation expenses, as calculated for accounting purposes, were overstated or understated.

This does not mean that book values should always be ignored when passing judgment on an equipment replacement proposal. If an after-tax analysis is made, the difference between the old equipment's book value and its market value will have to be considered. The reason is that tax laws may dictate treating such a difference as a gain or loss on disposal, and there may be tax consequences. In brief, the book value of the currently owned equipment will be a relevant datum when ascertaining the tax effects of an investment decision but not when ascertaining the present investment in the equipment.

This is an appropriate point at which to mention another mistake which is sometimes made. The mistake is to take the old equipment's market value into account twice instead of once. To explain, let us suppose that the old equipment has a market value of $2,000 and that the new equipment has a first cost of $40,000. Since the total investment in the old equipment is $2,000 and the total investment in the new equipment is $40,000, the new equipment calls for an additional investment of $38,000. Yet, there are analysts who will correctly state the total investment in the old equipment is its market value of $2,000, but who will incorrectly state that the total investment in the new equipment is its first cost of $40,000 minus the $2,000 that will be realized from the sale of the old equipment, or $38,000. The difference between this $38,000 and the old equipment's market value of $2,000, or $36,000, is then treated as the extra investment required by the new equipment. This is a mistake of double-counting, because the $2,000 is first shown as a cost for the one alternative and then as a revenue for the second alternative. It is interesting to note that these same analysts would not argue that the total investment in the old equipment is its market value of $2,000 minus the $40,000 which will not be spent on new equipment if the old equipment is retained.

Equal Time Periods

There are times when the mistake made in an equipment replacement analysis is the failure to consider the same number of years when comparing the alternatives. It is easier to make this mistake when the uniform annual cost method is being applied than when either the present worth method or the discounted cash flow method is being applied.

With the discounted cash flow method, a determination of the prospective return on the extra investment in new equipment must be preceded by a year-by-year comparison of the alternatives' respective cash flow patterns; so if the old equipment has an estimated remaining life of 3 years and the new equipment has an estimated life of 15 years, the analyst cannot determine cash flow differences unless he or she takes into account what might happen after the third year if the old equipment is retained.

With the present worth method, a total cost is calculated for each alternative; if the analyst were to obtain the sum of the costs that would be incurred during a 3-year period with the old equipment and the sum of the costs that would be incurred during a 15-year period with the new equipment, he or she would probably recognize that two such totals cannot be compared.

In the uniform annual cost method, however, average annual costs are computed, and it may not be evident what is occurring when averages for two periods of different duration are compared. For example, the average cost, during the next 3 years with the old equipment, may be found to be $69,000 per year and, during the next 15 years with the new equipment, to be $58,000. It might appear that each of these values is expressed in the same unit, that is, in terms of dollars per year and, therefore, that the two values can be compared. But, actually, the units are not the same. The unity, in one case, is *dollars per year for 3 years*, and, in the other, *dollars per year for 15 years*. This precludes any meaningful comparison of the two values. In spite of this, such incorrect comparisons are often made.

A mistake of another kind is occasionally made in the course of satisfying the equal-time-period requirement. This is to increase the service life estimate for the old equipment to a value equal to the expected service life of the new equipment or to decrease the service life estimate for the new equipment to a value equal to the expected remaining service life of the old equipment. With either approach, one of the alternatives is being burdened with an uneconomic service life which may result in an incorrect decision. That this is so is something most people recognize intuitively. To illustrate, we all know that it might be prudent to keep an old automobile one more year but not four more years; similarly, it might be prudent to buy a new automobile if we shall keep it for four years but not if we shall keep it for only one year.

Recurring Costs

Let us now turn to a mistake of another kind. Included in the cash flow pattern of each alternative will be an estimate of the series of annual recurring expenses which will be experienced with the equipment involved. Some of these expenses will be direct costs, such as direct labor and direct materials. Others will be indirect costs, such as supervision, utilities, supplies, and maintenance.

The difficulty of estimating these recurring expenses is alleviated to a significant degree by the fact that those that are unaffected by the choice of alternative are irrelevant and can be ignored. Nevertheless, a number will remain; some of these will be direct costs, and some will be indirect costs. The nature of indirect costs is such that they pose much greater problems of estimation than do direct costs. Some indication of the truth of this is that it may not be feasible to determine what they currently are for a specific activity in which the company engages; that is why it is common practice, for accounting purposes, to allocate total indirect costs among various activities with the use of burden, or overhead, rates. This is done as a matter of convenience and with the realization that the amount of overhead expense attributed to a certain activity may or may not be accurate.

The difficulty of estimating indirect costs when making an equipment replacement study may cause an organization to make an avoidable mistake in the process of applying a sound method of analysis. The organization will begin with an estimate of the values of the annual direct costs that will be affected by a decision to retain or to replace the old equipment. But it will then go on to generate estimates for the indirect expenses for each alternative with the use of the burden rates developed for accounting purposes. This is inadvisable. The application of such rates will yield estimates which reflect an assumption that all indirect costs will be affected by the choice of equipment and estimates which do not reflect the exact manner in which the relevant indirect costs will be affected. The resultant errors are almost certain to exceed those that would be made in the course of estimating what the actual relevant indirect costs will be with each alternative.

Cost Of Money

Mistakes are also made by firms when they are determining their cost of money, which becomes their minimum rate-of-return requirement. It will be recalled that the cost of money is the average of the interest expense to be associated with the debt portion of the total investment in an asset and the opportunity cost to be associated with the equity portion. Insofar as the opportunity cost is concerned, this cost is represented by the rate of return which could be realized by investing the equity capital elsewhere. But elsewhere is to be understood as being another investment which entails the same degree of risk as will be assumed if an investment is made in the new equipment. From this, it follows that riskier investments in new equipment generate higher opportunity costs, and, therefore, those investments should be expected to yield a higher rate of return. Yet, there are firms that use the same minimum rate-of-return requirement in all their equipment replacement studies, even though the amount of risk varies from case to case.

The rate which represents an organization's cost of money will also be affected by whether a before-tax or an after-tax analysis is being made of an equipment replacement proposal. In a before-tax study, the minimum rate-of-return requirement will be the required return before taxes, and, hence, the organization's cost of money is computed on a before-tax basis. Specifically, the interest expense used in this computation is the cost of borrowed funds before taxes, and the opportunity cost is the return that can be realized before taxes by investing equity funds elsewhere.

In an after-tax study, the minimum rate-of-return requirement becomes the required return after taxes, and it is obtained by computing the average cost of money after taxes. Hence, the opportunity cost becomes the after-tax return that could be realized by investing equity funds elsewhere, and the interest expense is the cost of borrowed money after taxes. It is in the determination of the after-tax cost of borrowed money that many organizations make a serious mistake.

The mistake stems from the fact that the cost of borrowed money is a tax deductible expense. Because it is, a company that is subject to an effective tax rate of 50 percent and borrows money at 16 percent may be inclined to use 8 percent as its after-tax interest expense. If it does, it will be making a mistake, because it will then conclude that an after-tax return of 8 percent on the debt portion of its investment will suffice to cover the cost of borrowing that money. But if the company earns a number of dollars, after taxes, which is equivalent to a return of 8 percent, that number of dollars will not enable it to repay the number of dollars it owes in interest as a result of having borrowed money at 16 percent. To cover that expense, it will have to realize an after-tax return of 16 percent on the debt portion of the investment.

Therefore, when calculating its average cost of money, an organization should use a lower rate for its opportunity cost in an after-tax analysis than it uses in a before-tax analysis. But it should use the same rate for the interest expense in both types of analyses. This does not mean that the fact that interest is a tax deductible expense is ignored in an after-tax study. That is taken into account when the firm is ascertaining how its income taxes will be affected by a decision to replace a currently owned asset.

A Summation

This brings us to the close of our discussion of the pitfalls to be avoided in equipment replacement analyses. Naturally, this has not been an exhaustive treatment of the topic, because there are innumerable ways of doing something incorrectly. Nevertheless, the mistakes considered are the major ones which are commonly made. As we saw, some of them are related to the adoption of a faulty method of analysis, and some are related to the improper application of a sound method of analysis. Avoiding them will not completely eliminate the risk of error in an equipment replacement decision, but it will certainly reduce that risk significantly.

Bibliography

[1] Canada, J. R., and J. White, *Capital Investment Decision Analysis for Management and Engineering*, Prentice-Hall, Inc., Englewood Cliffs, N. J., 1980.

[2] DeGarmo, E. P. *Engineering Economy*, 6th ed., Macmillan Publishing Co., Inc., New York, 1979.

[3] Grant, E. L., W. G. Ireson, and R. S. Leavenworth, *Principles of Engineering Economy*, 6th ed., John Wiley & Sons, Inc., New York, 1976.

[4] Mayer, R. R., *Capital Expenditure Analysis*, Waveland Press, Inc., Prospect Heights, Ill., 1978.

[5] Newman, D. G., *Engineering Economic Analysis*, rev. ed., Engineering Press, Inc., San Jose, Calif., 1980.

[6] Riggs, J. L., *Engineering Economics*, McGraw-Hill Book Company, New York, 1977.

[7] Thuesen, H., W. Fabrycky, and G. Thuesen, *Engineering Economy*, 5th ed., Prentice-Hall, Inc., Englewood Cliffs, N. J., 1977.

[8] White, J. A., M. H. Agee, and K. E. Case, *Principles of Engineering Economic Analysis*, John Wiley & Sons, Inc., New York, 1977.

Reprinted from **Production Engineering,** *March 1978.*

To Buy Or Not To Buy?

Weighing capital investments

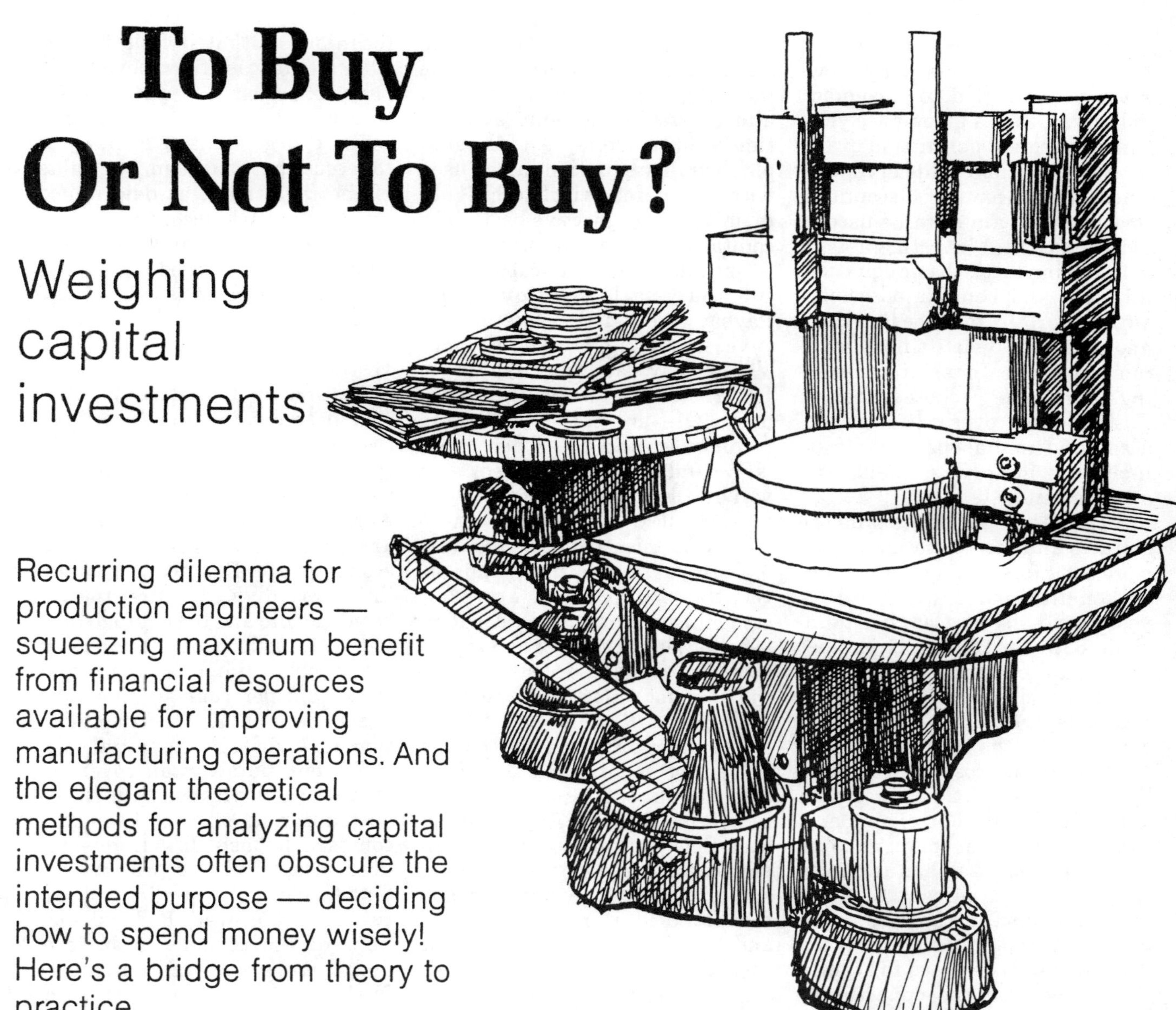

Recurring dilemma for production engineers — squeezing maximum benefit from financial resources available for improving manufacturing operations. And the elegant theoretical methods for analyzing capital investments often obscure the intended purpose — deciding how to spend money wisely! Here's a bridge from theory to practice.

By HENRY A. TOMBARI
Assistant Professor,
Management Sciences
California State University, Hayward
Hayward, Calif.

Capital investment evaluation begins with a qualitative assessment of how well the proposed expenditure fits into corporate financial and marketing plans. And if you are a typical production engineer, you probably have no shortage of projects that meet this criterion. More often than not, there isn't enough capital to do all the things that need to be done, and the question becomes one of deciding which projects to support.

A quantitative appraisal of the financial consequences of each project provides the basis for an objective go/no-go decision vis-a-vis supporting that project. And this data can also, if necessary, be presented to top management as justification for the capital expenditure.

A simple equation determines the savings or income that a project will generate each year:

Cash Flow = Benefits − Costs,

where benefits are the projected annual savings or the revenues from sales; costs are the recurring operating costs; and cash flow is the net income after taxes, plus noncash charges, or the net annual savings.

The key to this simple concept is that cash flow is the criterion for decision making. And, the annual cash flow projected for the life of the project or estimated life of a piece of equipment must be translated into valid economic terms. That is, annual dollar cash flows must be converted to a common dollar value for a given base year. The adjustment permits an evaluation for the time value of capital. Thus, the concepts of discounting and present value can be used to examine capital investments.

In setting up the basic equation, all costs and benefits associated with the project must be identified and included. Benefits include income from sales, or savings resulting from some proposed change. Costs usually cover initial costs of assets and annual operating costs, including materials, fuels, and other expenses.

A critical part of developing the benefits and costs associated with a project is forecasting the changes in benefits and costs in future years. And any forecast has to include both the dollar change per year and the number of years the benefits and costs will continue.

One important cost that is often overlooked is the opportunity cost. An opportunity cost is income the firm foregoes from not using an asset differently. For example, if cash can earn ten percent after taxes, the cash itself has an opportunity cost of ten percent. Hence, if an equipment purchase is made for cash, the opportunity cost is ten percent of the cash used to acquire the items. And the opportunity cost of equipment in use is the profit lost due to the unavailability of the equipment for other purposes. But if the equipment is idle, its opportunity cost is zero.

Funds shown in different years cannot be compared directly because time gives them dissimilar values. Therefore, the cash flow stream is converted to a base year, effectively assigning a time value to the capital. The time value of money put into an expenditure is called the opportunity cost of capital, or the marginal rate of return. This cost of capital is what the company would pay to utilize the capital on an alternative purchase. The opportunity cost of capital reflects the rate the firm decides it can be fairly sure of obtaining by utilizing the funds in an alternative project.

The economic criterion best suited for evaluation of a project or purchase is cash flow. Cash flow is a concept that permits an analysis including both comparison with alternative investments and the time value of money. And the simple cash flow equation can be used with many analytical techniques to provide the objective basis for a capital investment decision.

The four most practical and useful analytical techniques are: Payback Period (PP); Return on Investment (ROI); Net Present Value (NPV); and Internal Rate of Return (IRR). Two of these—PP and ROI—are called screening techniques, and the other two—NPV and IRR—are called time value techniques.

SCREENING TECHNIQUES can be used to perform a quick evaluation of capital investment opportunities. In general, however, screening methods should not be used for justification of major projects or purchases, because screening does not reflect the time value of money.

The information needed for the screening techniques is initial cost, annual operating cost, benefits, estimated life, and annual cash flow. Initial cost is the estimated dollar cost of the project or equipment under evaluation and the cost to put it in place. The cost and benefit items determine the cash flow and the estimated life factor permits some estimate of life expectancy of the project.

Payback period (PP) is the initial cost divided by the annual cash flow, with each annual cash flow being equal. If the annual cash flows are not equal, the calculations are somewhat more difficult, see the box entitled "The arithmetic is easy."

Comparing the payback period with the expected lifetime of the project provides a "ballpark" evaluation as to its potential for recoupment. As a rough rule of thumb, a payback period of less than one-half the lifetime of an investment would generally be considered a suitable expenditure where the lifetime is ten years or less.

This technique is obviously crude at best and only gives an indication of the time needed to recoup an investment. The major shortcoming of the payback technique is that it fails to consider annual cash flows beyond the payback period. Because an objective decision maker would not commit large sums of money just to recoup the initial sum, the value of the investment as determined by the payback technique is not conclusive. The payback period should be considered as a time constraint to be satisfied, instead of as an indicator of the full-time value to the company of the investment.

Return on investment (ROI) is a better screening technique than PP, because it takes into account the depreciation of the investment over its economic life. And the proposed project's benefits are also averaged over its economic life. Depreciation is the initial cost divided by the project's economic life and the average benefit is the total benefit divided by the economic life. ROI—expressed in percent per year—is the average benefit minus the depreciation, all divided by the initial cost.

ROI has the advantage, over PP, of putting capital expenditures with different expected lives on a comparable basis. It is frequently used to determine a project's potential due to its simplicity of calculation. However, ROI calculations do not consider cash flows and, therefore, are of limited usefulness in evaluating long-term, large-scale investments. As a general rule, if the ROI is less than 20%, time value techniques should be used.

TIME VALUE TECHNIQUES are those which incorporate an allowance for the time value of money, usually in the form of a discount rate. There are usually several different opportunities for capital investments to accomplish the same objective. And because these capital investment opportunities differ, the dollar held today is worth more than a dollar held in some future time period. Hence, the discount rate—a subjective factor established by the company—may differ widely between industries and even among firms within the same industry. The discount rate is typically 10-20 percent. Setting a discount rate is equivalent to stating that such a return can be realized elsewhere. That is, the

The arithmetic is easy

Here's how the calculations go. As an example, consider a capital investment for energy conservation. The equipment will cost $50,000 to purchase and install but will involve no new annual operating costs. The annual savings are expected to decrease from 22,500 MBTU of natural gas to 5,000 MBTU over the four-year life predicted for the equipment. The company sets a discount rate of 10 percent as the minimum return expected on its investments. The data are:

Initial cost (IC) = $50,000
Annual operating cost (AOC) = Zero
Annual fuel savings (AFS) = 22,500 MBTU (year one)
22,500 MBTU (year two)
15,000 MBTU (year three)
5,000 MBTU (year four)
Estimated life (EL) = 4 years
Discount rate = 10%
Estimated Fuel Price (EFP) = $1.00/MBTU
Benefits (B) = (AFS × EFP)

And the cash flow is:

Cash flow (CF) = Benefits − Costs
= (AFS × EFP) − AOC
= (22,500 MBTU × $1/MBTU) − 0
= $22,500 (year one)
= $22,500 (year two)
= $15,000 (year three)
= $5,000 (year four)

The four analytical techniques yield the following results:

PAYBACK PERIOD

PP = IC/CF with equal annual cash flows. For unequal annual cash flows, add the annual cash flows beginning with year one, and determine the number of years required to recoup the initial outlay. In this case:

$22,500 (year one)
\+ 22,500 (year two)
\+ 5,000 (⅓ of year three)
= $50,000 (Initial cost)

Hence, PP = 2⅓ years.

The rule of thumb is if PP is less than one-half the equipment lifetime, the purchase is acceptable. This is not the case, therefore the proposal should be rejected.

RETURN ON INVESTMENT

$$\text{Depreciation (D)} = \frac{IC}{EL} = \frac{\$50{,}000}{4\text{ yr}} = \$12{,}500/\text{yr}$$

Average Benefits (B) = Average annual benefits for estimated life of project.

$$\text{Average B} = \frac{\$22{,}500 + \$22{,}500 + \$15{,}000 + \$5{,}000}{4\text{ yr}}$$

$$= \frac{\$65{,}000}{4\text{ years}} = \$16{,}250/\text{yr}$$

$$\text{ROI (\%/yr)} = \frac{\text{Average B-D}}{IC} \times 100$$

$$= \frac{\$16{,}250/\text{yr} - \$12{,}500/\text{yr}}{\$50{,}000} \times 100\%$$

$$= 7.5\%$$

This is less than the minimum return on its investments set by the company, therefore the proposal should be rejected.

NET PRESENT VALUE

Year	Benefits	Costs	Cash Flow	PV* of $1 @ 10%	Discounted Cash Flow
0	$ 0	$(50,000)	$(50,000)	1,000	$(50,000)
0-1	22,500	0	22,500	0.952	21,400
1-2	22,500	0	22,500	0.861	19,400
2-3	15,000	0	15,000	0.779	11,700
3-4	5,000	0	5,000	0.705	3,500
Total	$65,000	$(50,000)	$ 15,000		$ 6,000 NPV

*Interest tables

The proposed purchase recovers all costs, the 10% opportunity cost of the firm's capital is realized, and the investment will yield an additional $6,000 return, therefore the proposal should be accepted.

INTERNAL RATE OF RETURN

Year	Cash Flow	PV* of $1 @ 10%	Discount Cash Flow	PV* of $1 @ 18%	Discounted Cash Flow
0	$(50,000)	1,000	$(50,000)	1,000	$(50,000)
0-1	22,500	0.952	21,400	0.915	20,600
1-2	22,500	0.861	19,400	0.764	17,200
2-3	15,000	0.779	11,700	0.639	9,600
3-4	5,000	0.705	3,500	0.533	2,600
Total	$ 5,000		$ 6,000 NPV		$ 0 NPV

*Interest tables

The IRR is that interest rate at which the cash flow stream equals all the costs of the proposed project and is determined by trial and error. In this case, the IRR is 18%—greater than the desired 10% minimum rate of return, therefore the proposal should be accepted.

The two screening techniques indicate rejection of the proposal while the two time value techniques indicate acceptance. Hence, it is apparent that without a full evaluation of the project, based on the opportunity costs of capital over the estimated life of the proposal, the signal would be to reject the project. However, it is evident that when the opportunity cost *is* included in the calculation, the proposal more than meets the company's minimum return requirement. Thus the proposal should be accepted.

The time value techniques adjust the cash flows to base year zero, so that all funds are on a common denomination. Hence, cash flow and rate of return are determined over the total life of the project.

Also, the four analytical techniques were applied with the tacit assumption that the proposed capital investment is being evaluated in a risk-free environment. However, some risk is inherent in most large investments. Hence, a subjective risk appraisal should be made prior to making the final decision to purchase or not to purchase.

discount rate is simply the opportunity cost of capital, and the project or purchase is justified if the return is greater than the discount rate.

Net Present Value (NPV) is determined by comparing the present value of the cash flows generated by a given investment, with the initial and annual operating costs associated with the investment. The present value of the expected cash flow discounts that flow for each year of the project's estimated lifetime, making the estimates equivalent in time. Summing the discounted cash flows for each year yields the net present value. The NPV is a very useful method for it considers all benefits and costs, including the opportunity cost of capital. The general rule in comparing alternative projects is to select the alternative that results in the highest net present value.

Internal Rate of Return (IRR) is the discount rate which reduces the stream of cash flows associated with the project to a present value of zero. Although IRR is a meaningful technique it is, unfortunately, more difficult to calculate than NPV. The IRR technique requires an iterative approach converging on the solution. In essence, the decision maker finds—by trial and error—the maximum rate of interest the project could pay on the investment and break even.

IRR permits an explicit comparison of the proposed project's return with the discount rate appropriate for the company in justifying the investment. And like NPV, IRR is useful when comparing expected rates of return for alternative capital expenditures.

It can be argued that IRR is just an extended version of NPV. However, each has its own advantages. NPV uses a given corporate discount rate to determine the annual cash flow present value. Total NPV is calculated and a decision can be made if this NPV is greater than zero. On the other hand, IRR ignores the firm's discount rate and determines the rate of return on the total cash flows. Generally, the two techniques complement each other. Either one, however, is a better analytic device than the two screening methods.

Reprinted from **Production Engineering,** *March 1978.*

A Hard Look At The Numbers

Weighing capital investments

By LAWRENCE HACKAMACK
and
BEATRICE HACKAMACK
Northern Illinois University
DeKalb, Ill.

How can you prove the need for replacement of major machinery? This question has been of constant concern to engineers in industry and government for years, and one has only to try to justify new machines to know the problems. Each methodology has its own purpose; yet when compared side by side, as we do in this article, do we see the problems and the lack of certain critical elements?

Thus, if one examines the field of replacement theories, we find Markov Chains, M.A.P.I., Pay-Off, Pay-Back, Flow of Funds, Present Worth Annual Cost, Productivity Criteria Quotient, Rate of Return, and very elaborate statistical machine replacement models all used. Certainly some are more frequently found in use than others. The major types that are most common are Pay-Off or Rate of Return, Present Method, Annual Cost Method, and Productivity Criteria Quotient.

To show how these methods compare under "service conditions," let us trace the same two metal cutting lathes through these major theories. We'll get a clearer indication of what prob-

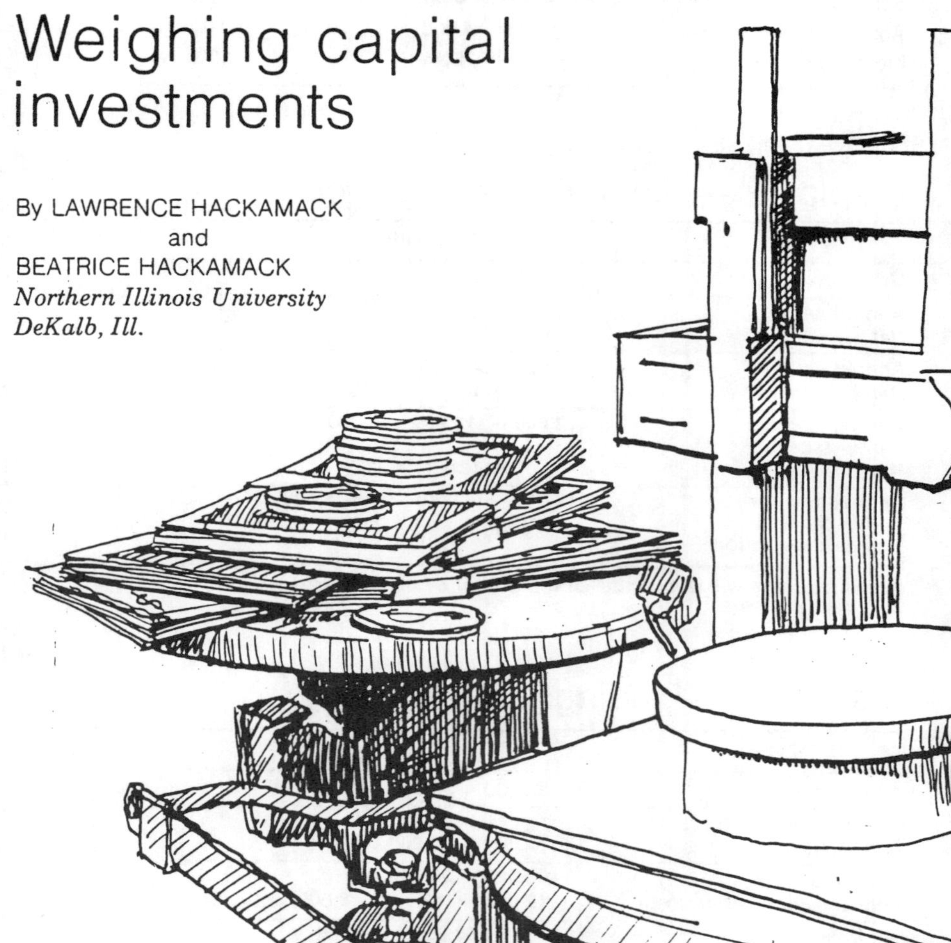

Display 1 — Problem statement for example 1

	Present machine	Proposed machine
Installed cost	$40,000	$80,000
Present book value	24,000	———
Present salvage value	20,000	———
Salvage value at end of life	———	4,000
Average annual operating cost	32,000	20,000
Service life	1 year	8 year
Annual depreciation	4,000	10,000
Income tax rate	50%	50%

Display 2 — Finding pay-off period

Method		Calculation		Period
Gross investment / Annual savings	=	80,000 / 32,000 – 20,000	=	6.66 Year
Net investment / Annual savings	=	80,000 – 20,000 / 32,000 – 20,000	=	5 Year
Net investment / Annual savings after taxes	=	80,000 – 20,000 / 12,000 – (12,000 – 8,000) (.50)	=	6 Year
Net investment / Annual savings after taxes and depreciation	=	80,000 – 20,000 / 12,000 – (4,000) (.50) – 8,000*	=	30 Year

*(New machine will depreciate at the rate of $8,000 per year.)

Display 3 — Finding rate of return

Method		Calculation		Percent
Annual savings / Gross investment	=	32,000 – 20,000 / 80,000	=	15%
Annual savings / Net investment	=	32,000 – 20,000 / 80,000 – 20,000	=	20%
Annual savings after taxes / Net investment	=	12,000 – (12,000 – 8,000) × (0.50) / 80.000 – 20,000	=	16.66%
Annual savings after taxes and depreciation / Net investment	=	12,000 – (4,000) × (0.50) – 8,000 / 80,000 – 20,000	=	3.33%
Average annual savings / Average book value	=	96,000/8 / (80,000 – 4,000/2)	=	31.57%
Average annual savings / Book value	=	96,000/8 / 80,000	=	15%

lems occur when justifying machine tool replacement.

Example I

Conventional vertical turret lathe purchased new in 1972 at a price of $40,000 and used at the two shift level since that time.

Example II

Conventional vertical turret lathe purchased new in 1964 at a price of $20,000 and used at the two shift level since that time.

What happens when we apply the various theories to these machine examples?

The pay-off theory is closely related to the rate of return method. This method judges the relationship of original cost, depreciation, interest rate, service life, and annual operating costs to attempt to discover whether the machine will pay for itself and turn a profit.

In this first problem, used in the following example, is whether a new machine costing $80,000 should replace an obsolete machine whose present book value is $24,000. The problem has been given in a more simplified form so that the service and tax lives are the same, and there is no salvage value. Straight-line depreciation is used, as well as a 50-percent income-tax rate and the annual costs remain constant, Display 1. Base methods for determining the pay-off period are shown in Display 2.

We assume that a new machine has been purchased for $80,000 to replace a machine with a salvage value of $20,000. This means that there has been a net investment of $60,000. In a given year the new machine costs $20,000 to operate, and the old machine costs $32,000 to operate, which yields a savings of $12,000 per year; the machine will depreciate at a rate of $8,000 per year.

In the rate of return analysis, we find a method which involves setting up a minimum rate of return. This is usually established by top management to prove through financial paper work that machine replacement will yield an acceptable return. Although this method has limitations when used for evaluating equipment replacements, its principal value is in ranking replacement projects.

In the previous example, a new machine was purchased to replace an old machine that had a salvage value of $20,000. The figures from that example will be used to illustrate the rate of return method, Display 3.

This method answers a very basic question: What annual interest rate must be applied to a deposit or investment of $60,000, and subsequent balances on that deposit, if one expects to withdraw $12,000 at the end of each year, including principal and interest, particularly after the interest has been applied each year?

The concept of present worth, also known as discounted cash flow, has many uses in evaluating new equipment. However, the principal objection to its use is that the results are difficult to interpret, because managers are accustomed to thinking in terms of annual cost and find it difficult to realize that a small change in the interest rate might dictate another course of action.

Following is an example which shows the present worth method as applied to the sample problem. The starred (*) numerical factors were obtained from a table that lists Present Worth factors available in most engineering handbooks, where both the old and the new machines are assumed to be in operation for 8 year. In the beginning the new machine would have a market or residual value of $20,000, but would have no value in 8 year. Furthermore, the time value of money to the firm is assumed to be 10 percent per year.

Old machine

Present worth of an expenditure for operating expenses of $32,000 for 8 years:

$32,000 × 5.335* = $170,720

New machine

Present worth of net expenditure for new machine:

$80,000 − $20,000 = $60,000

Present worth of an expenditure for operating expenses of $20,000 per year for 8 years:

$20,000 × 5.335* = $106,700

Present worth of $4,000 received 8 years from now:

$4,000 × 4.665* = $18,660

New machine total	$185,360
Old machine total	−170,720
	$ 14,640

New machine has an advantage of $14,640 over the present equipment.

The annual cost method (also called the capital recovery method) is one in which the annual cost for obtaining service from different sources is developed. The annual cost consists of the cost of getting the original investment back, the cost of getting a return on that investment, and the cost of operating under that particular alternative. When a salvage value is involved, the interest that could have been earned on that amount each year, had it been used otherwise, is also an annual cost of capital.

Two steps are involved when figuring the annual cost method. First is capital recovery—the amount that will just repay the initial investment at a stipulated interest over the economic life of the equipment. To this must be added the annual operating cost—all relevant operating costs converted to an annual basis and then simply added to the annual cost of capital recovery. Operating costs are those normally considered in any cost analysis—such as labor, tooling, maintenance, and scrap. In most cases the average of the series of annual disbursements is used as an approximation.

This method is used to compare mutually exclusive alternative; it is practical for use in estimating the economic life of existing or proposed equipment. The annual cost method forces the individual to recognize that the capital cost of an existing asset is its market value, not its book value. Furthermore, the annual cost method does not require the same termination date for each alternative.

The following example shows how the sample machine problem is applied to the annual cost method. Again, it uses numerical factors obtainable from standard engineering references.

Old machine

ACCR = (P−L) × (CRF) + L × i
= (20,000 − 0) × 0.18744* + [0 × (0.10)]
= $3,748.80

Operating Cost = ($32,000)

Total Average Annual Cost = $35,748.80

Proposed machine

ACCR = (P−L) × (CRF) + L × i
= (80,000 − 4,000) × 0.18744* + (4,000 × 0.10)
= ($76,000) × 0.18744* + 400 = $14,645.44

Operating Cost = $20,000

Total Average Annual Cost = $34,645.44

The results indicate that the proposed machine has a lower annual cost than the present machine. The proposed machine yields $1,103.36 before taxes beyond the 10 percent rate of return.

Old Machine	$35,748.80
Proposed Machine	− 34,645.44
	$ 1,103.36

Productivity Criteria Quotient process relies on the basic PCQ index of machine tool improvements to guide you in establishing when to change machines. And an allied index, the Productivity Return Rate, tells you how fast the new machine can pay for itself.

The PCQ process, completely computerized for easy analysis, has been in service for over a decade. You can use PCQ either for working up data on a single machine, or for a macro-analysis of an entire plant. For more information on PCQ, contact the author.

Reprinted from ***Industrial Engineering,*** *July 1981.*

Material Handling

Justify An Automated Storage & Retrieval System

AS/R systems may improve floor space utilization by as much as 300 to 500% over conventional methods.

By Ole B. Rygh
Munck Systems Inc.

The first step in the justification process of an automated storage and retrieval system is to establish the benefits. Some benefits are obvious, while others are more difficult to ascertain. They also vary from application to application. The most common benefits can include improved space utilization, lower operating costs, absolute inventory accountability, reduced pilferage, lower energy consumption and reduced product damage.

Improved space utilization

An AS/RS can improve floor space utilization by three to five times compared to conventional methods. The AS/RS technology allows inventories to be compressed into high-density, narrow aisle, high-rise storage.

Furthermore, AS/R systems not only fully utilize the air rights up to 100 feet and over, but also provide for greater cube utilization simply by better management of the space. These characteristics obviously convert to savings in real estate and building costs. In addition, rack-supported buildings are classified as equipment or machinery shelters and are, therefore, allowed an accelerated depreciation on the same basis as the rest of the equipment. For the same reason, they often qualify for investment tax credits and sales tax exemption.

Lower operating costs

The operating characteristics of an AS/RS can substantially reduce the size of the material handling work force resulting in direct labor savings. A work force reduction of two-thirds or more compared to conventional methods is often realized. The reduction can include fork truck operators, order pickers and others who are directly engaged in receiving, handling, storing, retrieving and delivering functions.

As the number of operating personnel is reduced, savings are also often realized in indirect labor such as clerical, inventory control and data processing personnel, as well as in personnel associated with supervision, expediting, administration, facilities, security, etc.

An AS/RS also reduces the need for forklifts and other peripheral material handling equipment. This equipment reduction also needs to be considered in the justification equation.

Absolute inventory

The cost of carrying inventory is real. A recent study published by the National Council of Physical Distribution Management indicated that the annual cost to carry inventory can be as high as 43% of the value of the inventory. The same study indicated an average annual cost of 30% for the companies surveyed.

Consider the fact that an automated storage/retrieval system today offers very close to absolute inventory accuracy. This means that an AS/RS effectively becomes a tool to reduce excess inventory.

The following are samples of such excess inventories:

- ☐ Inventories purchased or manufactured considerably in advance of need.
- ☐ Inventories pre-kitted or physically separated to support manufacturing operations far in advance of need.
- ☐ Inventories which are known to have been received but cannot be

used because they cannot be located.

☐ Inventories which are not lost but cannot be used because the paperwork has not been processed.

☐ Inventory quantities ordered over and above required quantities to insure against inventory shrinkage.

☐ Inventories of partially completed products which cannot be finished or shipped because of parts or material shortages.

Obviously, an AS/RS cannot eliminate all excess inventories, but it can eliminate most of the reasons for excess inventories.

Reduced pilferage

Industrial pilferage is a very real cost to business and was recently estimated to exceed $20 billion per year in the United States. An AS/RS obviously encourages less pilferage because the loads are generally inaccessible to operating personnel, since machines rather than people have access to inventories in the storage area.

Lower energy consumption

By virtue of an AS/RS's superior cube utilization and the reduction in the amount of operating equipment, many users have experienced it is evident that an AS/RS consumes less energy than a conventional system. Furthermore, no operating personnel are required to be inside an AS/RS. This reduces the need for light, heat, ventilation, etc. Many companies are experiencing energy savings of two to two-and-one-half times over conventional systems.

Reduced product damage

An AS/RS virtually eliminates product damage because of the gentle and controlled handling by the AS/RS equipment. This can become a significant benefit particularly where fragile or expensive items are handled.

Table 1. Example of Initial Investment for Two Systems

Conventional	
Building & Foundations (110,000 sq ft × $28.00)	= $3,080,000.
Land (110,000 sq ft × $1.00)	= 110,000.
Racks (10,000 openings × $50.00)	= 500,000.
Forklifts (15 × $20,000)	= 300,000.
Sub-Total	$3,990,000.
Less 10% Investment Tax Credit on Equipment	= 80,000.
Total	= $3,910,000.
AS/RS	
Building & Foundations	= $1,200,000.
Land (30,000 sq ft × $1.00)	= 30,000.
AS/RS	= 3,800,000.
Forklifts (4 × $20,000)	= 80,000.
Sub-Total	= $5,110,000.
Less 10% Investment Tax Credit	= 500,000.
Total	= $4,610,000.

It is more difficult to put a value on the less tangible benefits. These are, therefore, frequently eliminated from the justification equation. However, some users today are willing to do their homework in examining these areas and will soon have a technological advantage over their competitors. The most common of these less tangible benefits can include:

☐ *Improved working conditons*—An AS/RS normally offers a controlled, safe, clean and challenging environment compared to a conventional system where the jobs are often physically demanding, dull or routine, and, therefore, usually attract a lower caliber of personnel.

AS/RS users are experiencing a lower turnover rate and have little problems in filling vacant positions. The cost of recruiting and training is a real cost which should be part of the justification equation.

☐ *Easier housekeeping*—Since equipment and personnel, to a very high degree, are kept outside the AS/RS, the system is easier to keep clean and neat. Therefore, the cost associated with housekeeping is a measureable factor, particularly in large operations.

☐ *Less equipment damage*—Unlike forklifts and similar devices which are controlled by an operator, the AS/RS equipment is automatically controlled and has built-in devices to prevent damage to itself and adjacent equipment.

The cost of damage done to racks, pallets, containers, building structures, etc, by forklifts and their operators is very real and must be accounted for.

☐ *Improved customer services*—Most people will recognize the impact of lost business because of late deliveries of products or parts. In fact, one of the most common justification factors for an AS/RS in distribution or spare parts depots is improved customer service, which is translatable into new business.

What would it mean for your company to be able to process orders and make shipments in hours rather than days? Sometimes, however, an AS/RS which makes this happen will not get any credit for the increased profits generated by this new business. Another way of looking at it is: think of the business that the company would lose if it did not improve its ability to make prompt deliveries.

□ *Better management control*—Better management decisions based on accurate and timely inventory information are a benefit of an AS/RS. The effect on overall manufacturing efficiency as it relates to improved material control is difficult to measure; but, most companies can identify the effect on production efficiency of parts and raw material shortages even if the actual time delays are not recorded.

The cost of poor management decisions based on inaccurate or untimely inventory information may be even more difficult to determine. It is, however, obvious that decisions regarding major shifts in production schedules can be better made with the assurance that materials are available and major production delays will not occur. With such assurances, decisions to accept new orders with tight delivery schedules can be made with more confidence.

Planning for design changes and introducing new models is extremely dependent on reliable inventory information, and accurate inventory information is also essential for day-to-day production planning and scheduling.

Justification process

It is obvious that there are many substantial benefits to be derived from an AS/RS and that the benefits vary widely from situation to situation.

Many automated storage/retrieval systems do not need to go through a long justification process because there are no real alternatives.

Most systems are, however, justified based on careful analysis. The following example is for illustration purposes and is not an attempt to establish a detailed justification process. To do so would be futile since each company has a unique and different set of ground rules regarding this subject. This example, while hypothetical, is based on several real system justifications, and should be representative of an automated storage and retrieval system directly supporting manufacturing operations.

The system is rack-supported and is 75 ft tall. It provides storage space for 10,000 loads and is serviced by five automatic S/R machines. The storage system is computer controlled and has an automatic conveyor system transporting pallets between shipping/receiving, storage and the manufacturing area, and to and from an order-picking area.

A conventional system of similar capacity would have 110,000 sq ft of floor space and be equipped with 15 forklifts. The initial investment for both systems would appear as shown in Table 1.

It is interesting to note the significant initial advantage gained by the AS/RS in the 10% investment tax credit category. The conventional system only has an offsetting tax credit of $80,000, while the entire AS/RS receives credit as a system and warrants a tax credit of $500,000. This produces an immediate cash saving of $450,000 and reduces the cost difference to $700,000.

The breakdown of the cost comparison figures shown in Table 2 includes:

□ *Manpower (direct)*—This includes the direct material handlers such as forklift operators and order pickers, and includes a total of 25 people in the conventional system; 10 in the AS/RS.

□ *Manpower (indirect)*—This includes clerical personnel, checkers, supervisors, expeditors, security, etc., and has 16 people in the conventional system and 12 in the AS/RS.

□ *Maintenance costs*—Generally, an AS/RS requires less maintenance. However, the maintenance personnel are generally of a higher caliber and the AS/RS maintenance cost

Table 2. Cost Comparison of Two Systems

Cost Classification	Conventional	AS/RS
Manpower—Direct	$ 625,000.	$ 250,000.
Manpower—Indirect	375,000.	300,000.
Utilities	150,000.	80,000.
Maintenance	70,000.	70,000.
Insurance & Taxes	50,000.	60,000.
Interest	—	110,000.
Depreciation	168,000.	423,000.
Excess Inventories	125,000.	—
Miscellaneous	260,000.	37,000.
Total	$1,823,000.	$1,330,000.

may, therefore, be equal to the conventional system.

☐ *Insurance and taxes*—The higher value of an AS/RS reflects the higher cost of insurance and property taxes.

☐ *Depreciation*—Again, the entire AS/RS can be depreciated at the same rate as the equipment, which creates a much more attractive cash flow compared to the conventional system.

☐ *Interest*—The cost of money for the *additional* capital outlay for an automated storage/retrieval system is factored into the equation.

☐ *Excess inventory*—By eliminating excess inventories, I have estimated that this company can reduce its inventory by $500,000 per year. At an annual inventory carrying cost of 25%, which is very conservative, this should result in a cost savings of $125,000 per year.

☐ *Miscellaneous*—Under the miscellaneous column, I have factored in some cost for the following items: excess floor space due to excess inventories, pilferage, product damage, labor turnover and training, and housekeeping.

This represents a yearly cost savings in favor of the AS/RS of $493,000 or:

$$\text{Payback} = \frac{\$700,000}{\$493,000} = 1.4 \text{ years.}$$

An investment in an automated storage/retrieval system is not unlike the investment in a computerized management information system. Few company presidents would argue that the benefits of such a system do not justify the investment; however, I suspect that few companies could justify the investment in an MIS if they did not give considerable weight to the so-called intangibles, such as better informed and more timely management decisions. If we, therefore, factor in some of the less tangible benefits, our cost comparison equation would then look like Table 3. This shows a yearly cost savings of $893,000, or:

$$\text{Payback} = \frac{\$700,000}{\$893,000} = 0.8 \text{ year.}$$

Table 3. Cost Comparison Including Less Tangible Benefits

Cost Classification	Conventional	AS/RS
From previous Illustration	$1,823,000.	$1,330,000.
Profit on Lost Sales	100,000.	—
Poor Management Decisions	300,000.	—
Total	$2,223,000.	$1,330,000.

Room for improvement

More and more companies understand the true benefits of an AS/RS and are, therefore, able to justify such systems. But, there is also room for improvement, and the following will highlight three areas in which substantial improvements can be made.

Controls

The rapid technological advances in the mini- and microcomputer technology have greatly contributed to the growth of the AS/RS industry and will continue to do so. The computerization of automated storage/retrieval systems has become a major justification factor. The fact that the AS/RS computer frequently extends into areas of manufacturing and distribution very often becomes a determining justification factor.

Further economy can be realized by the new trend of providing a single source, totally integrated control package, where the same design philosophy and similar components are used throughout the system. Obviously, this represents tremendous implementation and maintenance advantages over a control package, which is comprised of many different subsystems.

Hybrid systems

The early AS/RS systems were traditionally used for warehousing finished goods at the distribution center or for storing raw materials at the factory, where a high volume of materials needed to be stored and retrieved. This picture is changing. Many companies are looking at AS/RS for their entire inventory of materials. The way to do this is to combine more conventional methods with an AS/RS where the common link is the computer control system. For instance, the computer can track and control materials stored in more conventional storage configurations and thereby include the company's entire inventory in an overall system. This not only solves a common problem for many companies, but it also gives the AS/RS a broader base for justification.

Implementation

We have made tremendous technological advances during the past decade. However, we have not advanced as rapidly in the area of implementation of these systems; it's an area that leaves much room for improvement. The implementation deficiencies are also more noticeable and costly as systems become larger and more complex. We must all learn to do a better job in implementing these systems. Both the systems supplier and the user can contribute to this area.

After all, a lot of the users' resources go into designing and implementing an AS/RS. There is a further economy in having the user and supplier work close together in regard to implementation. It is extremely important for the user to be involved throughout planning, design, manufacturing, installation, start-up and training; and training covers the entire organization, all the way from the top executive to the operators.

Some might say that this defeats the turnkey principle, perhaps, but as far as I am concerned, that term does not really apply to our industry. There really is no such thing as a turnkey AS/RS. I would rather like to think in terms of total systems contractors possessing the unique combination of talents and expertise so necessary to put together an automated storage/retrieval system.

An automated storage/retrieval system has two major elements—hardware and implementation. Unfortunately, the implementation end of the system is always underestimated. So much so, that some early customers were not even willing to pay for it.

To better illustrate what I am talking about, I would like to share with you what I call "the systems iceberg." Here, the hardware is the immediate visible portion and the implementation portion is under the water line. This, I believe, fairly represents the proportions involved. I will, however, make a few comments:

□ *Project management*—A good and experienced project manager is essential to successful project implementation. However, it is a trained project team having all the combined talents required to put together an AS/RS that makes the difference between a modest and successful system.

□ *Installation*—An automated storage/retrieval system demands precision installation. The site manager therefore plays a key role on the project team. However, a good site manager has no impact unless considerations have been given to all installation aspects during the planning and design phase. Some companies have the tendency to hope for the site manager to perform miracles; however, I have seen very few situations where this actually occurs.

□ *Training*—Everyone agrees that training is extremely important, but when it comes to committing personnel and funds for this crucial activity, there is very often a shortage of both. I see a distinct relationship between the amount of training given and the amount of time it takes to start up a system and get it into operation. It is, therefore, hard to believe that training is so frequently short-changed.

The objectives and framework of the training program must be established in the contract documents. However, the detailed program should be developed during the design phase with user input. The proper training program has both classroom and hands-on instruction.

One of the most common reasons for short-changing training is lack of funds. A firm may even pay for instruction but neglect to set aside funds to pay salaries for its own personnel. It costs money to take people out of production to be trained, and this cost must be part of the overall project cost.

□ *Start-up*—The start-up phase takes place after the system has been accepted. At that time it is gradually taken into operation. This is the time when the user takes over and the changeover can, unless properly planned for, be a painful experience. A proper plan must be developed to detail exactly how the system is to be loaded and how it is to practically interact with the rest of the operation. This plan must also stipulate how the system will operate while the conversion from the old to new system is made.

Considerations must be given to pallet requirements, spare parts provisioning, operating procedures, staffing, preventive maintenance, warranty, systems audit, etc.

I have talked about the major system elements and described the most common system configurations available. I trust you will agree that automated storage/retrieval systems are more versatile and have more practical applications than most people realize. More and more companies can today justify automated storage/retrieval systems because they offer many proven and substantial benefits, but there is even room for additional improvement. This is evidenced by the substantial growth of the industry and its positive trends. IE

Ole B. Rygh has been in the AS/RS industry for the past 16 years. He has held positions as a systems engineer, systems consultant, project manager, division manager and is presently the senior vice president of Munck Systems Inc. Rygh is currently chairman of the membership committee of the Material Handling Institute and is a member of its product liability steering committee.

A Cookbook Approach to Robotics and Automation Justification

abstract

A "how to" approach for economic justification of robot and automation systems is presented in a step by step manner. Direct cost savings areas are identified and quantified, as well as methods for quantifying indirect savings areas such as, inventory savings, material handling, eliminated operations, reduced scrap and rework, operator training, and efficiency improvements are detailed. The step-by-step sequence leads the reader through the methods and calculations. After the numbers are generated, the reader is shown how to organize data, interpret it, and compute the return on investment (ROI) and the payback, using discounted cash flow methods.

author

Ronald J. Meyer
Senior Manufacturing Project Engineer
Rockwell International
Troy, Michigan

conference

Reprinted from the Robots VI Conference Proceedings
March 2-4, 1982
Detroit, Michigan

index terms

Robots
Automation
Computation
Efficiency

I. Introduction

Today's enthusiasm and interest in robot and automation technology for improving our productivity is long overdue. The excitement and level of interest for the first plant installation of a robot system usually is quite high. Based on some of the industry survey results and my own experience, management seems willing to relax the financial justification requirements for that "first installation". However, after the honeymoon, the realities of financially justifying the future systems require us to put some "hard numbers" on all those wonderful benefits.

There are basically two major reasons for this in most companies: 1) competition for capital with various new product programs, and 2) competition for capital with machine tool replacements and other plant and facilities equipment. In some companies, capital set aside for potential acquisition, or new plant expansion, creates additional demands on the company's limited financial resources.

The ability to prove convincingly a robot or automation system is a good investment, and prioritizing its capital needs within the constraints of a limited capital budget seems to require the talents of a "super engineer".

The proposed system may double or even triple current production rates. One, two, or maybe three direct labor operators may be replaced by the system. Additional benefits may include better shop relations with the Production Control Department due to more consistent production output. Also, Quality Assurance may like the 100% parts' inspection, if automatic gauging is installed as part of the system. We know this is all true, but when we attempt to put dollar figures on these facts, the numbers do not seem to show the benefits which exist.

The reason may be some of the savings areas are being neglected or omitted in the justification analysis. These dollar savings may require some digging to quantify. They are not the easily quantifiable direct savings, but are the harder to quantify indirect savings, i.e., reduced costs associated with inventories, reduced scrap and rework, greater machine efficiency and utilization, reduced operator training time, eliminated secondary operations, reduced inspection costs, etc.

Some accounting and finance people consider these savings to be of minor importance, and discount their potential for additional savings. However, indirect savings can have substantial effects on ROI calculations and payback periods. Therefore, careful scrutiny of all potential indirect savings should be considered.

II. Inventory Savings

Accounting and finance people get nervous when manufacturing engineers try to quantify and incorporate inventory savings into their justification methods. Placement of dollar figures on this category of savings requires a certain amount of investigative effort and assistance from Production Control and Accounting.

The major premise regarding inventory savings is the following: producing piece parts at a faster, more consistent rate shortens the manufacturing cycle. The manufacturing cycle starts when the raw material is received, inspected, and issued to Production on a manufacturing shop order. Cost gets added to the raw material cost due to receiving, inspection, and issuing of the material. As the material moves through the shop, "value is added". However, value is only added when the material is being machined, stamped, formed, heat treated, plated, etc. Cost is added by material handling, waiting in queue, inspection, placing into inventory, etc.

When a robot or automation system produces parts faster, eliminates secondary operations, and moves the material through production faster, it lowers added costs. Since the produced part is at a lower cost, "working capital" needed to finance "work-in-process" (WIP) and finished goods inventories is reduced.

III. Indirect Savings

The other categories of Indirect Savings also require some good investigative work. In most cases, the hard part of the investigative work is due to a lack of adequate, detailed records. Therefore, "educated estimates" become the only attainable numbers. Some pretty stubborn resistance to these estimates may be put up by accounting and finance people. However, stand your ground! These estimates should not be omitted due to poor records. The savings do exist, and if they are disallowed, a value of zero has been assigned to them. This would certainly be more incorrect than the "educated estimates" would be. The most important thing to note is - make sure your estimates are reasonable, and can be audited! They should be calculated after certain facts, data, and assumptions are prepared and stated. Some compromises may have to be made with Accounting and Finance to gain their concurrence, however, be prepared to defend your numbers.

Regarding an audit of the system's performance after startup, the purpose of the audit should be for information to assist in preparing a capital appropriations request, for the next robot or automation system. This will permit you to gain experience with each additional installation, and will increase your credibility with Accounting and Finance.

IV. Justification Report

In order to assist the reader in working through a system justification, an example is presented in "Justification Report Format". Depending on the type of robot or automation system, certain savings categories could be omitted or expanded upon. Realistically, a single robot loading/unloading a diecast machine probably would not require the extensive analysis presented in this example. However, a system consisting of a flexible machining system, consisting of several machine tools and robots, may require further elaboration upon the various savings categories in the example. The Justification Report is intended to be submitted as backup to the capital appropriations request.

The example is divided into three sections. Section I details the Direct Cost savings calculations. Section II demonstrates the Indirect Cost savings calculations. The last section tabulates the combined dollar savings and shows how to convert them into a measure of profitability. The Discounted Cash Flow method is utilized for this conversion.

The example proposes a flexible manufacturing cell. The cell is comprised of a numerical control mill and center machine, two numerical control lathes, an automatic outside diameter gauging system, a robot with dual gripper, incoming/outgoing conveyors with part positioners, a programmable controller, and appropriate electrical sensors. The machine tools are assumed to be previously purchased. The robot, conveyors, gauging system, programmable controller, and electrical sensors are assumed to cost $225,000. Associated tooling, machine rearrangement, programming, and startup costs are an additional $25,000. Sections I, II and III on Page 1 of the Justification Report summarize this information.

The flexible manufacturing cell is to replace the current production method, i.e., one man operates the mill and center machine, one man operates the two numerical control lathes and inspection is manual based on an inspection frequency of one per twenty parts produced.

IVA. Direct Savings

Page 2JR summarizes the direct savings calculations. New and old methods are described and compared in regard to direct labor. The proposed flexible manufacturing cell requires one minute versus three minutes for the old manual method. Assuming the old method operates on a three shift basis at 6,000 annual hours, 2,000 hours per shift, the new method uses only 2,000 hours to produce the same output. A productivity advantage of three to one is realized.

The hourly direct labor rate plus variable burden costs (includes fringe benefits) is $20 per hour. Consequently, $80,000 in direct labor is saved by the new method. If any overtime were eliminated by the new method, it should be added to the direct labor savings calculation. However, for this example overtime is not a factor.

CHART 1

SOURCES OF DIRECT SAVINGS FROM ROBOTICS AND AUTOMATION TECHNOLOGY

CYCLE	Consistent Machine Cycle Times Predetermined and Optimized Floor-to-Floor Times No In-Process Manual Gauging Automatic Non-Stop Processing
SET-UP	Automatic Preplanning Reduced Tool Changing More Accurate Machine Response
EFFICIENCY	Less Downtime to Change Programs No Operator Fatigue Reduced Job Learning Time No Machine Time Waste
SECONDARY OPERATIONS	If operations are saved, the entire cost of operations, not just man-hour cost, is saved.

	Measure by Time Study	Measure Partially By Time Study	No Measure By Time Study
CYCLE	1. Floor-To-Floor Times Consistent	1. Optimized Machine Cycles	1. No Manual In-Process Gauging 2. Non-Stop Process Sequence
SET-UP		1. Reduced Tool Changing 2. More Accurate Machine Response	1. Automatic Preplanning
EFFICIENCY			1. Less Down Time to Change Tools 2. No Operator Fatigue 3. Reduced Job Learning Time 4. No Machine Time Waste

First year estimated system maintenance costs of $5,000 are subtracted from direct labor savings yielding a first year savings of $75,000.

IVB. Indirect Savings

The interesting part of the analysis, Indirect Savings, is now to be addressed. Indirect savings has been a largely neglected area in most justification attempts. Savings are more difficult to identify and quantify, therefore they have been ignored. However, as stated previously, indirect savings can be significant!

Page 3JR of the Justification Report summarizes the indirect savings dollars. As indicated, inventory expense savings and other miscellaneous indirect savings are totaled.

Page 4JR details the calculations for inventory savings from consistent floor-to-floor times. They can be calculated by breaking them down into three elements, inventory savings from:

A. Reduced Process Lead Time
B. Less Piece Cost
C. More Efficient Production Scheduling

Inventory savings evolve from the more efficient process of the manufacturing cell, which produces parts faster and more consistently than the old. manual method. The parts have a lower cost due to the new method, thus the work-in-process (WIP) inventory is of lower cost throughout the ten week manufacturing cycle. (See Figure 1) Additional savings are generated because the new manufacturing cell shortens the manufacturing cycle.

In any calculation of inventory carrying costs, three conditions require consideration. Inventory value is one. The length of time (days, weeks, months, etc.) the inventory is carried, i.e., inventory turns per year is second. The third consideration includes the costs of money (interest rate charges), storage, handling, damage, loss, obsolescence, and taxes. These costs can range from 12 to 30 percent of the value of inventory per year. For this example, 2% per month or 24% per year is used.

Page 4JR details the calculation due to reduced process lead time. One week of new method production replaces three weeks of old method production. Figure 1 graphically displays this result. The vertical axis represents dollars of cost starting with the raw material cost of $1.75 per piece. The incremental costs associated with receiving the material, inspecting it, and placing it into inventory, adds $0.25 per piece. Therefore, total material cost is $2.00 per piece. It should be noted that faster, more consistent manufacturing cycle times reduce the need to carry extensive raw material inventories. The Japanese KANBAN system is evidence of this. However, for simplicity's sake in this analysis, these savings are not calculated. They will be the subject of a future technical paper.

CHART 2

SOURCES OF INDIRECT SAVINGS FROM ROBOTICS AND AUTOMATION TECHNOLOGY

SAVING	SOURCES OF SAVING
IN-PROCESS INVENTORY	Faster Process Cycle Reduced/Combined Operations
FINISHED INVENTORY	Reduced Cost of Inventory Smaller Lot Sizes Faster Material/Process Cycle - Better Forecasting
SCRAP AND REWORK	Automatic Program - Less Operator Error "Proved" Program After First Run Reduced Scrap and Rework
INSPECTION	Automatic In-Line 100% Parts Check Higher Quality Parts
OPERATOR COST	Less Training Time to Reach Efficiency
TRUCKING	Fewer Process Operations - Fewer Moves No Inspection Area
FLOOR SPACE	Fewer Machines, Less In-Process Material Less Room for Incoming-Outgoing Jobs Less Waiting to Be Inspected Fewer Operations - Better Work Flow
GENERAL RISE IN DEPARTMENT EFFICIENCY	Renewed Operator and Foreman Enthusiasm New Look at Methods
MORE PROFITABLE BUSINESS	Better Delivery to Customer Better Quality Parts More Flexibility to Alter Standard Designs
LESS MACHINE OBSOLESCENCE	More Universal and Flexible than Special or Semi-Special Machines

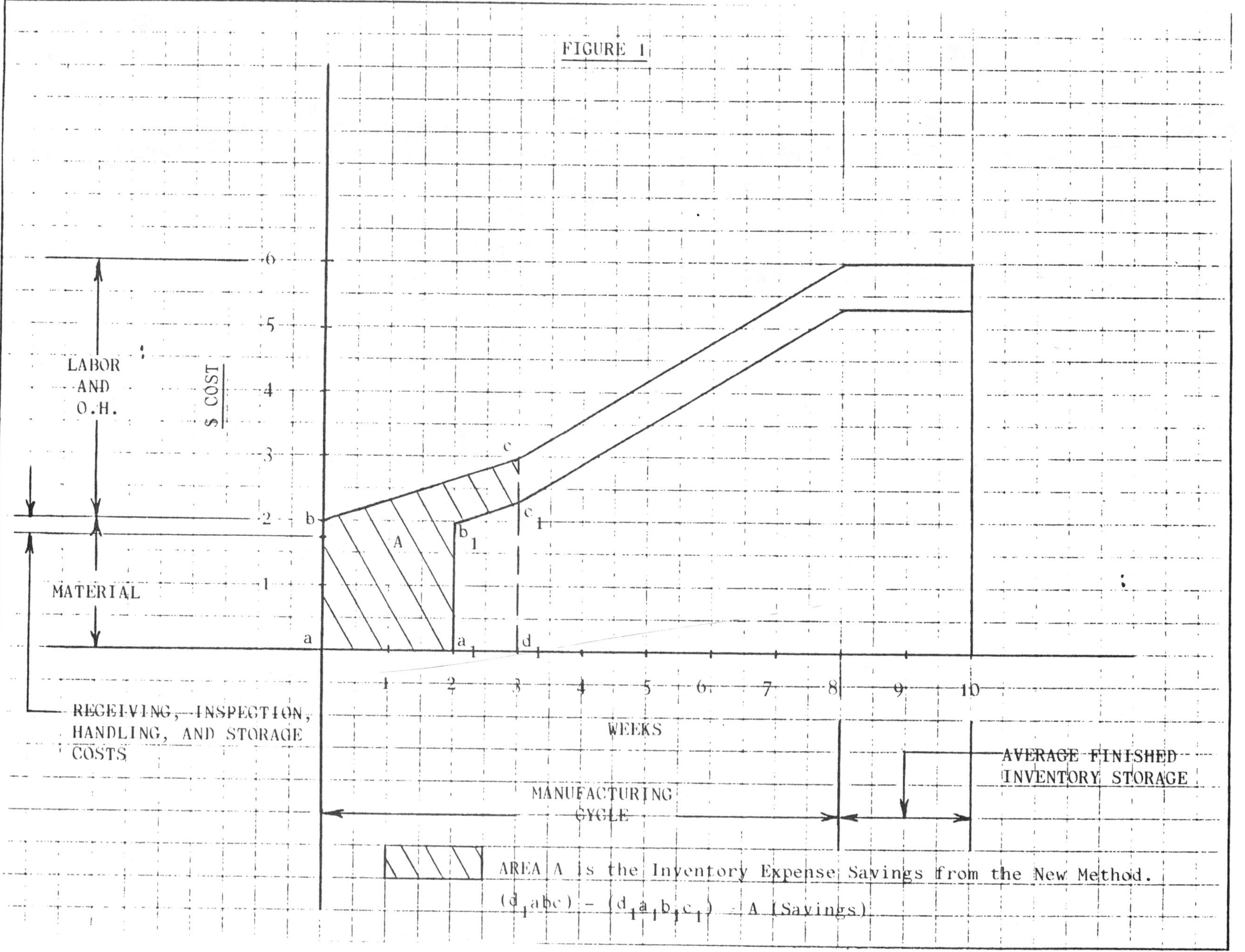
FIGURE 1
LABOR AND O.H.
MATERIAL
RECEIVING, INSPECTION, HANDLING, AND STORAGE COSTS
$ COST
6
5
4
3
2
1
a
b
c
b_1
c_1
A
a_1
d_1
1
2
3
4
5
6
7
8
9
10
WEEKS
MANUFACTURING CYCLE
AVERAGE FINISHED INVENTORY STORAGE
AREA A is the Inventory Expense Savings from the New Method.
$(d_1abc) - (d_1a_1b_1c_1) = A$ (Savings)

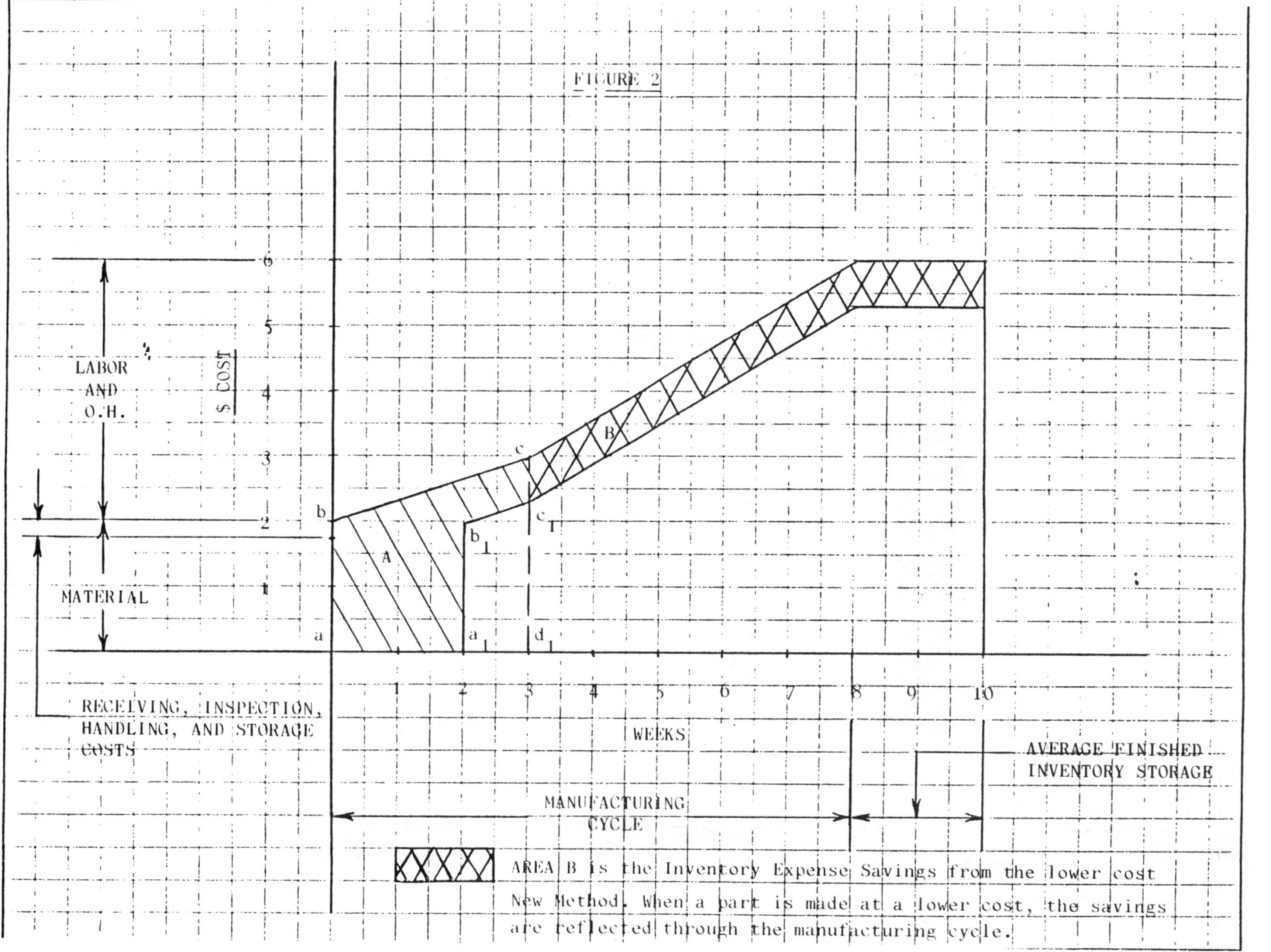
FIGURE 2
LABOR
AND
O.H.
$ COST
MATERIAL
RECEIVING, INSPECTION,
HANDLING, AND STORAGE
COSTS
6
5
4
3
2
1
a
b
A
B
c
c_1
b_1
a_1
d_1
1
2
3
4
5
6
7
8
9
10
WEEKS
MANUFACTURING
CYCLE
AVERAGE FINISHED
INVENTORY STORAGE
AREA B is the Inventory Expense Savings from the lower cost
New Method. When a part is made at a lower cost, the savings
are reflected through the manufacturing cycle.

FIGURE 3

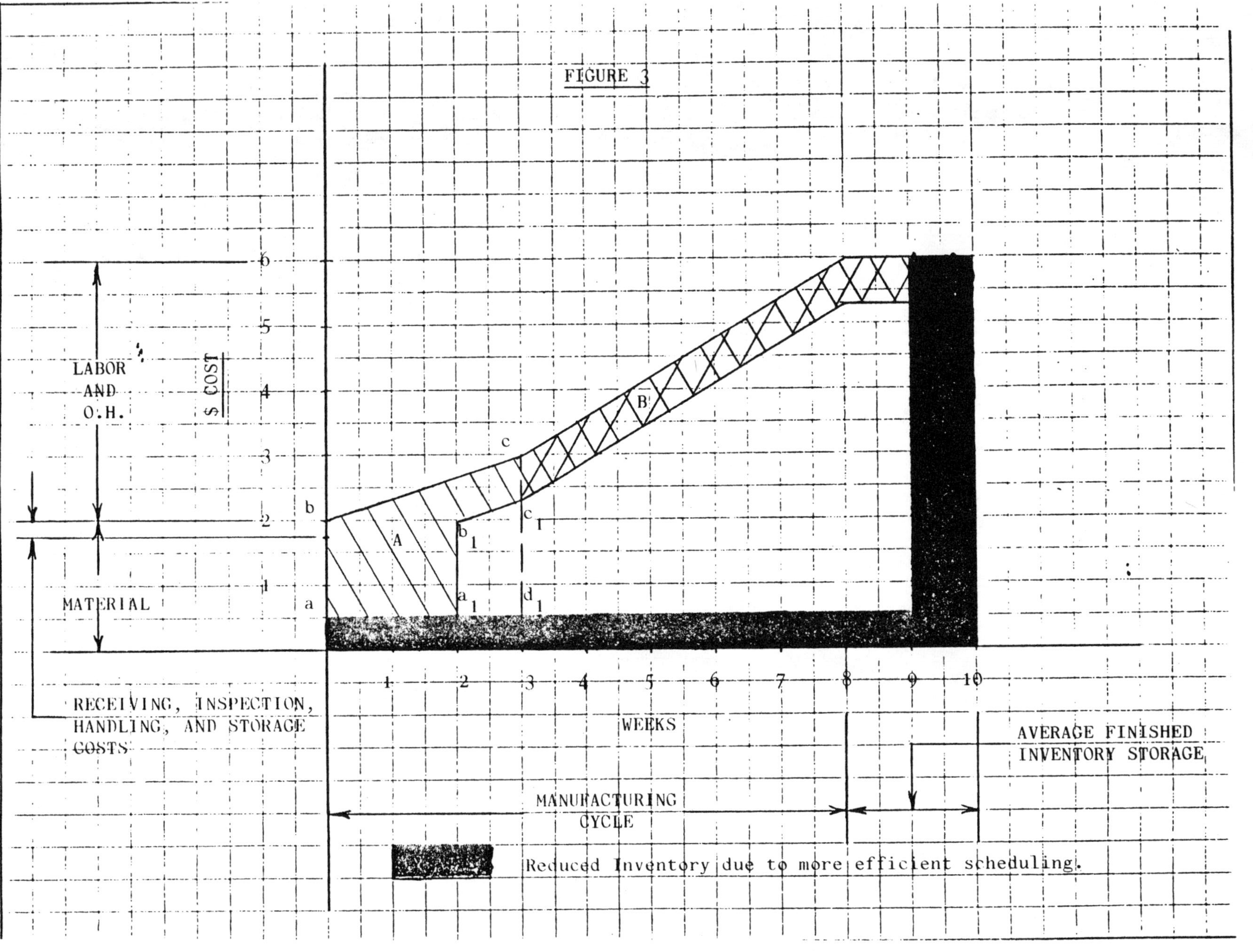

LABOR
AND
O.H.
$ COST
MATERIAL
RECEIVING, INSPECTION,
HANDLING, AND STORAGE
COSTS
6
5
4
3
2
1
b
a
A
B
c
b_1
c_1
a_1
d_1
1
2
3
4
5
6
7
8
9
10
WEEKS
MANUFACTURING
CYCLE
AVERAGE FINISHED
INVENTORY STORAGE
Reduced Inventory due to more efficient scheduling.

As graphically displayed (see Figure 1), cost increases as the material progresses through subsequent operations in the manufacturing cycle. The vertical axis indicates this cost increase due to material receiving, inspection, handling, storage, labor and overhead costs. The horizontal axis represents the time in weeks to complete the cycle, and the average length of time spent as finished goods inventory until shipped. The entire ten week cycle consists of eight weeks processing time, and two weeks finished inventory storage time.

Page 5JR calculates the inventory cost savings after the parts are processed by the manufacturing cell, and proceed through the manufacturing cycle for subsequent operations. Figure 2 graphically displays this effect. The old method cost is represented by Point C. The new method cost is C_1.

The last element in the inventory carrying cost savings is due to more efficient production scheduling. These costs are associated with the additional "safety stock" Production Control builds into the manufacturing cycle due to the inherent inconsistencies of the manufacturing process.

The logic to support the calculations on Page 6JR is, if with the new manufacturing cell it only requires one week instead of three weeks to get parts after release of a manufacturing shop order, the protective "safety stock" normally factored into the production schedule can be reduced or eliminated. This is due to the more consistent production output of the manufacturing cell. Graphically displayed in Figure 3, the savings are shown as a reduction in storage time and inventory value. In order to calculate the numbers, Production Control will need to determine the percentage improvement or reduction in "safety stock". It should be noted, some salesmanship may be necessary to elicit their cooperation!

The computation is a percentage reduction in inventory expense after completion of the manufacturing cycle. It is a definite cost savings based on the parts added to the weekly inventory at the conclusion of the manufacturing process cycle. Material and associated labor costs for the remaining operations are included.

Quite frequently, a robot or automation system combines or eliminates the need of a secondary operation due to manual processing requirements. Secondary operations such as sizing/inspection, polishing, deburring, washing, rustproofing, etc., can be combined into the robot or automation system. This eliminates the labor and additional operation costs associated with them. In this example, these operations are not combined or eliminated. However, discussion and calculation descriptions are included in the Justification Report. Figure 4 graphically illustrates where in the cycle this might occur.

The direct labor associated with the secondary operations is a direct savings. The calculation is detailed on Page 7JR. The results get added to Line G on Page 2JR.

The indirect savings are the same as the inventory cost savings calculations used previously. They are detailed on Pages 7JR and 8JR respectively. Figure 5 graphically displays the savings, and Chart 3 illustrates the corresponding lead time savings calculation.

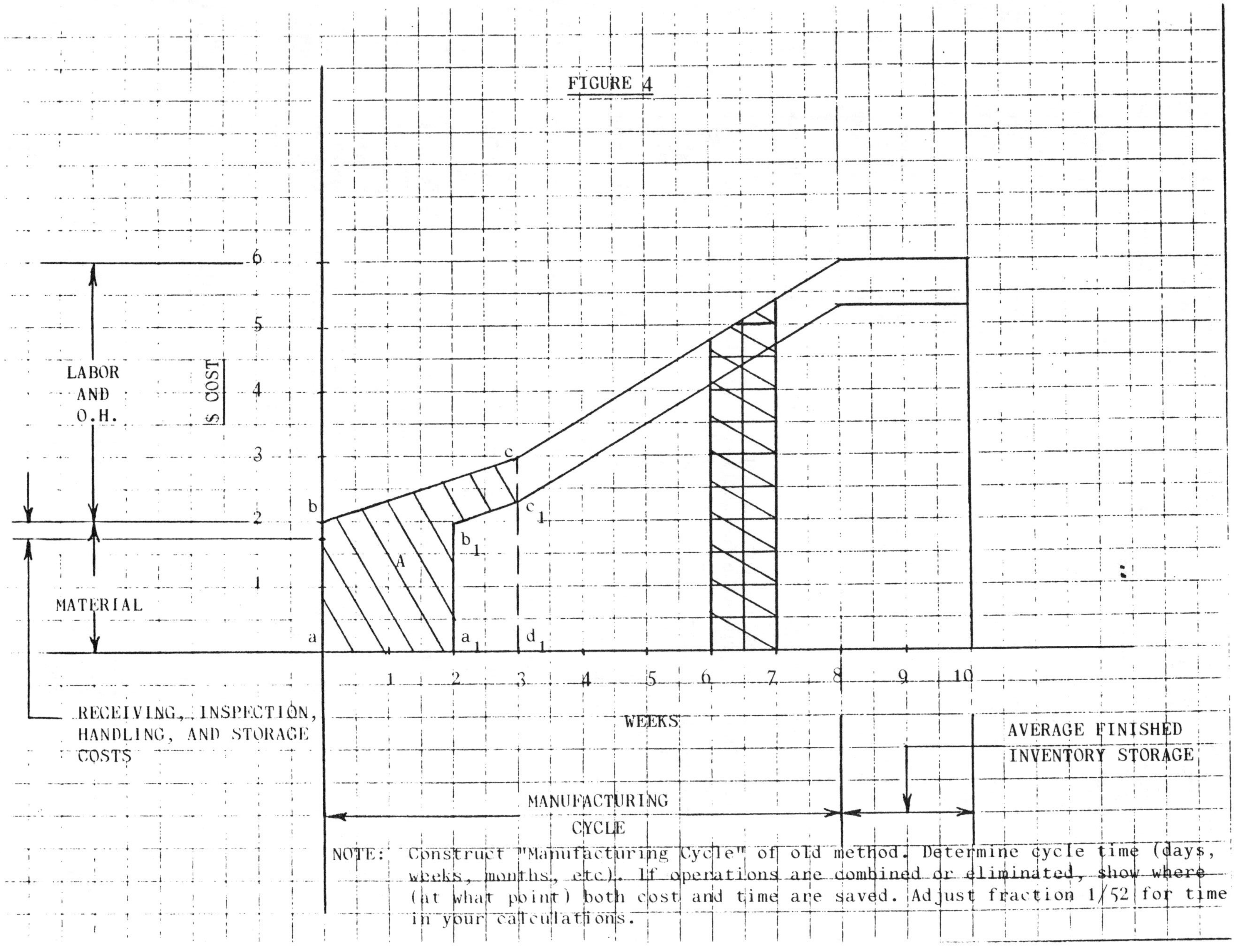
FIGURE 4
$ COST
6
5
4
3
2
1
LABOR
AND
O.H.
MATERIAL
RECEIVING, INSPECTION,
HANDLING, AND STORAGE
COSTS
a
b
c
A
a1
b1
c1
d1
1
2
3
4
5
6
7
8
9
10
WEEKS
MANUFACTURING
CYCLE
AVERAGE FINISHED
INVENTORY STORAGE
NOTE: Construct "Manufacturing Cycle" of old method. Determine cycle time (days, weeks, months, etc). If operations are combined or eliminated, show where (at what point) both cost and time are saved. Adjust fraction 1/52 for time in your calculations.

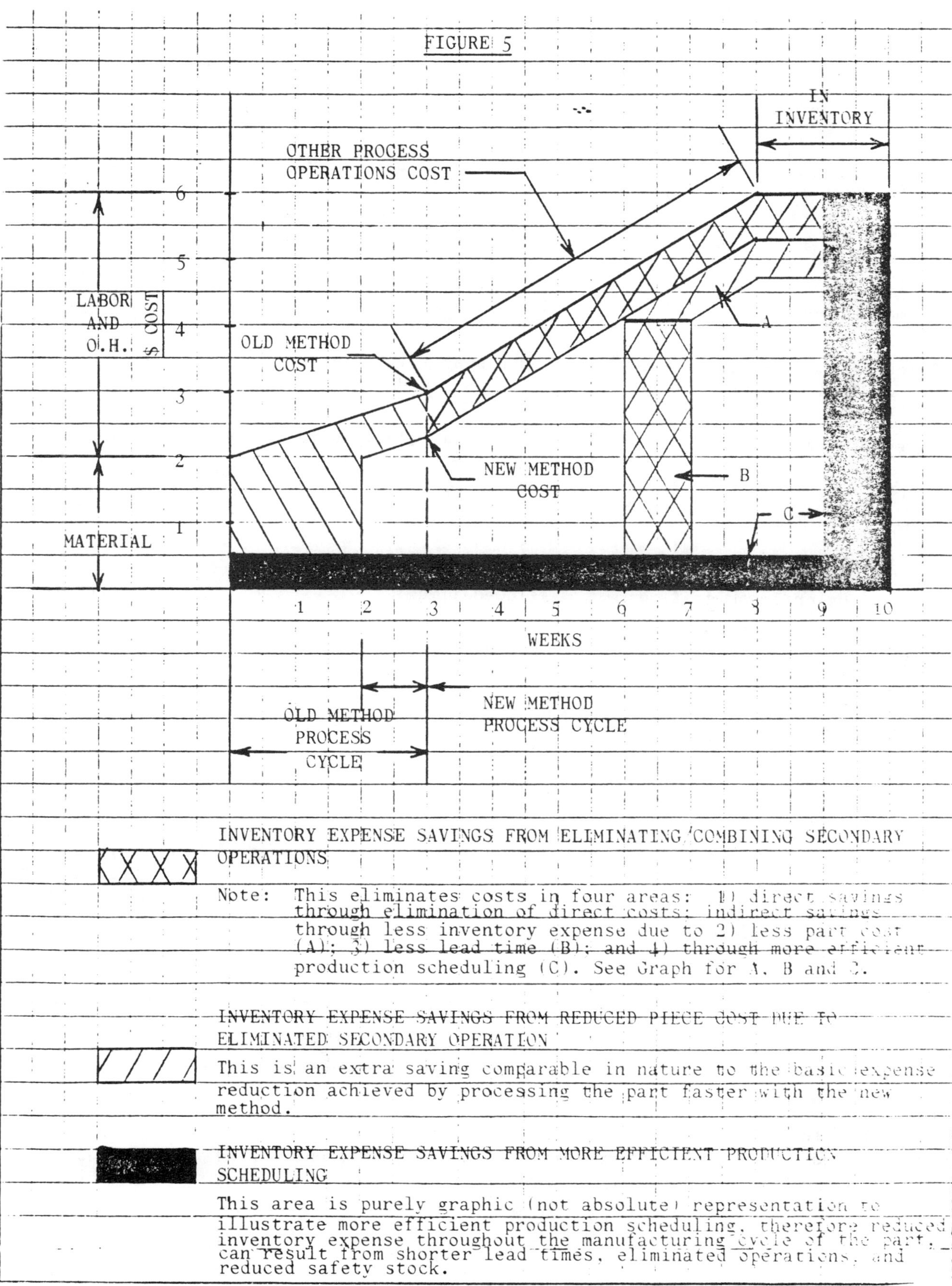

FIGURE 5
IN INVENTORY
OTHER PROCESS OPERATIONS COST
LABOR AND O.H.
$ COST
MATERIAL
6
5
4
3
2
1
OLD METHOD COST
NEW METHOD COST
A
B
C
1
2
3
4
5
6
7
8
9
10
WEEKS
OLD METHOD PROCESS CYCLE
NEW METHOD PROCESS CYCLE
INVENTORY EXPENSE SAVINGS FROM ELIMINATING/COMBINING SECONDARY OPERATIONS
Note: This eliminates costs in four areas: 1) direct savings through elimination of direct costs; indirect savings through less inventory expense due to 2) less part cost (A); 3) less lead time (B); and 4) through more efficient production scheduling (C). See Graph for A, B and C.
INVENTORY EXPENSE SAVINGS FROM REDUCED PIECE COST DUE TO ELIMINATED SECONDARY OPERATION
This is an extra saving comparable in nature to the basic expense reduction achieved by processing the part faster with the new method.
INVENTORY EXPENSE SAVINGS FROM MORE EFFICIENT PRODUCTION SCHEDULING
This area is purely graphic (not absolute) representation to illustrate more efficient production scheduling, therefore reduced inventory expense throughout the manufacturing cycle of the part, can result from shorter lead times, eliminated operations, and reduced safety stock.

CHART 3

LEAD TIME PER OPERATION (EXAMPLE)

Time Element	How Measure	Typical Time
Production Time on Whole Lot for Each Operation Saved	480 Pieces x 1.0 Minutes Each = 8.0 Hours	1.0 Shift
Queuing Time	Allowed Waiting Time to Get On Machine	2.0 Shifts
Inspection Time	Wait for Inspector to Approve and Solve Problems	0.5 Shift
Traveling or Trucking Time	Allowed Time	1.0 Shift
Waiting for Tools and Fixtures	Should Be Preplanned, But Average Delay	0.5 Shift
Unexpected Interferences	Machine Load, No Operator, Scheduling Goof, Lost Paperwork, Lost Parts, Excessive Scrap, Average Delay	3.0 Shifts
		8.0 Shifts

Total Lead Time Per Operation = 1 Week

(Assumes 1 Shift per Day Operation)

Use this Table as a guide in determining the approximate amount of lead time saved by eliminating/combining secondary operations. The fraction 1/52 in the equation on A1, Page 7JR, may be changed to read more or less than one over fifty-two.

IVC. Miscellaneous Indirect Savings

The final areas to be considered in the Justification Report are the miscellaneous indirect savings. Operator training, inspection, scrap and rework, material handling, floor space and department efficiency improvements are included. It should be noted that these miscellaneous areas are not an all inclusive list. Basically, if a robot or automation system reduces costs in indirect areas not listed in this example, by no means should they be overlooked. Remember, a value of zero is assigned to them if they are eliminated. The miscellaneous indirect savings shown in this example, frequently apply to robot and automation system installations.

1. Operator Training

Operator training is a continuous cost due to normal job turnover. It is assumed for this example that the job turnover rate is 50% per year. Turnover includes operator job changes, training required for relief and temporary fill-in operators, etc. The calculations are detailed on Page 9JR.

For the old method, training time requires approximately twelve weeks before the operators are performing at standard production rates. The average efficiency loss during training is about 20%.

For the new method, on the job training is estimated to require only four weeks to reach standard production rates. The benefits of the new method are only one operator requires training instead of two. Also, training is easier due to the automation of the process. The operator basically performs a supervisor function over the process. The system's production rate is predetermined by its design, thus requiring less direct input from the operator. The system determines production output, not the operator.

2. Inspection Savings

In most cases, when automatic gauging is installed as an integral part of a robot or automation system, the need for manual inspection of parts becomes eliminated. In this example, the parts receive a 100% continuous inspection by the automatic gauging system, thus eliminating the manual inspection requirements of the three machines. In this example, a yearly savings of $36,000 is realized. The calculation is detailed on Page 9JR.

3. Scrap and Rework

Scrap and rework costs associated with the old method should be greatly reduced. Since the parts are 100% inspected, when a part becomes out of tolerance, the system will flag the operator to make corrections. Only one part may require to be reworked or scrapped versus a whole production lot of parts with the old, manual method. Also, the costs associated with reinspecting out of tolerance parts are saved as part of the rework charges. As detailed on Page 10JR, these costs are significant. The old method scrap and rework rate of 5% versus the new method rate of only 1% shows an annual savings of $75,200.

4. Material Handling Savings

Material handling or trucking within a plant is one of the most inefficient and expensive operations. However, in most plants, forklift truck operations are not detailed on the process routing sheet. When a robot or automation system combines or eliminates operations, it reduces the material handling costs associated with moving parts from operation to operation. The forklift trucking savings are detailed for this example on Page 10JR. The $75 per hour forklift truck costs include the operator and all associated overhead costs to maintain the forklift operations.

Various industry studies indicate that an average part being processed in an average manufacturing plant spends somewhere between 5% to 15% of its total process time being machined, stamped, heat treated, etc. The remaining 85% to 95% of its time is being handled or trucked from operation to operation, or waiting in queue. Therefore, when a robot or automation system greatly reduces or eliminates forklift truck operations, the savings involved should become an integral portion of the justification economics.

5. Floor Space Savings

Floor space has an economic value in justification economics if less floor space is required by the new system. Its economic value can be classified as a first year savings. If more space is required, it becomes a penalty. In most cases, a robot or automated system uses less floor space. This is due to the consolidation of the machines and the reduction in floor space due to elimination of pallets of in-process material for each machine. The calculations are detailed on Page 11JR and amount to a first year saving of $7,500. This saving is realized in the first year of operation only! This is due to the assumption that the floor space saved will be put to productive use, and does not get saved throughout the life of the project.

6. Department Efficiency Increase

The final element considered in this example is the projected departmental efficiency increase. There are various methods for tracking this. Usually it can be measured as a reduction in non-standard hours and reflected as a percentage increase in efficiency. Some discussion with the department foreman is recommended to arrive at the percentage increase. The calculation is detailed on Page 11JR and for this example amounts to an annual savings of $6,000. The calculation shows that one operator per shift has been saved. The corresponding reduction in non-standard hours associated with that operator multiplied by three shifts, translates to a projected 5% efficiency increase.

V. Data Tabulation and Return on Investment Calculation

The following tabulation and calculations utilize the data compiled on Pages 1JR and 3JR of the Justification Report. Figures are rounded to the nearest dollar.

Annual Direct Savings (IVA)	\$ 75,000
Annual Indirect Savings (IVB)	151,253
1. Total Annual Savings (1st Year)	\$226,253
Annual Operating Costs:	
Maintenance	\$ 5,000
Taxes, Insurance	2,000
2. Total Annual Costs	\$ 7,000
3. Net Annual Savings (1 Minus 2)	\$219,253
4. Net Annual Savings After Taxes (52% of 3)	\$114,012
5. 10% Investment Tax Credit (10% of IIIF, Page 1JR)	\$ 25,000
6. Depreciation Allowance Credit (48% of \$83,333)	\$ 40,000

(Sum of Years Digits Used, based on 5 years write off 5/15 x \$250,000, 4/15 second year, 3/15 third, 2/15 fourth, 1/15 fifth year.)

7. Total 1st Year Savings	\$179,012
2nd Year Savings	138,512
3rd Year Savings	130,512
4th Year Savings	122,512
5th Year Savings	114,512

Note: Floor Space savings of \$7,500 is included in first year savings figure only.

8. Total Savings (Years 1 through 5)	\$685,060

Using the following equation:

$$\text{Present Value} = \sum_{1}^{N} \frac{\text{Total Average Annual Net Future Saving}}{(1 + r)^N}$$

$$\text{Total Investment Dollars} = \text{Total Average Annual Net Future Savings} \times \sum_{N=1}^{5} \frac{1}{(1 + r)^N}$$

$$\therefore \frac{\text{Total Investment Dollars}}{\text{(Total Annual Net Future Savings Years 1 through 5)}} = \sum_{N=1}^{5} \frac{1}{(1 + r)^N}$$

$$\frac{\$250,000}{\text{(Savings for Years 1 through 5)}} = \sum_{N=1}^{5} \frac{1}{(1 + r)^N}$$

Solving the equation for the present value of r equal to the total investment of $250,000.

$\therefore \quad r = .5258$

Return on Investment (ROI) = 52.58%

VI. Summary

In summary, this Justification Report has been developed in a cookbook approach format to assist the Manufacturing Engineer in the financial justification of Robot and Automation Systems. Depending on the type of system being justified, some additional savings categories may be added, whereas simpler systems may require deletion of some of the categories. Successful justification, i.e., resulting in an approved appropriations request, must contain all the potential direct and indirect cost savings benefits. If valid savings benefits are omitted or disallowed, the Return on Investment determination will be distorted.

The follow-up audit after the system is installed should not be neglected! It is your feedback mechanism to assist in your next system justification. Specific close follow-up on the indirect savings benefits will help "sell" their inclusion in future justifications. As shown in the example presented, they justify an investment that direct savings alone would leave questionable!

BIBLIOGRAPHY

A COOKBOOK APPROACH TO ROBOTICS AND AUTOMATION JUSTIFICATION

Anthony, R., "Management Accounting". Richard Irwin Company, 1965.

Fullerton, B. T., "Machine Replacement for the Shop Manager". Huebner Publications, Inc., 1965.

Harrington, J., "Computer Integrated Manufacturing". Industrial Press, 1973.

Tanner, W. R., " 'Selling' the Robot Justification for Robot Installations". Society of Manufacturing Engineers, 1978.

Warner and Swasey Company, "Introduction to the Concept of Total Measurement of Direct and Indirect Savings", Form 71003, 1971.

Woods and De Garmo, "Introduction to Engineering Economy". Mac-Millan Publishers, 1973.

ROBOTICS/AUTOMATION - JUSTIFICATION REPORT

Prepared By Ronald J. Meyer

Attention Division General Manager

I. PROJECT NUMBER RJM-8234R6

II. PROPOSED EQUIPMENT

(1) Robot with Dual Gripper

Incoming/Outgoing Conveyors

(1) Programmable Controller,

Part Positioning Tooling, Electrical Sensors

III. COST

A.	EQUIPMENT (ROBOT, P.C., CONVEYORS)	$ 225,000	
B.	INSTALLATION	$ 10,000	
C.	SPECIAL TOOLING COSTS	$ 5,000	
D.	PLUS ONE TIME COSTS Startup Training/Plant Rearrangement	$ 10,000	
E.	LESS RESALE VALUE/SALVAGE OLD EQUIPMENT (IF ANY)	$ 000	
F.	NET INSTALLED COST (A + B + C + D − E)		$ 250,000

IV. ESTIMATED FIRST YEAR SAVINGS

A.	(See Page 2) DIRECT	$ 80,000	
B.	(See Page 3) INDIRECT	$ 151,253	$ 231,253

V. GROSS FIRST YEAR RETURN

(Divide IV by III F)

92.5%

Submitted By Ronald J. Meyer

Date 3/4/82

DIRECT SAVINGS - FIRST YEAR SUMMARY

PRODUCTION INCREASE - REPLACEMENT BY DIFFERENT EQUIPMENT OR METHOD

A. Describe the new and old method very briefly. Point out the main advantages of the new method.

Old: Current mode of operation is one man operates mill and center, one man operates the two numerical control lathes, inspection is manual, one part per twenty produced.

New: The mill and center, two numerical control lathes, and 100% parts inspection by automatic gauging system, will be grouped with a robot in a flexible manufacturing cell.

Advantages: Current production is constrained, faster, more efficient operation, improved quality, consistent production, reduced and stable costs.

B. From production estimates on typical pieces, or other information, determine the time needed, in standard minutes, to produce by:

a. New Method 1.0 Standard Minutes

b. Old Method 3.0 Standard Minutes

Enter (b ÷ a) = 3 (Productivity Advantage)

C. Show number of standard minutes new method will operate per year at normal shop load. — 2,000 Standard Minutes

D. Multiply (B x C) to show equivalent standard minutes with old method. — 6,000 Standard Minutes

E. Subtract new minutes (Line C) from old minutes (Line D) to get total annual minutes saved. — 4,000 Standard Minutes

F. Divide by 60 minutes/hour and multiply (Line E) by hourly rate* to get value of hours saved. (* Direct Labor plus Variable Burden $20/Hour) — $ 80,000

G. List other significant annual savings because of the method improvement. (See Line 7, Page 7JR)

$ 0

H. Subtract any significant additional annual operating costs because of new equipment.

(Maintenance Costs) — $ 5,000

I. Total first year direct savings. — $ 75,000

(Post to IVA, Page 1JR)

INDIRECT SAVINGS - FIRST YEAR SUMMARY

A. INVENTORY EXPENSE SAVINGS FROM FASTER FLOOR-TO-FLOOR TIMES

1. From Reduced Process Lead Time	$ 15,550	
2. From Less Piece Cost	2,215	
3. From More Efficient Production Scheduling	1,828	
		$ 19,593

B. INVENTORY EXPENSE SAVINGS FROM ELIMINATING SECONDARY OPERATIONS

1. From Reduced Lead Time	$ 000	
2. From Less Piece Cost	000	
3. From More Efficient Production Scheduling	000	
		$ 000

C. MISCELLANEOUS INDIRECT SAVINGS

1. Operator Training	$ 4,800	
2. Inspection	36,000	
3. Scrap and Rework	75,200	
4. Trucking	2,160	
5. Floor Space	7,500	
6. Department Efficiency	6,000	
		$ 131,660
	TOTAL	$ 151,253

(Post to IVB, Page 1JR)

MEASURING INVENTORY EXPENSE SAVINGS FROM CONSISTENT FLOOR-TO-FLOOR TIMES

A. Reduced Process Lead Time

1. New Method One Week of Production, 120 Hours x $ 20 /Hour Department Cost = $ 2,400

2. Replaces 3 Weeks Prior Method, 360 Hours x $ 20 /Hour Department Cost = $ 7,200

3. Average Number Pieces Processed with New Method per Week = 14,400

4. Average Material Cost per Piece $ 2 x 14,400 (Line 3) = $ 28,800 Inventory per Week

Calculation:

Old Method

5. $\left[\$28{,}800 \text{ (Line 4) plus } \frac{\$\ 7{,}200}{2^{**}} \text{ (Line 2)} \right] \times$

$\frac{3}{52}$ Weeks (Line 2) x 24%* = $ 449 Weekly Expense
Calendar Weeks

New Method

6. $\left[\$28{,}800 \text{ (Line 4) plus } \frac{\$\ 2{,}400}{2^{**}} \text{ (Line 1)} \right] \times$

$\frac{1}{52}$ x 24%* = $ 138 Weekly Expense

Line 5 Less Line 6 = $ 311 Weekly Savings x 50 (Number of Operating Weeks/Year)

= $15,550 Annual Savings
(Post to A1, Page 3JR)

* 24% Assumes 2% per Month Inventory Carrying Cost (May Be Different for Each Plant)

** Process Cost Gradually Accumulates During Period. Average is one-half completed.

MEASURING INVENTORY EXPENSE SAVINGS FROM CONSISTENT FLOOR-TO-FLOOR TIMES

(continued)

B. Less Piece Cost

1. One Week of New Method Production _120_ Hours x $_20_/Hour Department Cost = $_2,400_
2. Replaces _3_ Weeks Old Method, _360_ Hours x $_20_/Hour Department Cost = $_7,200_
3. Line 2 Less Line 1, Reduced Inventory Cost for Each New Method Production Week = $_4,800_
4. After Process Operations, How Long Before the Average Piece is Shipped?
   10 Weeks - _8 Weeks_ = _2 Weeks_

Calculation:

$_4,800_ (Line 3) x $\frac{2}{52 \text{ Weeks}}$ (Line 4) x 24%* = $_44_ Weekly Savings x

_50_ (Number of Operating Weeks/Year)

$_2,215_ Annual Savings
(Post to A2, Page 3JR)

* In this example, the carrying cost of inventory stated at 2% per month includes storage, which in the case of these examples may not be regarded as saved.

MEASURING INVENTORY EXPENSE SAVINGS FROM CONSISTENT FLOOR-TO-FLOOR TIMES

(continued)

C. More Efficient Production Scheduling

1. From Line 4, Page 4JR, Material Cost per Week = $ 28,800
2. From Line 2, Page 5JR, Prior Operating Cost Equivalent to New Method Production Week = $ 7,200
3. Estimated Total Cost, Remaining Operations on Same Parts (If Any) = $ 14,400 (Line 3, Page 4JR) x $ 3 Remaining Operations Standard Cost Per Part) = $ 43,200
4. Add 1 Plus 2 Plus 3 to Get One Week Total Inventory Cost = $ 79,200
5. Estimate Average Length of Time Parts Will Be In Finished Inventory = 2 Weeks
6. Estimate % Efficiency Improvement in Production Scheduling $\frac{0.5 \text{ Week}}{10 \text{ Weeks}}$ = 5% .

Calculation:

$ 79,200 (Line 4) x $\frac{2}{52}$ (Line 5) x 24% = $ 731 x 50 (Number of Operating Weeks) = $ 36,554 x .05 (Line 6) = $ 1,828 Annual Savings (Post to A3, Page 3JR)

MEASURING INVENTORY EXPENSE SAVINGS FROM ELIMINATING OR COMBINING SECONDARY OPERATIONS

Frequently, Robotics/Automation will eliminate or combine entire operations. For example, using a manufacturing cell approach, several operations may be combined. The cell production output will be determined by the operation with the longest cycle time. You must calculate the savings in inventory expense by:

1. Measuring the Reduced Lead Time Savings - See A, Page 7JR for calculation.

2. Measuring the Reduced Piece Cost Savings - See B, Page 8JR for calculation.

3. Measuring the Additional Savings from 1 and 2 Due to Better Scheduling Decision - See C, Page 8JR for calculation.

Note: The elimination of an operation is also a direct saving and should be included on Line G, Page 2JR.

---

1. How Many of the Following Secondary Operations Will Be Saved on Each Work Piece: (This amount can be a fraction of one.)

   Washing — Sizing
   Deburring — Polishing - Burnishing
   Rustproofing

   ________ Number of Operations

2. Total Annual Part Production New Method ________

3. Line 1 Times Line 2 = Total Number Operations Saved ________

4. Average Production Per Hour for the Eliminated Operations ________

5. Average Departmental Cost Per Hour Eliminated Operations $________

6. Line 5 ÷ Line 4 = Cost Per Piece Eliminated Operation $________

7. Line 6 Times Line 3 = Annual Direct Cost Reduction $________
   (Include Line 7 in Summary, Line G, Page 2JR)

A. Calculation of Lead Time Savings

   1. Total Operations Saved (Line 3 Above) ________ x Prior Material and Manufacturing Cost ________ = ________ Total Prior Cost Annually.

MEASURING INVENTORY EXPENSE SAVINGS FROM
ELIMINATING OR COMBINING SECONDARY OPERATIONS

(continued)

B. Calculation of Less Piece Cost Savings

1. After the Secondary Operation, How Long Before the Average Piece is Shipped?

____ Weeks - ____ Weeks = ______ Weeks (See Figure 4)

Calculation:

$______ (Line 7, Last Page) x $\frac{\_\_\_\_}{52}$ (Line 1) x 24%* = $______ Annual Savings

(Post to B2, Page 3JR)

C. More Efficient Production Scheduling

1. Total Annual Cost Affected Parts (A1, Page 7JR) $__________

2. Estimate Average Length of Time Parts Will Be In Finished Inventory ______ Weeks

3. Estimate % Efficiency Improvement in Production Scheduling ________

Calculation:

$______ (Line 1) x $\frac{\_\_\_\_}{52}$ (Line 2) x 24% = $______

= $______ x ______ (Line 3) = $________ Annual Savings

(Post to B3, Page 3JR)

MEASURING MISCELLANEOUS INDIRECT SAVINGS

A. Measuring Savings In Operator Training Cost (On the Job)

Training Cost Old Method

1. Rate of Annual Operator Turnover 50 %
2. Number of Operators Required Old Method 2 Operators x 3 Shifts/Day = 6
3. Line 1 x Line 2 = 3 Operators to Train Annually
4. Training Time in Weeks 12
5. Average Efficiency Loss During Training 20 %

Calculation:

12 Weeks (Line 4) x 3 Operators (Line 3) x 20 % (Line 5) = 7.2 Weeks Loss of Production

7.2 Weeks Loss x 40 Hours per Week x \$ 20 Cost per Hour = \$ 5,760 Training Cost

Training Cost (New Method)

Annual Turnover 50 % x 3 Shifts x 4 Weeks to Train x 20 % Efficiency Loss x 40 Hours x \$ 20 per Hour = \$ 960 Training Cost

Old Method Training Less New Method Training = \$ 4,800 Annual Savings
(Post to C1, Page 3JR)

B. Inspection Cost Savings Due to In-Line or Automatic Gauging

1. New Method Combines/Eliminates 3 Machines
2. In Prior Method, One Inspector Handled 10 Machines
3. Line 1 Divided by Line 2 = 0.3 Inspector Prior Machines
4. Calculation $\frac{0.3}{0.3}$ (Line 3) x 3 Shifts x 2000* Hours/Shift x \$ 20 Hour = \$ 36,000 Inspection Cost
5. Estimate % Saving with In-Line/Automatic Inspection 100 %

   Line 4 x Line 5 = \$ 36,000 Annual Savings
   (Post to C2, Page 3JR)

*2000 Hours/Shift is used as an average. Actual hours depends on plant hours per year.

MEASURING MISCELLANEOUS INDIRECT SAVINGS

(continued)

C. Scrap and Rework

1. New Method Combines/Eliminates 3 Prior Machines
2. Average Number of Pieces Processed per Week New Method $ 14,400 (Line 3, Page 4JR)
3. Average Material Cost per Piece $ 2.00 x $ 14,400 (Line 2) = $ 28,800 Material per Week
4. Average % of Scrap and Rework Old Method 5 %
5. Average % of Scrap and Rework New Method 1 %

Old Method Scrap and Rework Cost:

6. 3 (Line 1) x 2000* x 3 Shifts x $ 20 /Hour = $360,000 Cost of Production
7. $ 28,800 (Line 3) x 50 Number of Operating Weeks = $1,440,000 Cost of Material
8. $1,800,000 (Line 6 Plus Line 7) x 5 % (Line 4) = $ 90,000 Scrap Loss Old Method

New Method Scrap and Rework Cost:

9. 2000 x 1 Shift x $ 20 Hour = $ 40,000 Cost of Production
10. $ 28,800 (Line 3) x 50 Number of Operating Weeks = $1,440,000 Cost of Material
11. $1,480,000 (Line 9 Plus Line 10) x 1 % (Line 5) = $ 14,800 Scrap Loss New Method

Subtract Line 11 from Line 8 = $ 75,200 Annual Savings (Post to 3C)

*2000 Hours/Shift is used as an average. Actual hours depends on plant hours per year.

MEASURING MISCELLANEOUS INDIRECT SAVINGS

(continued)

D. Trucking

1. Estimate Number of Different Parts on Which Operations Will Be Eliminated/Combined ___12___
2. Average Number of Times Each Part is Run per Year ___6___
3. Average Number of Operations Eliminated/Combined on Each Part ___1___
4. Number of Trucking Moves Required to Move the Average Job ___4___
5. Average Number of Moves per Hour per Truck ___10___
6. Trucking Cost per Hour $___75.00___

Calculation:

7. Line 1 x 2 x 3 x 4 = ___288___ Moves per Year Eliminated
8. Line 6 Divided by Line 5 = $___7.50___ Cost per Move

   Line 7 x Line 8 = $___2,160___ Annual Savings (Post to 3C, Page 3JR)

E. Floor Space

1. New Method Replaces/Combines ___3___ Prior Machines
2. Space Required for Machine and Parts Prior Machine ___150___ Square Feet x ___3___ = ___450___ Square Feet
3. Space Required for New Method and Parts ___300___ Square Feet
4. Line 2 Less Line 3 = Square Feet Saved $ x $___50.00___ Annual Cost =

   $___7,500___ Annual Savings (Post to 3C, Page 3JR)

F. Department Efficiency Increase

1. ___1___ People x ___2,000___ Yearly Hours x ___3___ Shifts x

   $___20___/Hour Department Cost = $___120,000___ Annual Cost
2. ___5___% Improvement in Efficiency x Line 1 = $___6,000___ Annual Savings

   (Post to 3C, Page 3JR)

Reprinted from ***American Machinist,*** *September 1983.*

How to justify multimachine systems

The economics of operating a flexible manufacturing system are sometimes apparent, sometimes hidden. Here are 14 costs to consider

OPERATING EFFICIENCIES achieved by grouping similar part families are now being expanded into multimachine groups for producing part families. Group technology, when applied to the production floor, yields great savings in operating costs.

To correctly justify the acquisition and installation of these multimachine manufacturing systems, a user must identify and assign dollar values to areas of operating efficiency. Fourteen critical costs can be defined to aid any justification study, regardless of the investment-ranking tool finally used to measure the proposed action. We'll take a look at each one of those costs; first, however, some broad observations about the economic-justification process itself.

The justification cycle has four stages. In the first, the operational evaluation, in which elements of savings are defined, the manufacturing-engineering function has the primary responsibility. During this stage, comparisons are made by considering operational differences due to the proposed changes in equipment.

In the second stage, basic economic assessments are made by the manufacturing function and by cost-accounting personnel. Here, dollar estimates are assigned to elements of the operational differences.

In the third, or financial-measurement, stage, the corporate finance division evaluates depreciation, special tax credits, and corporate-tax structure for the basic cost savings.

Finally, it is management's primary responsibility to rank the various investment opportunities. This is usually done by applying return-on-investment, discounted-cash-flow, or other formulas and strategies to measure the possibilities.

The first and second stages in the cycle are of primary concern if the proper information is to be provided to those who must determine the soundness of a new program.

In evaluation of two dissimilar manufacturing approaches, a schematic of each approach is useful. Schematics can also be meaningful in the formal justification presentation, assisting communication between manufacturing, design engineering, plant maintenance, and other branches of factory management. Fig 1 and 2 illustrate the use of such schematics.

In Fig 1, the conventional approach, production flows in a straight line from one stand-alone machine to the next, with manual transfer. Production of 21,000 parts per year comprises 19 different part numbers within five distinct part families. For such a system, we have found that equipment utilization is typically 100% of the available production year.

Fig 2, by contrast, shows an NC lineup of machines with random processing of parts and automatic transfer via asynchronous subsystem. Production in such a system would be identical to the present system in Fig 1, but the equipment would be utilized only 89% of the available production year.

By H. Thomas Klahorst, special-products manager
Kearney & Trecker Corp, Milwaukee

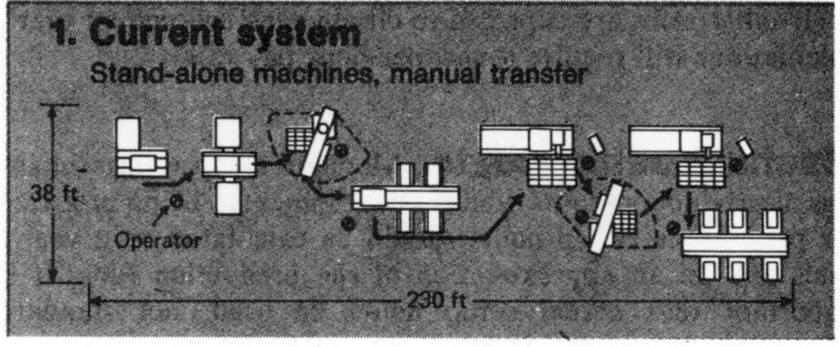

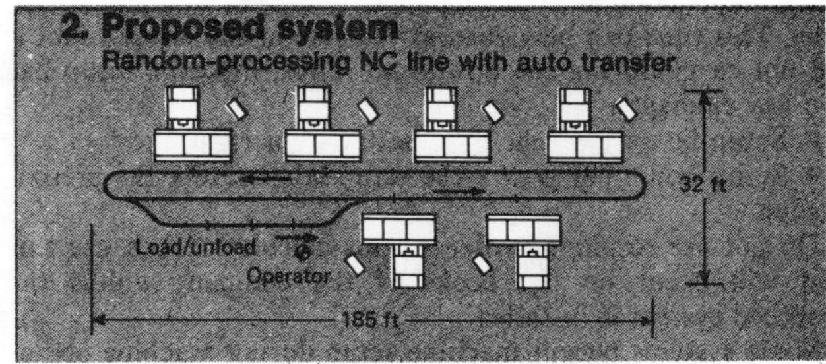

Modified burden rates

Many critical manufacturing costs are not directly proportional to the labor content in a manufacturing system. For this reason, Kearney & Trecker (Milwaukee), a machine-tool builder and systems supplier, recommends that a modified burden rate be used to give the proposed system a fair evaluation.

Examining key components of the burden separately and subtracting their effects from the overall shop burden permits a more accurate justification of any proposed project.

Direct-labor costs. Several elements are essential in calculating direct-labor savings: a productivity ratio, the residual burden rate, labor content, and the labor rates as projected for the period of the justification.

Use of a productivity ratio allows the system's productive capacity to be expressed as a common denominator by defining equal production requirements and evaluating machine load. In the two layout schematics, for example, dividing the current 100% utilization by the proposed utilization of 89% yields a 1.12 ratio.

Utilization estimates can be based on different efficiency ratings. A two-shift year (4000 hours) at 70% efficiency, for instance, yields 2800 available hours per year; at 80% efficiency, 3200 available hours each year.

In figuring labor content, many direct-labor-hour calculations are based on standard hours per part multiplied by part production per year. Labor variances and production rates can complicate this calculation tremendously. We suggest, therefore, that a comprehensive approach be used in estimating labor content, at least for the initial study. Looking back at the schematics (Fig 1 and 2), we see that

Current direct labor	7 operators per shift
Proposed direct labor	1 operator per shift
Saving	6 operators per shift

This simpler, more comprehensive approach can be used as a start. From the simple examples, direct-labor savings can be estimated: 6 workers x 2000 hours per shift x \$4.92 per hr x 2 shifts x 1.12 productivity ratio x (1 + 2.7 residual burden) = \$489,000 saving for the first year.

(If labor rates are expected to climb, each succeeding year's calculations will require a new per-hour pay rate.)

Machine-setup costs. Machine setup results in the loss of production hours while the equipment is being prepared for production. Setup hours should be calculated on a yearly basis and on an approximation of the production runs. It is important that actual setup hours be used, not standard allowances.

In addition to the cost of setup labor during part changeover, a high-cost element is the opportunity cost of lost production time. This time can be valued at the rate of the burden that it did not carry during the downtime. Thus, machine setup has two key elements:

- Setup hours per year x the setup-labor rate.
- Setup hours per year x the setup-labor rate x the present burden.

Do not use residual burden in this calculation because this cost will occur on the books of the company unless the proposed system is installed.

Some systems permit machine setup during machine operation, as with pallet-shuttle machines. In such cases, evaluate setup costs only at the setup hours, not including lost opportunity. Burden need not be calculated since the equipment absorbs its full share of overhead.

Tooling costs. Three attributes of tooling enter the justification framework: tool setup for production runs, tool storage, and tool maintenance. Tool setup at a machine tool, resulting in machine downtime, should be considered part of machine setup and not tooling. Tooling setup is defined as an off-line activity.

3. Sample tally list for tool maintenance

Tool type	Quantity in system	Frequency*	Standard rate (hr)	Resistant labor (hr)
Small mill	6	125	0.7	525
Large mill	6	83	0.5	249
Twist drill	461	55	0.1	2,535
Tap	392	31	0.2	2,430
C'bore	20	110	0.3	660
End mill	71	85	0.3	1,180
Reamer	7	65	0.7	318
Total tool-maintenance hours per year				8,527

* 4000 hours between changes

First, there is initial tool setup. Tool assembly for batch production must relate to the quantity of tools to be set and the frequency of setups during a production year. Itemize the differences in preparing tooling for production runs. Usually, an inverse relationship exists between tool-setup labor and machine-tool flexibility. Random production systems will not have extensive tool setup.

Calculate tool-setup costs with this formula: [(tool-setup hours per lot) x (setup-labor rate) + (setup-labor rate) x (burden rate)] x lots per year = setup costs for the system.

Next, tool storage. What are the average quantity and the typical types of tooling stored. Pay close attention to multispindle heads since they require floor area. A portion of the general tool crib should be allotted to the system's tooling. Current plant costs per square foot range \$20-\$35 or more.

Lastly, maintenance. A tally list, as in Fig 3, provides an estimate of costs for the FMS in Fig 2.

A general formula can be used to calculate tool-maintenance saving: (present total maintenance) − (proposed total maintenance) x (tool-regrind-labor rates) x (1 + current burden rate of regrind department).

Materials-handling costs. There are two aspects of materials-handling costs: movement of parts between machine units and capital investment in new equipment to remedy maintenance problems with materials-handling equipment in current facilities.

If it becomes desirable to estimate whether to automate handling, a simple capitalization formula may be used to convert worker-hours to capital investment. Fig 4 shows one

4. Estimating value of materials-handling automation

Current direct-labor support per shift (workers)	7.0
Time spent in materials movement (%)	30
Factored labor rate for handling (workers)	2.1
Shifts of operation	2.0
Direct labor per day in handling (workers)	4.2
Corporate cost per worker per year (w/fringes)	\$15,000
Yearly cost avoidance	\$63,000
Corporate amortization period	8 years
Approx break-even point	\$504,000

5. Example of evaluating materials-handling costs

Lift operations per day, activities	63.0
Average cycle time for lift, hours	0.2
Labor hours per day	12.6
Working days per year	250.0
Labor hours per year	3150.0

3150 x \$7.50 labor cost per hr x (1 + 3.85 burden rate)
plus
150 hr of maintenance of cranes at \$7.50 = \$115,700

way of determining how much automation you can afford. If we are to automate handling in the proposed (Fig 2) system, we must do it for \$500,000 or less. This is a guideline to more-formal justification work.

Current materials-handling equipment may require replacement or updating within the period of the justification. In this case, the required expenditure may be used as a credit against the new investment.

Part-inspection costs. Batch startup usually requires in-depth checks of first-run parts. Semirandom and automated systems usually eliminate many inspection needs.

The amount of normal on-line process inspection required is in direct proportion to the number of direct-labor workers involved in producing parts. Systems that reduce direct labor also reduce the chance for human error, and the saving in on-line inspection can be analyzed as in Fig 6.

6. Initial-part inspection cost

Average parts checked in startup	4.0
Actual inspection cycle, hr	1.5
Inspection work per lot, hr	6.0
Production startups per year	46.0
Yearly startup inspection, hr	216.0

216 x \$5.73 per hour with fringes x (1 + 2.67 inspection dept. burden) = \$4540 saved in inspection costs

Equipment-maintenance costs. Costs may vary depending on the extent of the shakedown period, on whether preventive-maintenance programs exist, on the equipment's own characteristics, and on the number of machines in the system. Fig 7 provides factors for estimating maintenance hours.

7. Estimating hours of equipment maintenance

Maintenance type	Percent of annual production hours: Conventional machine	NC equipment
Planned	0.5	3.0
Unplanned	2.5	5.5
Planned with diagnostic system	0.5	1.5
Unplanned with diagnostic system	2.0	3.0

It is reasonable to expect that, from the time of installation, an NC machine tool will have a high downtime rate but that this rate will drop to 5-10% after 6-8 weeks. Once stabilized, downtime becomes an important measure of the maintenance department's effectiveness.

The cost of equipment maintenance can be estimated in two ways:

- (hours of maintenance) x (maintenance labor rate) x (1 + maintenance-department burden).
- (hours of maintenance + direct labor rate) x (1 + residual burden). The first element is the pure cost of maintenance; the second is the lost productive value due to machine downtime.

Shop-supervision costs. The cost of supervisors varies in direct proportion to the number involved in the supervision, and the type and number of machines play an important role. Our studies have shown that one supervisor is required per every 17 NC machines, one per 50 transfer stations, one per 18 conventional machines, and one per 18 direct-labor workers. From these ratios, you can develop estimates of how much supervision is involved in the proposed project.

Production-control costs. Shop scheduling (a front-office function) and part dispatching (a plant-control operation) are the basic elements to be used in estimating in production-control costs.

Scheduling costs can be estimated with this formula: (machine units) x (lots per year) x (scheduling pay rate).

Dispatching costs are based on the lead-time needed to complete a particular part lot. Because about 2% of lead-time is taken by parts dispatching, expediting can be estimated: (lot lead-time) x (lots per year) x 0.02 x (scheduling pay rate).

Manufacturing-engineering costs. You will need to review initial-part processing work and analyze how engineering and process changes can be accomplished. Initial-part processing is important when new alternatives are contemplated. Engineering and process-changing flexibilities are always important elements to analyze.

8. Hourly factors for engineering changes

General type of machinery	Minor (no new tools)	Average (some new tools)	Major (new sections to routings)
Conventional	0.25	6.0	40
NC	0.25	2.0	16
CAM	0.25	1.8	14

Estimate engineering-change costs by using the following formula: Frequency of average change x factor x engineering pay rate + frequency of major change x factor x engineering pay rate = cost to implement production changes during the year.

Plant-facilities costs. Multimachine systems can save substantial amounts of floor space. The main evaluation is the market value per square foot of space, which can be subtracted from the gross investment in plant and equipment since it is a one-time saving (Fig 9).

9. Floor-space disinvestment

Current system in sq ft (38 x 20)	8740
Proposed system in sq ft (32 x 185)	5920
Floor-space saving, sq ft	2820
Floor-space market value, sq ft	\$15.00
Value of disinvestment	\$43,300

Another factor directly related to floor space is shop-maintenance costs. Since these recur each year, treat them as a cost saving. Studies show that maintenance averages 50¢/sq ft per year. For the two alternatives in Figs 1 and 2: 2820 sq ft saved x 50¢ cost/sq ft = \$1410 annually.

Inventory cost. Figuring the reduction of inventory is similar to calculating the shop-floor area in that there is a disinvestment element and a cost-saving element.

Three kinds of inventory must be considered: raw material, in-process, and finished goods. Random systems tend to reduce in-process and finished-goods levels but increase the need for raw materials on hand. Batch systems have the opposite effect.

10. Evaluating savings due to inventory reduction

Item	Value
Finished goods:	
Current average stock on hand, units	3,500
Current value per finished unit	$ 115
Current inventory value	$402,500
Proposed stock on hand, units	1,500
Value per finished unit	$ 115
Proposed inventory value	$172,500
Saving in investment	$230,000
In-process goods:	
Current quantity in in-process queues, units	850
Proposed quantity required in process, units	150
Net saving in units	700
Weighted average value of processed parts	$ 85
Saving in investment	$59,500
Inventory space:	
Units storage reduction, units	2,700
Space per unit for storage, sq ft	1
Total space reduction, sq ft	2,700
Value per square foot	$ 15
Saving in investment	$40,500

Inventory reductions and space savings should be deducted from the investment in plant required to affect these reductions.

Associated with the finished-goods inventory are additional carrying costs, such as handling in and out of storage, deterioration and damage, taxes, insurance, and interest. We find that 14½% is a good multiplier in figuring these additional carrying costs. Thus, if an FMS saves 2000 units per year (each valued at $115, for a total value of $230,000) from being placed into and then taken out of inventory, the additional carrying cost saved per year is ($230,000 x 0.145 =) $33,350.

Fixturing costs. Storage and maintenance of production fixtures can create added manufacturing costs. Floor space used for storage should be calculated as it was for inventory floor space. Remaining elements include the list of items shown in Fig 11.

11. Cost elements of fixture storage

Item	Cost as percent of current market value
Physical maintenance, %	6.0
Handling and transportation	3.0
Taxes and interest	8.5
Total cost per year, %	17.5
Calculation example:	
Current fixturing value	$385,000
Processed fixturing value	205,000
Value difference	180,000
Carrying-cost multiplier	0.175
Yearly saving	$ 31,500

Prototype and new-part costs. More-automated production systems may lend themselves more readily to new-part introduction. The saving is directly proportional to the amount of machining on a workpiece and the type of equipment being used. Fig 12 assumes a gearbox-type part that requires about 1 hr of metal-removal time. The cost to produce a prototype part in each of the two systems in Fig 1 and 2 is given.

12. New-part production

	Conventional line		NC line	
Element	Engineer	Shop	Engineer	Shop
Process routing hours	92		40	
Engineer rate	$7.40		$ 7.40	
Engineer-process cost	$681		$296	
Manufacturing hours		55		3
Manufacturing-cost rate		$ 28		$28
Manufacturing cost		$1,540		$84
Frequency per year	0.75	0.75	0.75	0.75
Resultant costs	$511	$1,155	$222	$63

Rework and scrap costs. The more highly automated systems can reduce scrap levels 40-60%, but it is hard to generalize on formulas to calculate these elements. The most logical approach is to evaluate current scrap and rework cost and to eliminate a percentage based on the quantity of finished machines in the system and the amount of direct-labor content in part production.

Now do the justification

After investigating the many cost savings and completing the calculations, you are ready to assemble some numbers that represent the value of the project. These will tell you whether the project is justifiable, and several approaches may be used.

Payback divides the average yearly cost saving into the net investment. It does not calculate the time value of money, consider uneven annual savings per year, or look beyond the payback period to assign a value for further returns. On the other hand, it is simple and does assign a risk value to a project.

Basic ROI (return on investment) is the reciprocal of the payback formula and divides the average yearly savings by the net investment.

Current-value ROI uses the actual yearly cost saving and discounts it to the date of installation. The basic idea is to find the interest rate that equates anticipated cost savings to a current-investment amount.

An investment produces income over a period of time, and the fact is that money does not retain a constant value over the years. In an economy in which interest and investment opportunities exist, a dollar received today could be invested to earn money immediately. But a dollar received a year from now cannot be invested until it is received; therefore, it is less valuable. Discounted-cash flow (DCF) and net-present value (NPV) are varieties of current-value ROI methods.

There are other methods. Regardless of the approach, however, the strength of any justification rests in the proper evaluation of the operating differences that will occur. In evaluating an FMS, the changes that will reach throughout the company must be considered ■

*Reprinted from **1983 Fall Industrial Engineering Conference Proceedings.***

FLEXIBLE MANUFACTURING SYSTEMS: AN OVERVIEW

Edward J. Phillips, P.E.
Consulting Associate
Richard Muther & Associates, Inc.
Kansas City, MO 64113

ABSTRACT

The FMS concept has evolved over the last twenty years. Low cost microprocessor technology is today making FMS a reality. This paper discusses where we are and where we are going with FMS along with its opportunities and problems.

INTRODUCTION

FMS's or Flexible Manufacturing Systems have been part of the manufacturing scene for at least the last twenty years. The developed western nations, along with Japan, have made steady but relatively slow progress in advancing this outgrowth of technology.

Automatic transfer lines which were pioneered by the automotive industry and were really the birth of the FMS have probably been in use more than twenty years.

However, we should not become fixed in our view of FMS's particularly in regard to machining systems. The FMS has a much broader scope than just machining. I have had personal experience on the development of similar processing and testing systems in the electronic industry. Just two examples would be the electronic laser trimming of random hybrid circuits as they are tested on a real time basis in one system, and the automatic sizing of gold electrodes on random piezo-electric components of various sizes in another system. This is also being done on a real time basis. Needless to say, neither of these systems, nor any FMS in general for that matter, would be possible without the computer. You cannot speak about FMS intelligently without at least some basic knowledge of computers. However, rather than expound on computer and electronics applications, this paper will generally be centered around what are considered traditional FMS applications. That is to say, machining applications.

In most industries today, we can no longer afford to set up fixed production lines such as Henry Ford originally did. There are just too many competitors offering an abundance of different product features and variety. The important word here is variety. This is really the forcing function behind the FMS. That forcing function along with the rapid development of microprocessor technology have been the key elements in the advancement of FMS technology.

At the beginning of this introduction, I mentioned the relatively slow growth of FMS over the past 20 years. With the extremely rapid advancement in microprocessor technology over the past 5 or so years, FMS can be expected to become relatively wide-spread in at least the United States and Japan by the end of the decade (1)(9).

This paper will briefly describe the advantages and problems associated with the typical FMS. I will also try and give some insight into where we stand worldwide with FMS installations. Finally, I will try and cite some of the U.S. examples and will briefly go over one with which the writer is personally familiar. A short 16mm film presentation will serve to highlight some of these examples.

FMS DEFINED

There has been no one overall accepted definitior of FMS. My definition is shown in Figure 1.

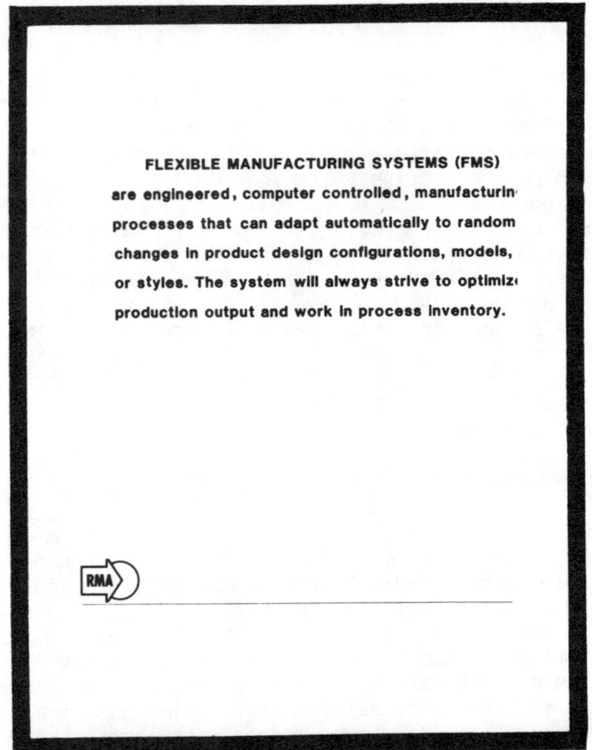

Figure 1. FMS Defined

In most conventional machining systems, workpieces are manually loaded and unloaded from unique fixtures on a particular machining center. This double handling at each particular machine during a series of processes is virtually eliminated with FMS. However, this is not true in all cases. Frequently, some additional set-up may be required within a cell, however, this situation is entirely dependent on the complexity of the workpiece.

It has been stated in the trade that some conventional machine shop's "door-to-door" times are 30 times greater than the actual machining "floor-to floor" times. FMS strives to narrow this gap.

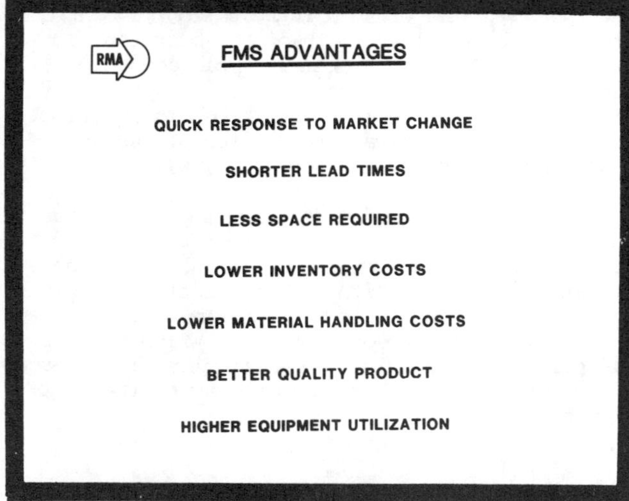

Figure 2. FMS Advantages

1. Quick Response to Market Change

Obviously, the faster our reaction time to a change in the "market", the faster our product will enter the marketplace. With FMS we don't have parts for model number x sitting on the floor until we break up the machine setup for model y. All parts are fed in a random sequence to the FMS. The computer directs the part to the most suitable machine. Setups and tooling tend to be more flexible and universal in nature, thereby reducing setup and tooling costs. Of course, the proper utilization of an FMS depends on a strategic commitment by all members of the individual companies involved. Production parts should be designed so that there is minimal effect on the FMS thereby offering maximum manufacturing advantage to the company.

2. Shorter Lead Times/Space Reductions

Some observers of FMS have estimated lead time reductions of up to 40% (3). The firm I am associated with, Richard Muther and Associates, Inc., is extremely interested in both this topic and also the "lower inventory" topic. Since two of our prime consulting fields of specialization are Facilities planning and materials handling, these lead time and inventory effects have tremendous impact for us. Long lead times equate with inefficient use of floor space and material handling problems in general. We have estimated that proper layouts of FMS departments can save up to 35% in floor space when compared to conventional machining areas.

3. Lower Inventory and Material Handling Costs

As previously noted, there is an abundance of wasted floor space in a typical machining area. That wasted space is usually full of production piece parts waiting for their next operation. Those piece parts usually translate into unneeded work in process inventory. This is one of the biggest space users in the typical machining area. Also, individual workpieces normally have to be loaded and unloaded manually at each conventional machine. Not only are the pieces double handled, but they are usually double handled four or five times in this process. Normally, in an FMS, parts are loaded and unloaded only once. Since the FMS will normally accept different parts in any order, work in process inventory is reduced to a minimum. There is no longer any need to queue up parts waiting for the next operation.

In addition, FMS usually creates a chain reaction effect along the entire inventory chain which includes both internal and external suppliers. The potential for cost savings in this one area is extremely high. Additionally, Material Handling costs are lowered throughout the system.

4. Better Quality Product

Primarily, from the reduction in parts handling throughout the system, FMS can provide far more consistency in resultant dimensions than conventional methods. This consistency is normally translated into improved product quality. Time savings in subsequent assembly operations may also be expected. Scrap levels are also generally reduced.

5. Higher Equipment Utilization

Another major advantage of the randomness features of FMS, equipment utilization can be expected to increase substantially. Again, this is an area in which we as facility planners are keenly interested. That increase in equipment utilization can usually be translated into a significant reduction in machine tool population. Of course, this in turn translates into the "freeing-up" of large amounts of existing floor space or a higher return on capital investment in new floor space required. Either way, it benefits our clients and can impact their P & L quite handsomely.

Figure 3 is a continuation of figure 2, and is shown on the following page.

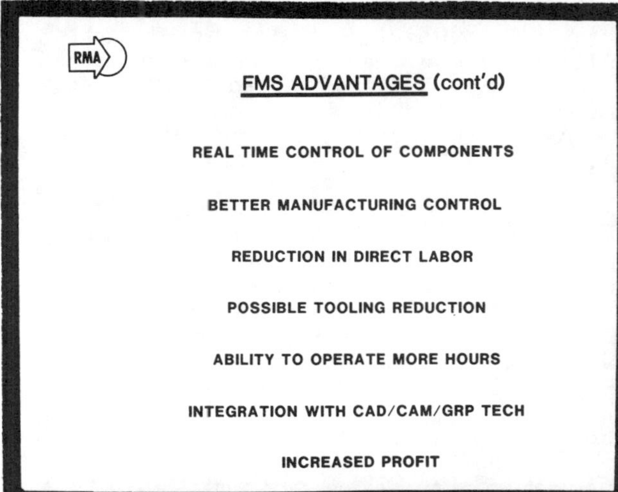

Figure 3. FMS Advantages (cont'd)

6. Real Time Control of Components

What is meant by "real time" control? In "real time" systems, we (and/or the computer) at any given time know exactly where each component production part is; the computer knows exactly what the status of each part is; it knows approximately, through sensor feed back circuits, how tool wear is affecting part quality; most importantly, it is preparing all machines within the system for the optimum next operation. Also, through inspection stations and self diagnostics, the computer will warn us if a problem is developing or it will shut portions of the system down and alert us if a critical problem has suddenly developed. The system will not only tell us what has gone awry, it will also offer us alternatives to get around and/or correct the problem. Of course all systems do not have all of these attributes today due to high cost, but they are available, and hopefully will be coming down in price during the next few years.

7. Better Manufacturing Control

This is almost a superfluous category and is an extension of all the other advantages. FMS clearly gives us better control. When tied to a well implemented MRP system, if there is such a thing, FMS offers an outstanding opportunity for both purchased materials improvements and scheduling improvements. Shop floor paperwork and inherent mistakes can be reduced to a minimum. Accounting and inventory procedures can obviously be improved. The systems manager is still able to override the computer's directions in the event the need arises either from urgent scheduling changes or machine breakdowns. An ideal FMS consists of a number of machining centers of the same type. If one breaks down, the host computer and/or the systems manager can redirect part movements to the optimum machine in the system.

8. Reduction in Direct Labor

Although reduction in direct labor costs are a primary target of FMS, they are not the only target. FMS can rarely be justified on labor savings alone. Lead time reductions, quality, and lowered inventory costs all play a key role in justifying FMS. Also, the space savings effects are not insignificant. Consider a proposed 50,000 square foot machining facility that could be reduced to 35,000 square feet in the planning stage by the use of FMS. At a conservative $30 a square foot building cost, that 15,000 square feet saved translates to a one time capital investment savings of $450,000, not to mention the ongoing upkeep costs associated with that square footage.

9. Possible Tooling Reduction

As noted earlier, FMS tooling tends to be much more universal in nature than conventional machine tooling. Robotic fixture changes within FMS cells tend to be expensive, but are effective for large systems. However, tooling is not always cheaper with FMS. Precise rotary type fixture tooling can be more expensive than conventional tooling. Some units have a full machine station dedicated to just rotating parts and/or fixtures in order to present the proper face of the part in the next machining operation. The "possible" tooling cost reduction stems from replacing many individual conventional fixtures with a much smaller number of FMS universal type fixtures.

10. Ability to Operate More Hours

For lunch breaks, shift changeovers, and coffee breaks it is only necessary to have an adequate supply of piece parts in queue for input to the FMS. The machine system should be able to run unattended during these normally short human work break periods. Of course, there should also be built-in safeguards and alarm systems. However, if we assume full three shift operation with 1/2 hour lunch and 1/4 hour breaks and shift changes the FMS can give us a theoretical 15% increase in machining time over conventional methods. Obviously, this isn't true in all cases, but the potential is there.

11. Integration with CAD/CAM & Group Technology

The FMS, when integrated with CAD/CAM and group technology, is clearly a component, or feature, of the factory of the future. Today, we are much more advanced in CAD than in CAM. At least the marriage of the two together has not really taken place. The forward movement of group technology with CAD may in fact be the go-between that ties CAD and CAM together. FMS can be thought of as the honeymoon vehicle for this marriage to take place. There is no doubt that computer coding systems for production parts and FMS go together. It is only a matter of time before the marriage takes place.

12. Increased Profit

All of the above attributes of the FMS lead, or will lead in the future, to increased profit. Rule of thumb numbers are thrown around the industry all of the time. Typical are 10% reductions in material cost, 50% reductions in direct labor costs, 50% cuts in lead time, 30% cuts in indirect labor and so on (3). FMS's are good, but it is highly doubtful that they are consistently that good! If that were true in general there would be a revolutionary rate of change in the United States to FMS. Why isn't there such a revolution? That leads in to the next topic.

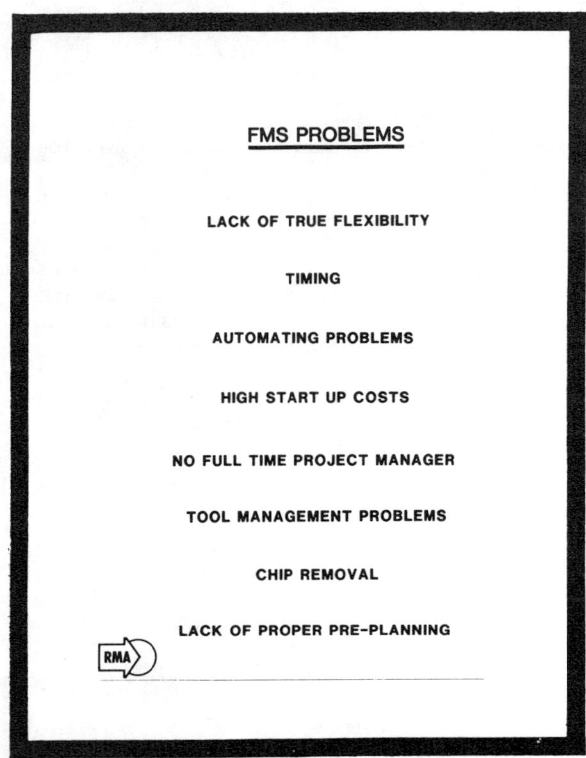

Figure 4. FMS Problems

1. Lack of true Flexibility

Flexibility in an FMS system is a function of the creativity of the equipment supplier and the user. That creative responsibility, however, is probably more needed by the user. The machinery in an FMS, that is the hardware itself, is usually not flexible at all. It is up to the planners to insure that flexibility is forced into the FMS during every phase of the planning.

One can't go out and buy an FMS "off-the-shelf" so to speak. The pre-planning, part selections, and planning for the future are key ingredients. This point can't be stressed enough. If one were to look at what appear to be FMS failures, they are typically failures in planning and not in hardware.

2. Timing

Typical FMS systems can easily take two to five years to implement. Again, planning is extremely important. The original parts that were thought to go first on the system may be obsolete by the time the system is fully implemented. Parts that haven't even been designed as yet today may have to be forced on the system calling for even longer delays.

3. Automating Problems

It is extremely important that we don't select parts for an FMS just because they give us so much of a manufacturing engineering problem today. These problems must be worked out before placing the parts in the system. Otherwise, one will only compound troubles and automate problems.

4. High Start Up Costs

Don't expect to plan, design, and install an FMS, "press the button" and watch it perform. There usually will be expensive changes and manufacturing engineering costs unplanned and not budgeted. It would be wiser to count on at least 6 months of de-bugging time.

5. No Full Time Project Manager

One study indicates that the part time assignment of planners for an FMS is almost a guarantee of failure. Almost all successful systems had a full time experienced project manager in charge of the project from the beginning until quite sometime after de-bug (3). This is absolutely necessary for the planning of a smooth operation.

6. Tool Management Problems

It is essential to reduce the cutting tool variety to an absolute minimum before going to an FMS. Looking at the typical NC machining center and the number of slots in its tool changer lulls us into a false sense of security. Why reduce the number of cutting tools required when we have so much capacity? There is no better method of reducing the flexibility of a machining cell than by not minimizing the number of cutting tools required. Some of the best planned FMS's have more than 4 days worth of machining capacity in each tool magazine. This leads to an exponential increase in the number of problems.

7. Chip Removal

This has become a particular problem with all automated machinery and not just FMS's. It is particularly true in Japan where manning is held very low. Titanium parts are almost bound to give problems at some time. Chips can cause major problems with subsequent part locations if they are not removed completely. Some FMS's are being planned today with at least one station devoted entirely to the cleaning of pallets and fixtures.

8. Lack of Proper Pre-Planning

Although touched upon earlier, one cannot overemphasize this potential problem area. Some larger companies feel they can just buy a system, place it on the floor, and if any problems develop, they can buy solutions. This course of action cannot be further from reality. Any such system purchased in this manner is certainly doomed to failure. The system must be planned to interface with the firm's present or planned data base and MRP system. A stand alone FMS would be a foolish investment. The entire hardware system and software system must be planned in detail. Simulation techniques should be used for optimization in the planning period. These simulations should preferably be performed by an independent third party (8). Do not rely on the equipment vendor's simulation study. Those have been optimized to suit what he is selling. The FMS user wants optimization on what he is buying. They are not necessarily one and the same. To give you an example of pre-planning, a study done by one consulting firm based in England cites a Japanese FMS consisting of 18 principal machines. That particular Japanese company spent 2 years and 100,000 hours in the pre-planning stage (3)! It is doubtful that western management would normally invest in such an up-front commitment in resources. However, if we don't invest the money up front, we will wind up investing much more in problem solving later on.

FMS WORLDWIDE

As of early this year there were 53 reported FMS's operating in Japan (6). It has been estimated that 30 or so of these are multicell with automated materials handling between cells. Forty-two of the systems were designed for prismatic production parts while only eleven were designed for rotational parts. It is difficult to separate the number of systems that are reported as FMS, but which in reality consist of only one robotic type cell.

It is also difficult to pinpoint the amount of investment and the benefits gained in monetary values. However, when one looks at the history of FMS installations in Japan over the last 10 years, it is quite clear that they are beginning to experience explosive growth in this technology (4)(6). See Figure 5. Japan's developmental push also appears to be zeroing in on wire guided vehicles for materials handling between FMS cells.

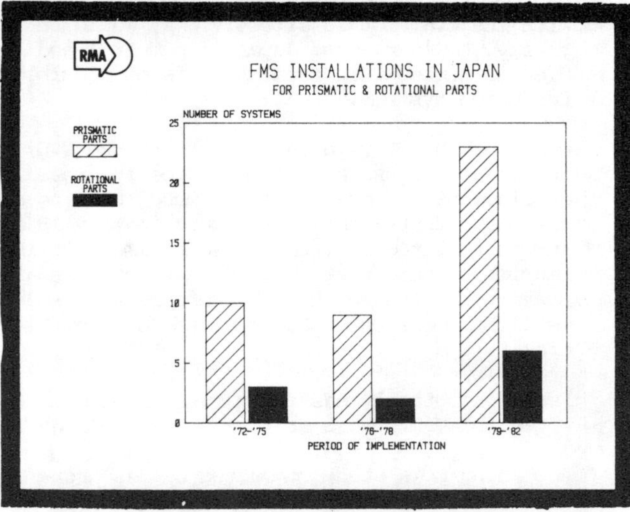

Figure 5. Japanese FMS Installations (6)

The largest FMS in Japan is reported to be the one operated by Fujitsu Fanuc (1). That one uses 30 machine cells with pallet carousels and robotic loading, along with an automated (AS/RS) warehouse. Robocarts are used for transportation between cells. Fanuc manufactures robots, EDM machines, and mini CNC machine tools. The FMS produces 450 different production parts (4). Funuc's future plans call for opening an FMS plant in Luxembourg that will be similar to the Fuji plant and almost double in size.

Incidentally, Yomazoki Machinery Co. of Japan which supposedly has the most advanced FMS in the world is planning a factory similarly equipped for Florence, Kentucky in the United States (1).

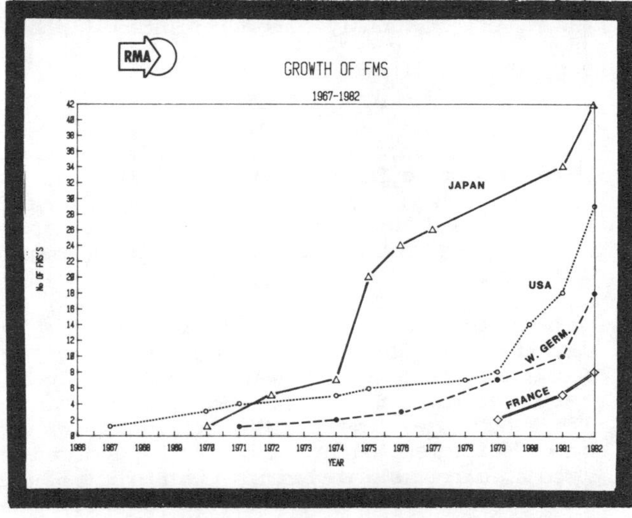

Figure 6. Growth of FMS

The United States has approximately 25 FMS's in operation which are multicelled and have some mode of automatic material handling between cells. Since the Japanese numbers include some stand-alone robotic type cells in their counts of FMS equipment, one cannot compare the Japanese totals with the U.S. totals directly. It would be fair to say, however, that Japan is number 1 and the U.S. is number 2 in total numbers of flexible manufacturing systems.

West Germany is relatively close to the U.S. in FMS installations and also seems to be spearheading development efforts for sheet metal working applications. West Germany has approximately 18 or so FMS installations at the present time. It is worth mentioning here that the U.S. Air Force is also heavily involved with the development of a sheet metal FMS cell under the auspices of the ICAM program (9).

France reportedly has eight systems in operation while the U.K. has 5 or 6, as shown in Figure 6.

According to a study conducted by researchers at the University of Strathclyde, Scotland, one of the biggest obstacles to FMS advancement in the U.K. stems from perceived risk (7). That study reports that roughly 75 percent of all metal parts made in Britain are produced in batches of less than fifty. This is certainly fertile ground for FMS, however, British industry has been slow to respond.

In my own experience the biggest hindrance to more FMS installations in the United States are the capital investment justification procedures used by U.S. industry. Japan takes a much longer view on capital investment payback. Apparently, their required payback periods are typically two to three times longer than the averages in U.S. industry. Where a typical investment in the U.S. would be acceptable with a two or three year payback, Japanese industry would normally accept a five to six year payback and even longer on strategic investments. However, we are becoming smarter, we are now placing monetary values on what were previously considered intangible benefits accruing from FMS installations.

SOME U.S. EXAMPLES

The largest users of FMS in the U.S., other than the automotive industry, have typically been heavy equipment manufacturers. John Deere, Caterpillar Company, and International Harvester are large users of FMS and helped pioneer these systems in the U.S.

The writer had the opportunity to work on a consulting assignment with FMC's Ordnance Division in California. My firm was engaged to help develop a long term facilities plan for that division. FMC presently uses a quasi-FMS installation for machining 7 different hulls, or bodies, of armored personnel carriers. We had the opportunity of helping develop facility plans for a new component manufacturing plant. The plans called for the inclusion of an FMS installation for machining component parts (8). That system is nearing installation and will use wire guided vehicles for automatic material handling between the FMS cells. A film produced by Cincinnati Milicron includes the FMC installations mentioned along with a few other examples (10).

WHERE ARE WE GOING WITH FMS?

I don't think there can be any question concerning the viability of the FMS concept in the "Factory of the Future". As a matter of fact, the FMS is fast becoming an integral part of that Future Factory concept. When FMS is fully integrated with computer aided design and group technology, we will be making a gigantic step forward. Actually, this integration of technology may be closer than we realize. Just a few months ago, Computervision Incorporated purchased the Organization for Industrial Research, Inc., who are one of the leaders in group technology services. It doesn't take a huge amount of mental prowess to figure out where that development is headed.

There are a number of complex technical advancements that are needed to further advance the FMS concept. This is particularly true in the area of flexible assembly systems. Reliable visual sensory devices incorporating character recognition and feedback definition will certainly be required in the assembly area. When combined with advances in force sensing and tactile grippers, we will be taking a huge step forward in lowering fixturing costs. This will also be beneficial in machining systems. In-process adaptive control of machining finishes or surface roughness still needs further development. All of these advancements are technically viable. Some are available now, although at a relatively high cost. The point to be made is, industry requires these advancements and there is a tremendous amount of current research and development efford aimed at producing them. We haven't even touched on the exponential advances being made today in microprocessor technology. To give you an example of how fast microprocessor technology is moving, consider Hewlett Packard's most advanced microprocessor chip. Approximately five years ago, it is my understanding, H.P. had approximately 40,000 transistors on their most advanced chip (5). Today they have over 450,000 transistors on that chip. That is more than a 10 fold increase in circuit density in five years! Although there is not a direct linear correlation between circuit density and computer size and cost, they definitely are related. Witness the widespread abundance of low cost computing power just with the advent of personal computers over the last 5 years.

Where are we going with FMS? There appears to be little doubt where we are headed. I firmly believe we are getting closer and closer to CIM or computer integrated manufacturing. The FMS will certainly play a key role in the factory of the 1990's. Whether or not that factory will be the true "Factory of the Future" as concepted by authorities today, is up to us.

BIBLIOGRAPHY

(1) Bylinsky, G., "The Race to the Automatic Factory", Fortune, February 21, 1983, p. 52.

(2) Hartley, J., "Japan Flexes It's FMS Muscle", Iron Age, January 21, 1983, p. 56.

(3) Ingersoll Engineers, "The FMS Report", IFS Publications Ltd., England, 1982.

(4) Knight, J.A.G., "The Latest Developments of FMS in Japan", "Proceedings of the 1st International Conference on Flexible Manufacturing Systems", Brighton, England, October, 1982.

(5) Mead, C., "Visionary Leading the Way in Microchips", Business Week, August 8, 1983, p. 66.

(6) Ohmi, T. and Ito, Y., and Yoshida, Y., "Flexible Manufacturing Systems in Japan - Present Status", Proceedings of the 1st International Conference on Flexible Manufacturing Systems, October, 1982.

(7) Parkinson, Dr. S.T., and Avlonitis, Dr. G.J., "Management Attitudes to Flexible Manufacturing Systems", Proceedings of the 1st International Conference on Flexible Manufacturing Systems, October, 1982.

(8) Redmond, G., "The Evaluation of a Flexible Manufacturing System - A Case Study, "Proceedings of the 1983 IIE Conference, Louisville, KY, May, 1983.

(9) Sania, A.M., "Air Force ICAM Project Paves the Way for the Factory of the Future", Iron Age, January 21, 1983, p. 48.

(10) Milacron Manufacturing Systems, a 16mm film demonstration supplied by Cincinnati Milacron, Cincinnati, OH.

BIOGRAPHICAL SKETCH

Edward J. Phillips is a consulting associate with Richard Muther & Associates, Inc. (RMA) located in Kansas City, MO. He is a registered professional engineer in Pennsylvania and Illinois. He earned his BME magna cum laude at Villanova University and his MBA at Widener College.

Mr. Phillips has substantial experience in manufacturing engineering and technical operations management within the aerospace, high technology electronics, high volume component manufacturing, and consumer appliance industries. He has been associated with firms such as Boeing, Smith-Corona, TRW, and Motorola. At Motorola's Franklin Park, IL facilities he advanced from mechanization engineering manager to manager of advanced manufacturing technology to technical operations manager. His responsibilities encompassed facilities planning, construction, and management along with associated plant engineering and support services. He has been involved in automation and customized materials handling equipment for most of his working career. With RMA, he is heavily involved in CAD/CAM and facilities planning for industrial clients.

He is a member of:
- Tau Beta Pi
- Robotics International of SME (senior member)
- American Society of Mechanical Engineers
- National Society of Professional Engineers

The use of a conceptual model to evaluate financially flexible manufacturing system projects

P L Primrose, MSc, CEng, MIMechE, MIProdE, MBIM and **R Leonard,** BSc, PhD, CEng, FIMechE, MIProdE, MBIM
Total Technology, University of Manchester Institute of Science and Technology

It is shown that conventional financial methodologies embody assumptions which are invalid when applied to major projects such as FMS. After critically discussing the assumptions normally made during financial evaluations, the technical problems associated with designing a computer program for FMS evaluation are described and it is shown that by devising a suitable conceptual model, these limitations can be overcome. The paper concludes by demonstrating how, by use of the model developed, together with the correct data input format for the interactive program, the problems normally considered inherent with conventional financial evaluation techniques can be readily overcome.

NOTATION

CNC	computer numerical control
DCF	discounted cash flow
IRR	internal rate of return
NPV	net present value
FMM	flexible manufacturing module
FMS	flexible manufacturing system
ZM	zero machine
WIP	work-in-progress

1 INTRODUCTION

Throughout the extensive literature associated with flexible manufacturing systems (FMS), a constantly repeating problem is stated, namely the inability of companies to determine analytically whether such systems are financially viable. To overcome this deficiency, a major research project is being conducted at UMIST (University of Manchester Institute of Science and Technology), with the dual objectives of establishing whether individual projects are viable and to devise a methodology for specifying the general range of conditions where FMSs can generate acceptable returns to justify their investment. Earlier papers (**1**) and (**2**) reported how a computer program could be used to evaluate financially single machine tool purchases and it was established that by using a correct methodology, the discounted cash flow (DCF) return from investing in advanced technology was considerably greater than that indicated by conventional techniques.

When investigating the attitudes of management towards FMS, Parkinson and Avlonitis (**3**) identified the following problems:

1. The lack of information to make sound estimates of future net returns.
2. The inability to quantify the numerous intangible benefits flowing from flexibility.
3. The accounting procedures which are not able, as yet, to tackle the new technology assessments.

The MS was received on 22 March 1984 and was accepted for publication on 15 August 1984.

During the current development of financial evaluation procedures, the difficulties which had to be overcome could be categorized as either conceptual limitations or accounting problems. This paper is primarily concerned with conceptual limitations, whilst a companion paper (**4**) is devoted to the accountancy area of FMS.

Conceptually, it is extremely difficult to evaluate an FMS comprehensively, with Sackett and Rathmill (**5**) suggesting that conventional costing methods could not be used when comparing existing practice with advanced technologies such as FMS. Within the present investigation it was decided to progress in controllable increments, with a computer program being initially devised for evaluating the returns on single machine tool purchases (**1**, **2**). The successful use of this program within several companies established the need for a new conceptual framework, and associated techniques, if the added complexities of FMS were to be encompassed correctly within an enhanced methodology. Therefore, the objective of the present work was to produce an accurate and practical methodology to achieve the following:

1. Enable companies to identify, prior to detailed design, areas in which FMS would be financially viable.
2. Readily interface with appropriate simulation techniques; thus ensuring that the final engineering design of a FMS was financially viable.
3. To act as a catalyst for directing the high research costs, associated with improved FMS, into areas which give the best return on investment.

Based on the experience of using the first program (**1**), (**2**) for selecting individual machine tools, a considerably more complex program, embodying a new conceptual framework, was written to evaluate financially FMS. The programs have been specifically designed to consider any type of machine tool or FMS, therefore, they are complex in terms of internal operation and comprise 900 and 1900 lines of Pascal source code respectively. It is clearly impractical within this paper to attempt a detailed critique of the program listing, thus the intention is to concentrate on the principles embodied in the FMS program.

2 THE CONCEPTUAL MODEL

When considering individual machines for purchase, it is intrinsically assumed that the basic operation of the company will remain unaffected. Likewise, when evolving from numerically controlled (NC) machines to computer numerically controlled (CNC), the effects of the change on the organization can be conceptually appreciated because of the clear relationship between the machine, its operator and the parts to be manufactured. With FMS, however, the whole operation of the company is potentially involved in the change and thus a comprehensive list of factors requires evaluation. In the previous work relating to single machine tools, thirty-one factors were identified as having a significant effect, hence these factors were incorporated within the program. When considering FMS, it was evident that if all incremental cash flows were to be identified correctly and quantified, this list of factors would need to be increased considerably.

A prime requirement of the work was to adhere strictly to DCF principles within the financial appraisal. This ensures that the results are acceptable to both the professional engineers and accountants in a company considering a possible FMS system. It should thus be noted that a DCF appraisal always carried the implication that a comparison is being made between two reasonable working alternatives, even if one of the alternatives is simply to do nothing and retain the 'status quo'. Therefore, in order to overcome the problems associated with defining an 'alternative' to the FMS, a novel conceptual model, designated the 'zero machine' (ZM), was devised. This conceptual model allowed the viability of the existing manufacturing processes to be evaluated against the ZM, thus correctly identifying all cash flows. The procedure then progresses to an evaluation of the proposed FMS against ZM. By creating the concept of ZM, the basic conceptual and technical problems normally associated with financially evaluating the viability of FMS were found to be capable of resolution.

Definition of the 'zero machine' (ZM):

'The zero machine has the capability to produce all the components currently being considered for the FMS system.
The zero machine will be commissioned at the end of year 0, with the corresponding work load being transferred in zero time.
The capital and installation cost of the zero machine is zero.
The zero machine occupies zero floor space.
Zero direct or indirect labour is required to operate the zero machine.
The zero machine incurs zero tooling and running costs.
The zero machine produces components in zero time, thus necessitating zero work-in-progress or stock.'

3 COMPUTER PROGRAM

The FMS program was written for use on a VAX 11/750 computer, with data being input via an interactive question/answer mode at a VDU terminal. The questions relate to all areas of potential cash flow change, with the wording of the question, together with the form of data input required, precluding the possibility of non-cash flow information, such as standard costs, fixed overhead or depreciation being wrongly included. To assist with the use of the program, two sheets are initially provided, corresponding to the data requested by the program. The first sheet relates to components, with twenty-five parameters being listed, whilst the second concerns sixty factors contributing to costs. Not all parameters are necessarily applicable to a specific FMS application, however, the format of a typical VDU question might be: 'Is any of the planned work for the zero machine currently being subcontracted out?' Type Y or N. An affirmative answer results in a series of secondary questions relating to such factors as the volume of work and the cost of subcontracting. Thus each of the sixty cost parameters has a corresponding cascade of related data items.

Although it may appear a sizeable task to input the necessary data to the program, reasonable initial estimates can be made to establish if an FMS is potentially viable. Should this prove positive, detailed data can subsequently be generated from system design and simulation models. When either estimates or accurate data have been input, the program generates a detailed print-out of the associated cash flow changes for each parameter considered, together with a statement of the overall internal rate of return (IRR) and net present value (NPV) for the project. To manipulate manually the volume of data required to calculate the DCF returns accurately would be completely impractical, however, by use of the program, a company can rapidly evaluate a range of possible FMS applications, thereby identifying those which potentially give the best financial returns.

4 TIMING OF CASHFLOWS

Existing literature generally assumes that the total expenditure on a new machine takes place at a single point in time, with full cashflow savings similarly being achieved. However, Darnell and Dale (**6**) correctly point out that in many cases, the costs of commissioning, and the loss of revenue during a period of run-up, seriously affect a project's financial viability. For example, the number of programs which have to be written for a CNC machine may cause a considerably time delay before full utilization is achieved. The present authors recognized this problem when developing the first program, with due provision being made for lower savings during the start-up period. For example, a simple straight line increase in utilization might be assumed for a CNC machine, while for less complicated machine tools, the short installation and commissioning time could enable the assumption to be made that expenditure did actually occur at a single point in time. For FMS, however, caution must be exercised regarding the start-up period and the timing of expenditure. There will be an extensive period of proving fixtures, programs, control software and system hardware. In addition, the level of manning will not reflect the time-scale of production build-up. Reports on existing FMS installations reveal periods of up to three years between the first major expenditure on a system and the com-

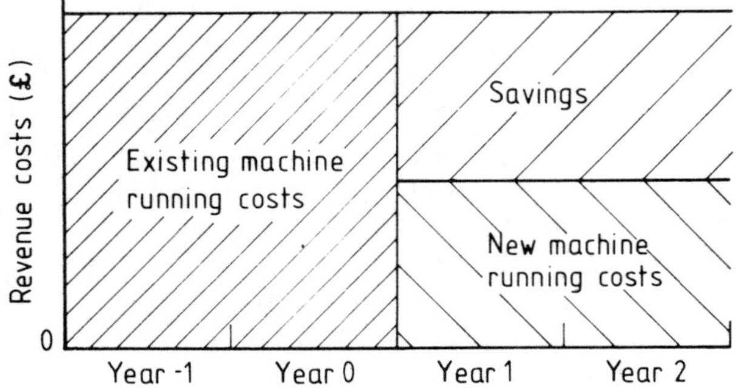

Fig. 1 Conventional assumption of cash flow changes

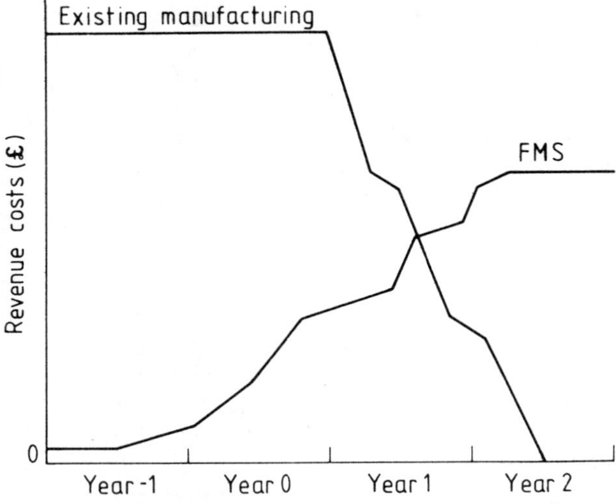

Fig. 2 Cash flow changes with FMS

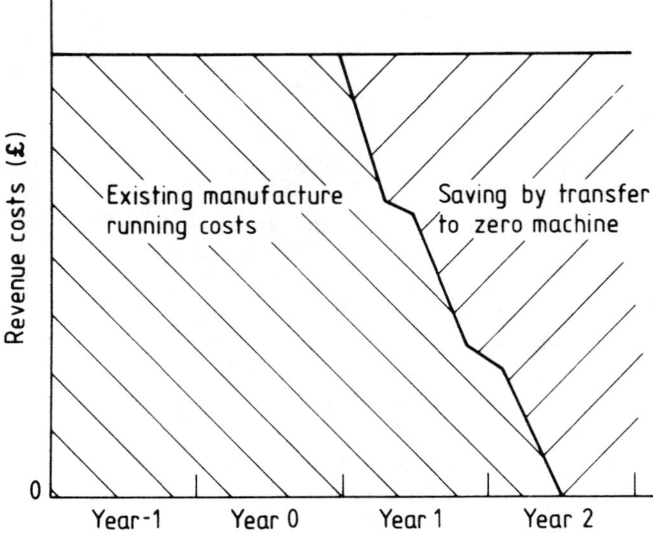

Fig. 3 Zero machine compared with existing methods

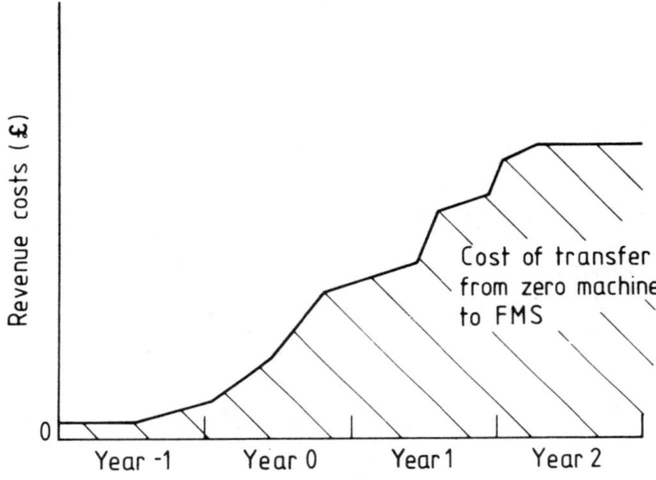

Fig. 4 FMS compared with zero machine

mencement of production, and even this delay may be followed by additional years before full benefits are achieved.

Conventional evaluations also assume that when a new machine is commissioned, the corresponding, outmoded facility is terminated, therefore, incremental cash flows, of both costs and savings, occupy the same time-scale. Unfortunately, the complexity of FMS renders this assumption invalid; for example, no specific point in time may actually exist where the FMS can be considered to be 'fully commissioned'. However, by adopting ZM as a conceptual framework, and by defining the date of ZM's commissioning as the end of year 0, a common time datum is established for running down the existing facility and building up of cash flows on the FMS.

Without the conceptual model, extreme difficulty exists when trying to relate two time-scales, with errors being easy to make and difficult to identify. In addition, because the time pattern of cash flows is of considerable importance with DCF, and with errors tending to relate to the early years of an FMS project, the viability of a potential system significantly relates to accurately identifying cash flows with chronological time. Figure 1 shows the simplistic assumptions incorporated within a conventional evaluation, whereas Figure 2 represents the much more complex way that cash flows actually change with time in an FMS. It is, therefore, apparent that the net cash flow can only be determined by evaluating the individual cash flows separately. Using a standard accounting convention, the time of commissioning the ZM is taken as the end of year 0, with the years -2, -1 and 0 giving a three year period during which system design and installation take place. Hence, year + onwards is assigned for the run-up to full production Figures 3 and 4 show that by first comparing the ZM with existing methods, and then separately to the FMS, the savings and costs can be evaluated in the same conceptual way as in Fig. 1.

5 SYSTEM DESIGN AND PROGRAMMING

In companies with an NC programming department, a new CNC machine will not necessarily result in the employment of an additional programmer as the extra work-load might be shared by existing staff. Therefore, the only incremental cash flow change will be for items such as training courses and software costs if computer aided programming is used. However, if an additional programmer must be employed, the cost of his wages should be identified so that cash flow changes can be calculated. A difficulty which arises concerns the proportion of the additional programmer's cost which should be attributed to the new machine. If it is calculated that all the required programs will be written and

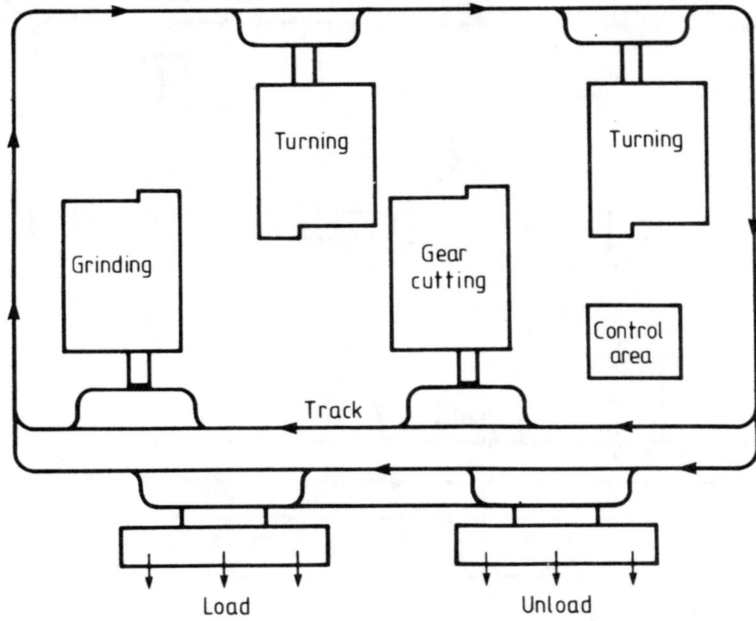

Fig. 5 Complementary type FMS

proved within two years, with little program maintenance being required, it seems reasonable to only charge programming costs to the new machine for two years, thus leaving the question unanswered regarding what the programmer will be working on after that time.

With the purchase of a single machine, it was shown (**1**) that the problem of support staff costs is not normally critical because the magnitude of the potential error is relatively small and companies will find suitable tasks for their employees to fulfil. However, the number, and cost, of staff engaged on FMS system design and programming will be considerable, with the previous discussion regarding CNC programmers, indicating the conceptual problem of defining which staff should be included as a cash flow change. Fortunately, by deploying the ZM concept, the costs associated with designing and programming the FMS can be readily stated. Thus, by considering the two systems separately, the overall cash flow changes can be correctly identified and embraced within the financial analysis.

The Department of Industry provides financial assistance towards the cost of external consultancy with respect to FMS. Thus when considering FMS, companies must decide to what extent extra staff should be employed for system design, or whether external consultants might be used because cash grants are available. Fortunately, the costs of these alternatives can be compared by use of the logic incorporated in the program, with only the long-term staff requirement and costs needing to be stated.

6 DIRECT LABOUR

When evaluating a single machine for purchase, various assumptions are implicitly contained in the technique used, yet the assessor may not be unaware that any assumptions have actually been made. Similarly, when considering an FMS, assumptions might intuitively be made which also have no validity. However, the high costs associated with FMS result in the need for all assumptions within the analysis to be identified and their degree of validity accurately assessed.

For example, with respect to direct labour, the total batch 'floor to floor time' may be calculated on both the old and new machines and hence the overall production ratio determined; specifically the times on a new CNC lathe may equate to the workload of 2.75 turret lathes. Therefore, if it is taken that one man operates one machine and that all machines are double shifted, it follows that an annual saving of 3.5 operators will result. For this situation, the implicit assumption has been made that the labour force is large, and its rate of turnover sufficient for the wages of 3.5 operators to be considered as a cash flow saving, with 1.75 operatives coming from each shift and two operators transferring directly to the new machine. It is also assumed that there is a direct relationship between machine hours and operator hours.

The introduction of FMS, however, will have a major impact on *both* operation and manning levels, therefore, it is necessary to consider separately the proposed reduction in existing labour, followed by the planned build-up of labour for FMS. This situation is further complicated by the need to have a separate labour force working on the FMS for training, commissioning and proving, before production is transferred and the existing operators become available. Fortunately, the concept of the ZM makes it possible to draw up separate lists and timetables for the respective run-down and build-up of labour requirements.

Another difficulty is that for conventional facilities, a direct relationship normally exists between the number of machines and operators; for FMS however, this is no longer valid. Within the FMS, machines will exist with widely different manning levels, worked on a multi-shift system which include periods of completely unmanned operation. Further complications are added by the

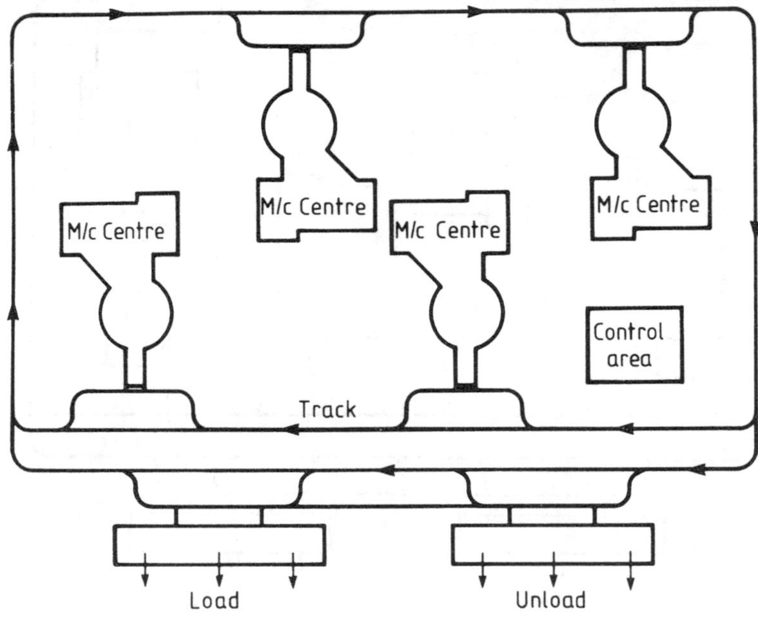

Fig. 6 Interchangeable type FMS

variations in utilization of the different machines within the total FMS arrangement. As a general statement, flexible manufacturing systems may either be classified as complementary or interchangeable, with Figs. 5 and 6 depicting the two types. Complementary systems comprise machines of different types, resulting in the high probability that each machine will have a different utilization, being independent upon a given product mix. Conversely, because the machines are basically 'identical' in an interchangeable system, the utilization of each element is likely to be the same.

7 INDIRECT LABOUR

No direct link exists between operator hours and machine hours within an FMS, therefore, the normal distinction between direct and indirect labour costs loses its importance, with jobs such as load/unload components, set tooling, system maintenance, fixture/program proving etc. perhaps being done by the same group of people. Therefore, the conventional approach of establishing indirect costs, by analysing each department, e.g. maintenance, and deciding if a charge for labour exists, is no longer appropriate. Fortunately, this difficulty is overcome by using the ZM concept; specifically, any reduction in existing labour can be assessed separately to the need to define the labour requirements for FMS.

8 SCRAP AND REWORK

Within an FMS it is necessary to adopt a different approach for calculating the cost of scrap and rework to that normally used for conventional manufacture. For example, with respect to the existing methods, the costs may be taken as:

scrap cost = raw material + 50 per cent (labour + specific overhead)
rework = labour + specific overhead

The labour cost in the above expression represents the basic operator wage plus any shift premium and overheads relating to employment costs such as national insurance, pension, welfare benefits etc. It is also normally assumed that rework is carried out on a similar machine to that for which the operation was originally planned. With FMS, however, no direct relationship exists between machine time and labour costs, thus it is not possible to cost scrap and rework in the above way. Similarly, the new cost cannot simply be subtracted from the old cost to obtain the financial saving in scrap and rework made by introducing the system. The method which should be adopted is as follows:

8.1 Scrap costs

If an FMS produces both good and scrap components, the cost of running the unit is not affected by the ratio of good to defective parts. However, for each defective item manufactured, a corresponding good unit is lost. As far as 'cash flow' is concerned, the cost of producing scrap is solely the value of raw material wasted. Traditionally, scrap is costed in an analogous manner to work-in-progress (WIP), with the assumption being made that 50 per cent of the required operations have been completed. Similarly with FMS, it is normally taken that 50 per cent of the total component manufacturing time has been consumed before the job is found to be defective, although it should be noted that the concept of a per piece time can be misleading when considering an FMS. Thus the capacity of the FMS is reduced by:

quantity of scrap × (50 per cent × total machine utilization time per component)

The assumption of 50 per cent work complete at the time of scrap needs to be reconsidered for each application; for example, in the complementary system shown in Fig. 5, once a component has been scheduled into the

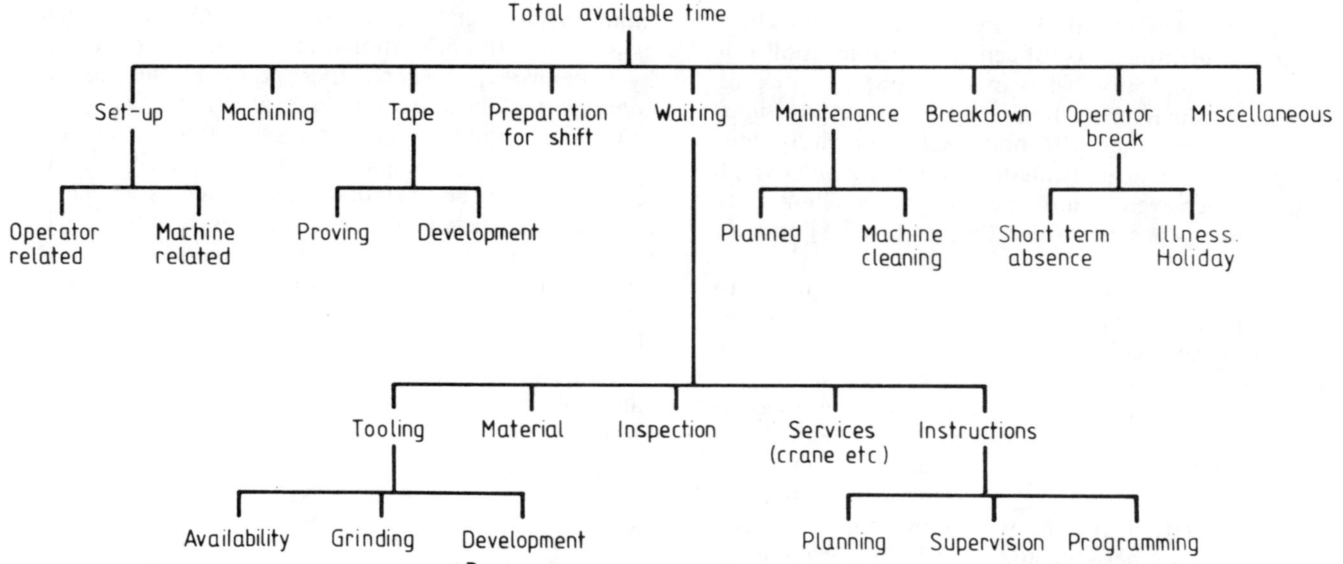

Fig. 7 Factors affecting machine utilization

system, the effect of removing a scrapped part is to lose 100 per cent of the capacity planned for that component. For this reason, the program allows the conventional 50 per cent value to be suitably altered.

8.2 Rework costs

Components, produced on an FMS, which require rework will, almost certainly, have the rectification performed on conventional machines. Hence the associated costing of rework relates to labour plus specific overheads on the conventional process, with the corresponding need to rework components not affecting the utilization of the FMS.

For conventional production, components normally experience a series of operations on a variety of machines. Therefore, when the requirement for rework arises, it will either be done after all the operations are completed or when the need is first identified, with subsequent operations being completed normally. Conversely, because an FMS may be considered as a single, complex machine, if the need for rework occurs, perhaps by the job halting part way through manufacture (e.g. due to faulty castings), rework, carried out on conventional machines, will consist of both correcting the fault and also, due to the impracticability of introducing part machined components into the system, having to finish off the outstanding machining operations manually.

9 SUBCONTRACT WORK

Work that was previously subcontracted and is now to be brought back 'in-house' has already been included in the evaluation as part of the savings made by replacing existing methods by the ZM; the cost saving is simply the number of hours involved × subcontract rate/hour. The actual cost of doing the previously subcontracted work on the FMS is zero; this occurs because the running cost of the new system is not altered simply as a result of the work coming from outside, with the costs having already been allowed for elsewhere within the computer program. However, it is necessary that due allowance is made for such work when defining the required capacity of the FMS. Conversely, if part of the work load for the FMS is to be components subcontracted from another company, a question arises with respect to deciding what rate to charge per hour. Obvious alternatives for charging include asking 'What the market will bear' or performing 'detailed cost calculations'.

10 CAPITAL EXPENDITURE

The program contains the assumption that, for taxation purposes, all capital expenditure will be classified as plant and machinery, rather than as land or buildings, and that suitable grants may be available from the Department of Industry. In addition, due to the cost and time-scale of the investment, with several years possibily elapsing between order and commissioning, provision exists for multiple payments.

The deployment of an FMS will normally result in a company avoiding the need to purchase a corresponding capacity of 'conventional machines'. In addition, any existing machines which thus become superfluous to requirements can be duly sold. It is, therefore, first necessary to identify the pattern of capital investment that would have taken place if the FMS had not been purchased, then the time-scale of investment and scale of assets with FMS must be separately established. When discussing machine tool disposal, most authors tend to assume that the residual value of a machine is of the order of 10 per cent of its initial capital cost. This, however, implies that the machine will either be sold in 'working order', or possibly for reconditioning. This assumption is clearly suspect with FMS and a significant probability exists that the large installation cost will prevent a system being sold for an external life. Thus the assumption is provisionally made that the system is disposed of as scrap at a residual value of only 1 per cent of initial cost.

11 UTILIZATION

A major factor affecting the viability of FMS is the utilization of existing machines and the projected utilization of the new system. Figure 7 shows factors which have been identified as affecting machine tool utilization. Although this list is extensive, apart from the per piece manufacturing times and initial setting times on existing machines, the only factor which is considered in the financial evaluation of FMS is the overall utilization achieved. Thus the existing machines are treated quite separately from the projected figures for FMS.

One objective of the financial program design was to enable a direct interface with any simulation packages used for system design. Therefore, the factors listed in Fig. 7 would be evaluated for the proposed FMS as part of the system design process, with the utilization figures of existing machines being obtained from either historical data or by observation and study. A number of investigations have been conducted with respect to machine utilization and, as was shown in (7), a good correlation exists between them. Therefore, established data are of sufficient accuracy to be incorporated in the initial evaluations to identify potential areas for detailed FMS study. For example, within (7) a simple program was shown to be capable of evaluating the viability of the first stage of involvement in flexible manufacturing, namely a flexible manufacturing module (FMM).

12 CONCLUSIONS

By using a computer program which incorporates a conceptual framework amenable to FMS evaluation, incremental changes in cash flows can be correctly identified such that the problems normally associated with the financial evaluation of such systems become resolvable. In addition, the design of the program, together with the format of the interactive 'question/answer' data input, ensures that the implicit assumptions normally associated with conventional methodology are replaced by a detailed procedure, appropriate to the cost and complexity of the system under evaluation.

The conceptual model developed in this work enables the difficulties of comparing two completely different technologies to be overcome. It also allows a financial evaluation to be made which adheres strictly to DCF principles thus creating a common and effective interface between engineers and accountants. When all the potential factors affecting cash flows are incorporated in the analysis, the intitial results from the program indicate that FMS is financially viable over a much wider range of applications than has been suggested by previous authors.

REFERENCES

1 **Primrose, P. L.** and **Leonard, R.** The development and application of a computer based appraisal technique for the structured evaluation of machine tool purchase. *Proc. Instn Mech. Engrs*, Part B, 1984, **198**, No. 10, 141–146.

2 **Primrose, P. L.** and **Leonard, R.** Optimizing the financial advantage of using CNC machine tools by use of an integrated suite of programs. *Proc. Instn Mech. Engrs*, Part B, 1984, **198**, No. 10, 147–151.

3 **Parkinson, S. J.** and **Avlonitis, G. J.** Management attitudes to flexible manufacturing systems. *Proc. 1st Int. Conf. on FMS*, 1982, pp 405–412 (IFS Publications Ltd.)

4 **Primrose P. L.** and **Leonard R.** The use of discounted cash flow to evaluate the 'intangible' benefits of FMS. *Proc. Instn Mech. Engrs*, Part B, 1985, **199**, this issue.

5 **Sackett, P. J.** and **Rathmill, K.** Manufacturing plant for 1985—developments and justification. *Proc. Instn Mech. Engrs*, Sept. 1982, **196**, 265–280.

6 **Darnell, H.** and **Dale, M. W.** *Total project management*, 1982 (British Institute of Management).

7 **Primrose, P. L.** and **Leonard, R.** The financial evaluation of flexible manufacturing modules (FMM). International Machine Tool Conference, June 1984.

ANALYTIC APPROACHES

This section describes the variety of analytic techniques that have been applied to the justification exercise. Included are scoring models, "value analysis," analytic hierarchy process, risk analysis, simulation, and expert systems. The more common approaches are described in detail and examples are included to illustrate their application. The less common techniques, such as expert systems, are described briefly.

In the first article, Sullivan gives an overview of the analytic approaches and briefly demonstrates their operation. Next, Frazelle presents a discussion and case study of applying the analytic hierarchy process to the analysis of alternative material handling systems.

In the third article, Keen describes a type of pilot study approach he terms "value analysis" that has been used successfully for justifying computerized information systems. Buck follows this with a description of the risk analysis approach, detailing the application of the deterministic "closed-form" style as well as the probabilistic simulation style.

Sullivan and Orr continue the simulation technique with three examples of Monte Carlo simulation applied to probabilistic environments. Last, Suresh and Meredith include a broad perspective of the justification problem in their detailed simulation/risk analysis study.

Reprinted from **Industrial Engineering**, March 1986.

Models IEs Can Use To Include Strategic, Non-Monetary Factors In Automation Decisions

By William G. Sullivan, P.E.
The University of Tennessee, Knoxville

Declining productivity and GNP in the United States is a much discussed issue. The search for the cause of this decline has covered many facets of business, including capital investment practices, union contracts, income tax policies, foreign competition and management support.

One factor which deserves more examination is the consequence of a lack of new capital investment in plant and equipment. It is reasonable to assume that if plant and equipment are not kept up to date technically and operationally, productivity will decline. Unless a firm is in a captive market, a decline in productivity will likely cause a decline in its competitive position. This could be disastrous to a firm whose competitors were aggressively pursuing the same market.

The manufacturing industry in the U.S. has recently experienced a rapid explosion of technology in an uncertain economic environment. This situation has made the justification of capital investments especially difficult. Rapidly changing technology and shortened product life cycles can cause capital equipment to become obsolete quite quickly.

For example, a few years ago a state-of-the-art line for fabricating silicon wafers into microchips could be purchased for $20 million. Today the type of line that is required to stay competitive costs approximately $60 million (Uttel, "For further reading"). Over the past few years, the economy has been in and out of a recessionary period, and the demand for these microchips has fluctuated wildly.

The combination of increased capital expenditure and uncertain markets is not limited to the semiconductor industry, but pervades every manufacturing industry. For instance, replacing conventional machine tools with a numerically controlled machining center can greatly increase a company's productivity and capacity, but such equipment can cost more than $250,000 and hence add significantly to the company's financial and operating risk. Although capital investment is assuming a more crucial role in the success or failure of a business, the decision to make such investments is becoming ever more complex.

Factory automation, which includes computer-aided design (CAD) and computer-aided manufacturing (CAM), will assume increased importance in productivity improvement programs in manufacturing industries. These systems are a substitution of capital for labor and are ideal for environments where *flexibility* can be economically traded off for efficiency (see Ayers and Miller).

A subtle benefit of flexibility is that it encourages technological change and innovation. However, flexibility is initially destructive of capital—whether in the form of labor skills, technological processes or capital equipment—as it tends to create rapid obsolescence and uncertainty regarding existing plant and equipment. In the face of these complexities, it's no wonder that a "wait and see" attitude develops when large investments in fixed capital assets are being considered.

Traditional engineering economy

Early capital investment studies performed by engineers dealt with the replacement or retirement of obsolete equipment in industrial and government organizations (see Grant). Today's engineering economy has evolved into a "bottom up" set of principles and techniques for evaluating the economic merits of proposed capital investments in view of numerous real-world complications, such as nonmonetary factors and uncertainty.

Engineering economy practitioners typically deal with tactical rather than strategic investment decisions. (A strategic decision is long-term in outlook and involves "top down" issues such as technology advancement, sourcing of capital, plant modernization and competitive position of the firm. A tactical decision is concerned mainly with the short-term position of the firm in such areas as cost avoidance, reduction of inventories and rework, reaction to product changes and replacement of relatively inexpensive capital assets.) Thus, we usually concern ourselves with the cost reduction aspects of the profit equation rather than the revenue creation part.

Critics of engineering economy have chided us for this orientation and repeatedly charged that we give "the narrowest possible interpretation" to our domain in concentrating on the "economic evaluation of minor investment projects" (Terborgh) and the rendering of "low-level advice for management decision making" (Radnor).

Such criticism may have been valid in the days when replacement analysis, for example, involved a comparison between a "defender" machine tool and its "challenger" which accomplished an identical function. However, criticism has recently grown even more vocal in view of our alleged inability to (1) deal realistically with multiple objectives; (2) include nonmonetary factors in the decision calculus; and (3) cope with environmental uncertainties and estimation inaccuracies (see Horowitz).

Justification studies involving the increased use of automation give rise to at least three considerations that upgrade them from a tactical to a strategic domain. First, innovations and flexibility in automated factories create a dynamic environment in which many traditional machine processes and manual operations are being redefined or eliminated.

Second, functions to be accomplished in successive corporate planning intervals are time-variant and become integrated over time. Third, the need to reduce unit costs through the adoption of new technology tends to focus on strategic business issues

Figure 1: Example Profile Chart for Two Alternatives and Simple Scoring System

	ALTERNATIVE A				ALTERNATIVE B			
	UNDESIRABLE		DESIRABLE		UNDESIRABLE		DESIRABLE	
DECISION FACTOR	-2	-1	+1	+2	-2	-1	+1	+2
EFFECT ON ANN. SALES			■			■	■	
MANPOWER AVAIL.		■	■		■			
PRODUCT QUALITY		■					■	■
INVESTMENT COST			■					■
INVENTORY REDUC.		■				■	■	■
⋮			■	■		■		

Table 1: Example of Symbolic Scorecard for Evaluation of CAD Systems

Decision factor	CAD Alternative			
	Vendor A	Vendor B	Vendor C	Reference (do nothing)
Cost of purchasing the system	$115,000	[illegible]	$32,000	$0
Reduction in design time	60%	67%	50%	[illegible]
Lead time	Excellent	Excellent	Good	[illegible]
Flexibility	Excellent	Excellent	Good	[illegible]
Inventory control	Excellent	Excellent	Excellent	[illegible]
Quality	Excellent	Excellent	Good	[illegible]
Market share	Excellent	Excellent	Good	[illegible]
Machine utilization	Excellent	Excellent	Good	[illegible]

Legend: green = best, yellow = intermediate, red = worst

and places pressure on engineers to broaden their understanding of corporate-level policy issues.

It is precisely because of these new dimensions that financial justification studies have assumed strategic importance and have forced industrial engineers to upgrade their analytical techniques.

Rebirth of engineering economy

According to Robert F. Huber, "Traditional financial justification procedures, based on internal rate of return or payback period, are quite likely the single greatest barrier to the utilization of new manufacturing technologies by U.S. manufacturing industries." (See "For further reading.")

Unfortunately, many organizations have come to accept these models as synonymous with *engineering economy* and have not made full use of several multiattributed techniques for including strategic (nonmonetary) factors in their analyses. The appropriate question may well be: "Have managers relied too heavily on conventional time value of money procedures for justifying proposed capital investments because they can all too easily hide behind the apparent rationality of such financial analyses while sidestepping the hard decisions . . ?" (Hayes and Garvin). The significance of this question is borne out by a 1984 survey of the National Electrical Manufacturers Association which revealed that 91% of responding business executives use conventional justification procedures as their *major* consideration in approving factory automation efforts.

Interestingly, it is generally *perceived* by top management that traditional measures of project worth, such as return on investment (ROI), payback and net present worth, are conceptually inadequate for judging the strategic merits of computer integrated manufacturing (CIM) technologies. However, these measures continue to be used for lack of a more acceptable approach. Whether they are theoretically correct or incorrect is apparently not the primary concern in coming to grips with how to justify manufacturing automation.

Throughout the literature, repeated reference is made to the need for new justification models for automation projects. However, the response has not been overwhelming. After all, what executive would approve an automation project whose benefits would not be fully realized for five years or more when he or she anticipated a major promotion in four years or less?

If engineering economy practitioners are to assist management in the appraisal of new technologies that are essential to a firm's survival, different models and methodologies must be developed (or resurrected) and applied to actual problems.

The remainder of this article describes several straightforward approaches for dealing with multiple, often conflicting objectives as well as the nonmonetary performance criteria that emerge when the strategic and tactical ramifications of automation are factored into justification studies.

Research into the methods discussed below, as well as other more advanced techniques, is being undertaken at the University of Tennessee's Center for Computer Integrated Engineering and Manufacturing (CIEM). The Center for CIEM is a multidisciplinary organization utilizing state-of-the-art hardware and software to facilitate information integration in the CIM environment. For instance, the center is currently seeking to make extensive application of GM's Manufacturing Automation Protocol and related technology.

Table 2: Example of a Ranking and Rating Linear Additive Model for Machine Tool Replacement

				j→	Alternative (1)	Alternative (2)
i	w_i	Rank	Decision factor		Keep Existing Machine Tool	Purchase a New Machine Tool
1	1.0	1	Annual cost of ownership (capital recovery cost)	Rank	1	2
				x_{i_j}	1.0	0.7
2	0.5	4	Flexibility in types of jobs scheduled	Rank	2	1
				x_{i_j}	0.8	1.0
3	0.8	2	Ease of training and operation	Rank	1	2
				x_{i_j}	1.0	0.5
4	0.7	3	Time savings per part produced	Rank	2	1
				x_{i_j}	0.7	1.0
				V_j*	2.69	2.30
				V_j (Normalized)	**1.00**	**0.86**

$${}^*V_j = \sum_{i=1}^{n} w_i x_{i_j}$$

Profile charts and scorecards

A basic problem in justification studies is choosing a small, but comprehensive set of mutually exclusive criteria with which to judge salient differences among alternatives. After these criteria are evaluated for each alternative, the results can be pre-

Figure 2: Benefit and Cost Hierarchies for Evaluation of an FMS

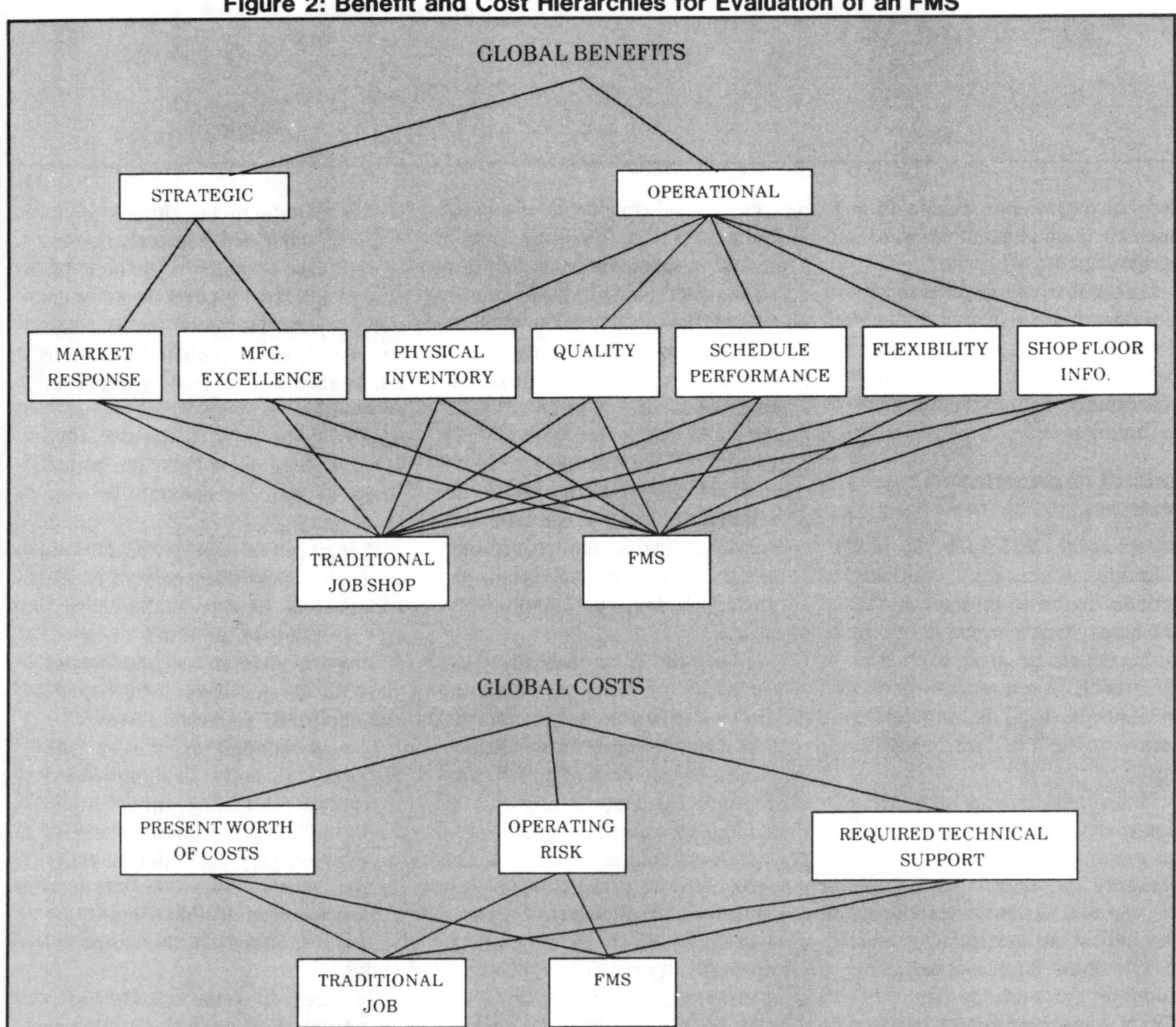

sented visually on a profile chart. Selection among alternatives that involve nonmonetary or "intangible" performance criteria can often be facilitated by graphical displays of relevant data. The outstanding features (strengths and weaknesses) of actions being considered can be highlighted so that careful attention will be given to them.

Figure 1 illustrates a type of profile chart that utilizes a simple scoring system to indicate how well a particular alternative meets each criterion. As an example of scoring, the criterion "Effect on Increased Annual Sales" might be evaluated as follows:

-2—Less than $100,000.
-1—$100,000 to $200,000.
+1—$200,000 to $400,000.
+2—More than $400,000.

Four levels of desirability are shown for each criterion that range from +2 to −2, but a greater or smaller number of levels could be utilized. Equilibrium (no change) status need not be plotted. Notice that uncertainty can be indicated by showing two or more values. Emphasis should not be placed on the numerical values assigned to each level, because the purpose of the profile is to indicate visually the differences in levels of desirability for each criterion from one alternative to another.

Most profile charts make no attempt to indicate the relative importance of the various criteria used. They merely provide the decision maker with a visual summary of information to be used in choosing between alternatives according to his or her subjective judgment as to the relative importance of the criteria.

Another type of profile chart is known as the symbolic scorecard (see Fleischer). In this chart, different colors or symbols are utilized to show the relative desirability of two or more alternatives for each criterion of interest.

Table 1 is an example using symbols for "best" and "worst" alternatives for implementing a CAD system. Other symbols might have been used, for example, to denote "next best" and "next worst" alternatives. Note that numbers are used to denote values (in any applicable units) for most criteria in this example, but numbers do not have to be used except where that level of detail is desired.

Several advantages of these two types of profile charts are apparent. First, the possible confusion and difficulty of formally weighting criteria (and sometimes of specifying a ratio scale for criterion performance) are avoided. Second, such charts focus on relevant differences among alternatives and permit the decision maker to envision the mental tradeoffs involved in making a selection. Third, the need for a defense of "precise" numerical estimates on the part of the analyst should be reduced.

It is sometimes difficult to decide how to assign ratings to alternatives for particularly ambiguous criteria. Also, one may be tempted to equate the best alternative with the one having the largest number of desirable characteristics. In either situation the real differences in importance among the criteria may be overlooked.

Linear additive models

The linear additive model is a decision tool that aggregates information from different, independent criteria in a linear fashion to arrive at an overall score for each course of action being evaluated. The alternative with the highest score is preferred. The general form of the model is:

$$V_j = \sum_{i=1}^{n} w_i x_{ij}$$

where V_j = the score for the jth alternative.

w_i = the weight assigned to the ith decision criterion ($1 \leq i \leq n$).

x_{ij} = the rating assigned to the ith criterion, which reflects the performance of alternative j relative to maximum attainment of the criterion.

The multiplication indicated in this model is based on the assumption that weights are measured on a ratio scale and ratings are measured on at least an interval scale (see Anderson, et al.).

Many different types of linear additive models have been developed and reviewed over the years (see Klee and also Moore and Baker). A simple and direct method is ranking and rating (see Morris). According to Morris, this method "has been found by decision makers to be easy to use, has led to evaluations which they found useful, and has produced rank orderings which seem to correspond closely to their actual ultimate choices." An application of this method to a simple machine tool

Figure 3: Typical Rule from the Rule Base of XVENTURE

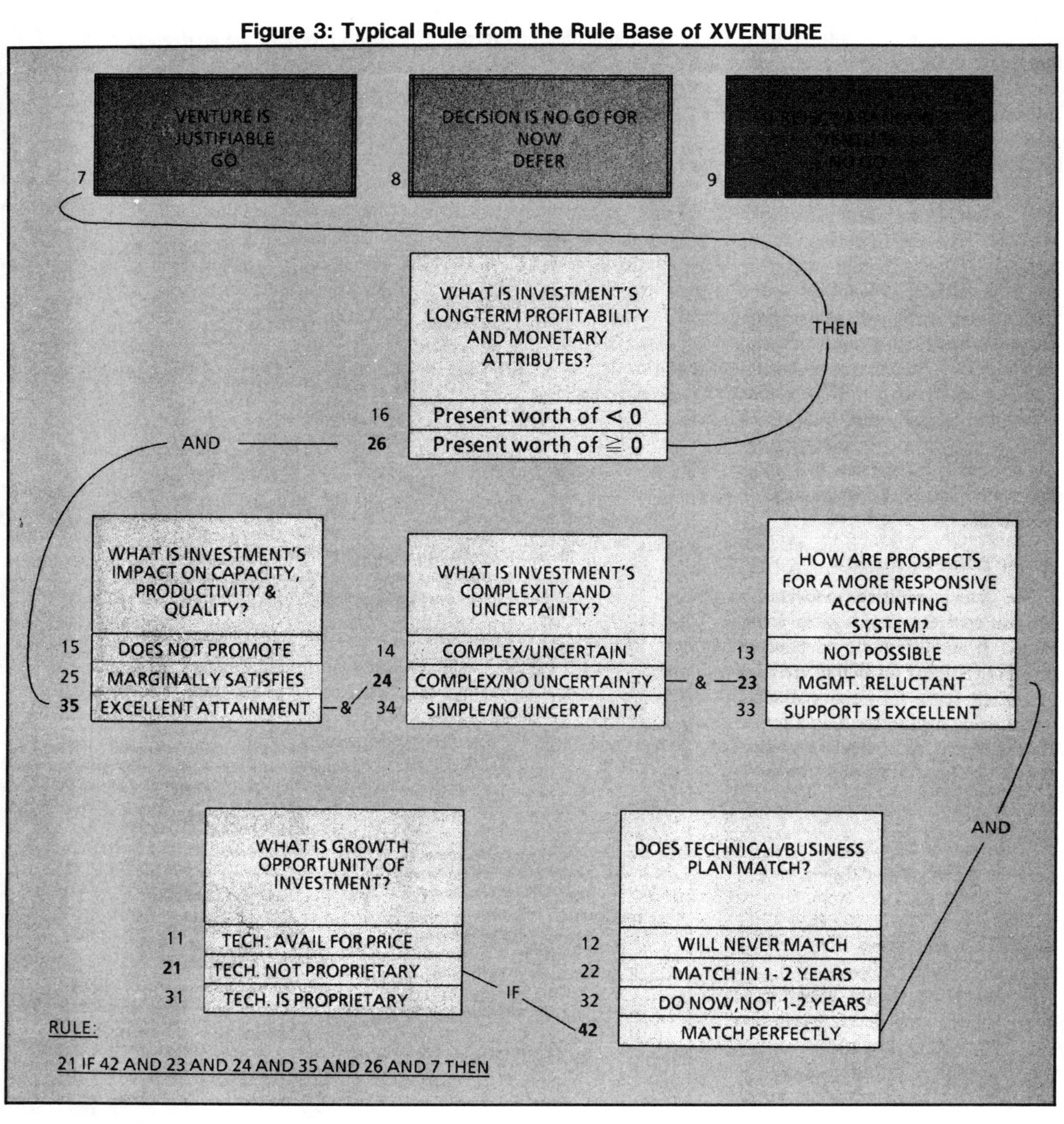

replacement decision is shown in Table 2 and described below:

1.) The decision maker lists all the factors that are important in evaluating whether he or she should replace the existing machine tool with a new one. One possible list might be:

a) Annual cost of ownership (capital recovery cost).

b) Flexibility in changeovers.

c) Ease of operation.

d) Time savings per part produced.

2.) The decision maker then verifies the independence of the above criteria, possibly combining some of them or redefining those that are not independent of each other.

3.) The independent criteria resulting from the first two steps are ranked by the decision maker in order of importance, with ties permitted.

4.) The decision maker assigns ratings to the decision criteria on a ratio scale from 0.0 to 1.0. The highest ranked criterion is rated 1.0, and the second ranked criterion is assigned a rating of 0.80 if it is perceived to be 80% as important as the highest ranked criterion. This same procedure is applied to the remaining criteria. The rating process is fairly subjective in that the decision maker is rating different criteria based on different units of measurement; e.g., dollars versus ease of operation.

5.) The next step is to rank and rate the available alternatives against each decision criterion. In the example of the machine tool, the decision maker would rank and rate the current machine tool compared to the new (proposed) machine tool for each of the four criteria defined in step 1. For instance, the current machine tool may be assigned a 1.0 for the annual cost of ownership criterion, while the new machine tool may be assigned a 0.70 rating because it is seven-tenths as desirable as the current asset in terms of this criterion.

6.) A score for each alternative can now be determined by using the general form of the model to aggregate the ranking and rating information obtained in steps 1 through 5, as illustrated in Table 2.

The score for each alternative is then interpreted as a prediction of the decision maker's preference for that alternative. Thus, alternative 1 is the recommended choice in the machine tool replacement example.

Analytic hierarchy process

The analytic hierarchy process (AHP) was developed in 1972 as a practical approach to solving relatively difficult decision problems. T.L. Saaty describes the AHP as follows:

> The analytic hierarchy process enables decision makers to represent the simultaneous interaction of many factors in complex, unstructured situations. It helps them to identify and set priorities on the basis of their objectives and their knowledge and experience of each problem. Our feelings and intuitive judgments are probably more representative of our thinking and behavior than our verbalizations of them. The new framework organizes feelings and intuitive judgments as well as logic so that we can map out complex situations as we perceive them. It reflects the simple, intuitive way we actually deal with problems, but it improves and streamlines the process by providing a structured approach to decision making. (Saaty, "For further reading.")

The general approach to using the analytic hierarchy process is:

1.) Develop the hierarchical structure for the decision problem of interest.

2.) Establish priorities among the decision attributes in the hierarchy.

3.) Synthesize the judgments to obtain a set of overall priorities.

4.) Check on the consistency of the judgments.

5.) Make a final decision based on the results.

The hierarchical structure applies very well to financial justification problems. An example is shown in Figure 2 for the justification of a flexible manufacturing system (FMS) (see Varney, et al.). Kamenetzky concisely describes the determination of the relative importance of decision attributes, the relative standing of each action with respect to each attribute, and the overall score of each action (see "For further reading").

Expert systems

An expert system (ES) is a computer program that attempts to solve difficult and complex problems normally solved only by applying human expertise. At present, expert system technology is regarded as an important subset of artificial intelligence having immediate commercial promise. Expert systems have wide application and have been designed, for instance, to configure computer systems, make medical diagnoses and discover the location of mineral deposits.

Typically, three components constitute an expert system: (1) a knowledge base which consists of the rules, facts, etc., utilized by an expert to solve a particular problem; (2) an inference engine which contains the line of reasoning used by a human expert to arrive at a conclusion in view of (1); and (3) a dialog system by which it is possible for users to communicate with the knowledge base through the inference engine. Expert systems accept task-specific data from a user, select and utilize a suitable inference logic for the knowledge base, and then reach a conclusion for a problem in the appropriate domain.

The number of operational expert systems for financial justification in manufacturing and service industries is increasing rapidly. A computer program called XVENTURE is representative of such expert systems (see Sullivan and LeClair). This expert system makes use of six issues that are vitally important to investment decisions that involve advanced manufacturing technology, then links these issues together into a rule base that represents the expertise of a single expert.

It should be noted that XVENTURE makes use of a traditional financial evaluation criterion (net present worth) in addition to five other "irreducibles" (see Figure 3) in the process of balancing benefits and costs for a proposed capital investment. Furthermore, the six issues need not be independent. Successful combinations of responses to the issues that lead to a "venture is justifiable" outcome are based on expert heuristics. An example rule from XVENTURE's rule base is shown in Figure 3.

The advantage of using an ES such as XVENTURE is that it affords a decision maker available expertise with above average ability to analyze strategic, nonquantifiable decision criteria that lend themselves well to heuristic analysis. It also preserves otherwise perishable human expertise and reduces the risks of poor human performance. Hence an ES represents a viable training aid from which "corporate culture" in areas such as financial justification can be quickly learned.

Hope springs eternal

Automation of manufacturing functions must be regarded as a vital ingredient of a firm's strategic plan as well as a significant commitment having far-reaching financial implications. Such investments need to be justified in a broad context relative to competitive positioning of a firm, and possibly even its *survival* in the long term.

To fill the widely acknowledged need for holistic financial justification analyses, the total impact of factory automation on a particular firm ought to be addressed. A broad view of tomorrow's manufacturing systems *must* be taken that permits synergies among their components to be considered in the strategic decisions of a company.

The relatively simple tools discussed in this article are obviously not all-inclusive, but they permit a comprehensive picture of the benefits and costs associated with manufacturing automation to be presented in terms that top managment can easily grasp. Industrial engineers must keep in mind that the financial justification techniques we utilize are selling tools for ideas and concepts that will position our companies for a more profitable future.

Clearly, many nonmonetary ("irreducible") considerations are involved in decisions regarding automation. The techniques discussed above encourage the explicit inclusion of these subjective, and often dominant, factors in the justification process. "Hope springs eternal" for the acceptance, refinement and application of multiattributed decision tools in industry.

For further reading:

Anderson, B.F., D.H. Deane, K.R. Hammond, and G.H. McClelland, *Concepts in Judgement and Decision Research,* Praeger Publishing Co., 1981, p. 253.

Ayers, Robert U. and Steven Miller, "Robotics, CAM, and Industrial Productivity," *National Productivity Review,* pp. 42-60, Winter 1981-82.

Fleischer, G.A., *Symbolic Scorecard Methodology,* I&SE Report 76-4, University of Southern California, Los Angeles, CA, 1976.

Grant, Eugene L., *Principles of Engineering Economy* (Third Edition). New York: The Ronald Press Co., 1950, Chapters 16 and 20.

Hayes, Robert H. and David A. Garvin, "Managing As If Tomorrow Mattered," *Harvard Business Review,* pp. 71-79, May-June 1982.

Horowitz, Ira, "Engineering Economy: An Economist's Perspective," *AIIE Transactions,* pp. 430-437, Vol. 8, No. 4, December 1976.

Huber, Robert F., "Justification: Barrier to Competitive Manufacturing," *Production,* September 1985, p. 46.

Hudson, C.A., "Computers in Manufacturing," *Science,* pp. 818-825, Vol. 215, No. 4534, February 12, 1982.

Kamenetzky, R.D., "The Relationship Between the Analytic Hierarchy Process and the Additive Value Function," *Decision Sciences,* Vol.13, 1982, pp. 702-713.

Klee, A.J., "The Role of Decision Models in the Evaluation of Competing Environmental Health Alternatives," *Management Science,* Vol. 18, October 1971, pp. 52-67.

Miller, J.H., "A Glimpse at Practice in Calculating and Using Return on Investment," *N.A.A. Bulletin,* June 1960, pp. 65-76.

Moore, J.R., and N.R. Baker, "Computational Analysis of Scoring for R&D Project Selection," *Management Science,* Vol. 16, December 1969, pp. 212-232.

Morris, W.T., *Decision Analysis,* Grid Publishing Co., 1977, p. 239.

National Research Council, *Computer Integration of Engineering Design and Production,* National Academy Press, November 1984, 62 pp.

Radnor, Michael, "A Critical Evaluation of the Field on Engineering Economy," *The Journal of Industrial Engineering,* pp. 133-141, May-June 1964.

Saaty, T.L., *Decision Making for Leaders,* Lifetime Learning Publications, 1982, pp. 12-13.

Sullivan, W.G., and S.R. LeClair, "Justification of Flexible Manufacturing Systems Using Expert System Technology," Proceedings of the Autofact 85 Conference, Detroit, MI, November 1985.

Terborgh, George, *Business Investment Policy—A M.A.P.I. Study and Manual,* Machine and Allied Products Institute, Washington, DC, 1958.

Uttel, Bro, "The Coming Glut of Semiconductors,"*Fortune,* Mar. 19, 1984, p. 125.

Varney, M.S., W.G. Sullivan, and J. Cochran, "Justification of Flexible Manufacturing Systems with the Analytic Hierarchy Process," Proceedings of the 1985 Annual Industrial Engineering Conference, Los Angeles, CA, May 1985.

William G. Sullivan, P.E., is professor of industrial engineering at the University of Tennessee, Knoxville. He is also director of the Center for Computer Integrated Engineering and Manufacturing in the College of Engineering. Sullivan received his PhD from the Georgia Institute of Technology and is author of three books and numerous articles. He is a registered professional engineer and a senior member of the Institute.

Reprinted from Industrial Engineering, February 1985

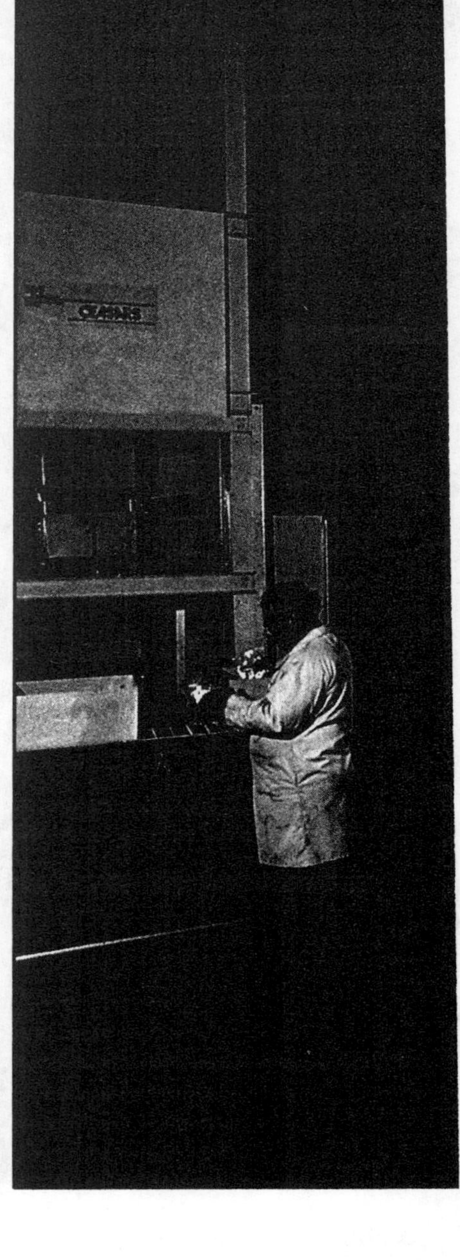

Suggested Techniques Enable Multi-Criteria Evaluation Of Material Handling Alternatives

By Ed Frazelle
North Carolina State University

An inappropriate material handling system selection based on an inappropriate evaluation can lead to losses in plant capacity, dramatic productivity losses, safety hazards and excessive market response time. The multitude of material handling equipment alternatives adds to the complexity of the evaluation process.

This article will describe general evaluation procedure guidelines and two multi-criteria evaluation techniques—the weighted evaluation technique and the analytical hierarchy process (AHP). Special emphasis will be placed on the latter, a relatively new evaluation tool.

All multi-criteria evaluation procedures share several common activities:

1.) *Identify the relevant, important system evaluation criteria.* Jim Apple, in *Material Handling Systems Design,* identifies 30 potential criteria upon which to evaluate material handling equipment and systems. These criteria can be aggregated into the following five major areas of concern for material handling systems/equipment evaluation:

A. Return on investment.
B. Flexibility.
C. Safety.
D. Compatibility.
E. Maintainability.

A. Return on investment (ROI)—ROI and a short-term outlook are the primary hangups in American management today. Yes, ROI is important! However, companies that continue to use ROI as the *sole* criterion for selecting and evaluating material handling systems stifle entrepreneurship and progressive thinking. Those that live exclusively by ROI will also die by it.

The relevant question in today's business environment is not what is the ROI now? The relevant question is, what will my business be like in five or ten years?

If the answer is that it will be a highly competitive business which produces a large variety of short-life-cycle, high quality parts, in an international marketplace, the following comparison should be made: Compare the cost to be competitive with conventional material handling equipment with the cost to be competitive with automated equipment.

The cost with conventional equipment will probably be in the form of significant losses in competitive position stemming from the inability to respond quickly or productively to market demand. The cost to be competitive with automated and more expensive material handling equipment is a low ROI, as we think of ROI today. However, the benefits will be improved competitive position stemming from an improved ability to respond to market demands with higher quality products and improved process control.

B. Flexibility—Flexibility is the capability to respond or conform to new situations easily. There is no doubt that our material handling equipment and systems must be flexible! They must be able to respond to new products, new product mixes and changes in volume for the same

An effective evaluation tool for proposed material handling systems takes into account possible biases or inconsistencies on the part of the evaluator. (Photo courtesy of SPS Technologies, Automated Systems Division)

product. Consequently, they must be able to operate at different speeds, with a variety of loads, over a variety of routings. Automated guided vehicle systems and tow line and power-and-free conveyor systems offer all these features.

C. Safety—According to the National Safety Council, over 60% of all industrial accidents involve material handling. Many of those accidents are associated with careless operation of fork lift trucks. In a transmission manufacturing plant in Indianapolis, the fork lift truck accident rate increases annually in May, the month of the annual running of the Indianapolis 500 auto race.

Conveyors, overhead equipment, industrial trucks and manual material handling all present safety challenges. Training in the use of this equipment and general safety procedures must be encouraged and improved.

D. Compatibility—For the integrated—much less the computer-integrated—factory to be achieved, material handling equipment must be compatible with existing or proposed machine tools, storage equipment, computer control hardware and software, and other system components. Compatibility requires reducing the number of equipment varieties (i.e., simplification) and reducing the number of models and makes of equipment (i.e., standardization). Herman Miller's Action Factory System is an excellent example of simplified, standardized, compatible material handling equipment for electronics manufacturing.

Compatibility is also relevant to the evaluation procedure chosen. It is critical that the evaluation technique chosen be compatible with the level and complexity of the system being evaluated. Certainly the procedures described here would not be appropriate for choosing between a two-wheel and a four-wheel hand cart.

E. Maintainability—With the advent of just-in-time and continuous flow manufacturing, the ability of our material handling equipment and systems to operate frequently, reliably and inexpensively will become increasingly important.

2.) *Determine the weight/importance of each criterion.* The weight or importance of each criterion should be determined by upper level management. The weights will vary depending on the type and state of the business. For example, a consumer electronics manufacturer in a highly competitive, weekly product life cycle business will place much more emphasis on flexibility than would a steel beam manufacturer.

Also consider a rapid growth, highly profitable firm operating in a booming economy. Such a firm can afford to overlook ROI much more than a questionable firm operating in a depressed economy can.

3.) *Evaluate each alternative with respect to each criterion.* The best way to make any evaluation process fruitful is by having a number of excellent alternatives to choose from. Reviewing the material handling literature, visiting other facilities, attending trade shows and confer-

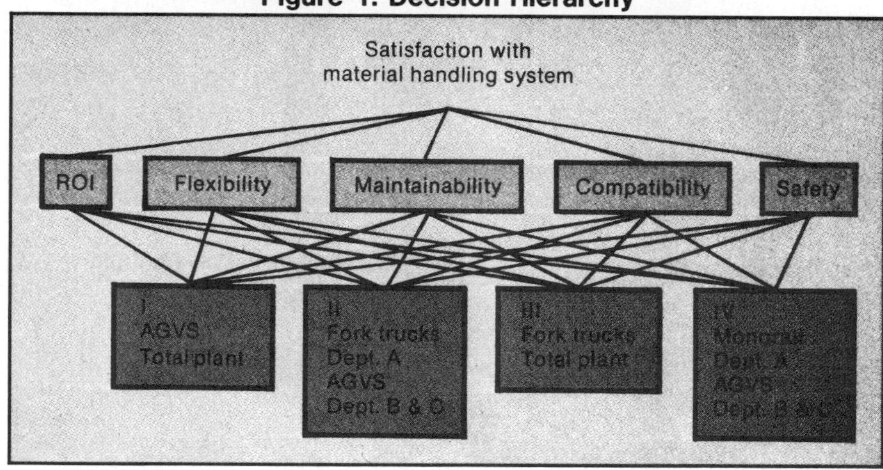

Figure 1: Decision Hierarchy

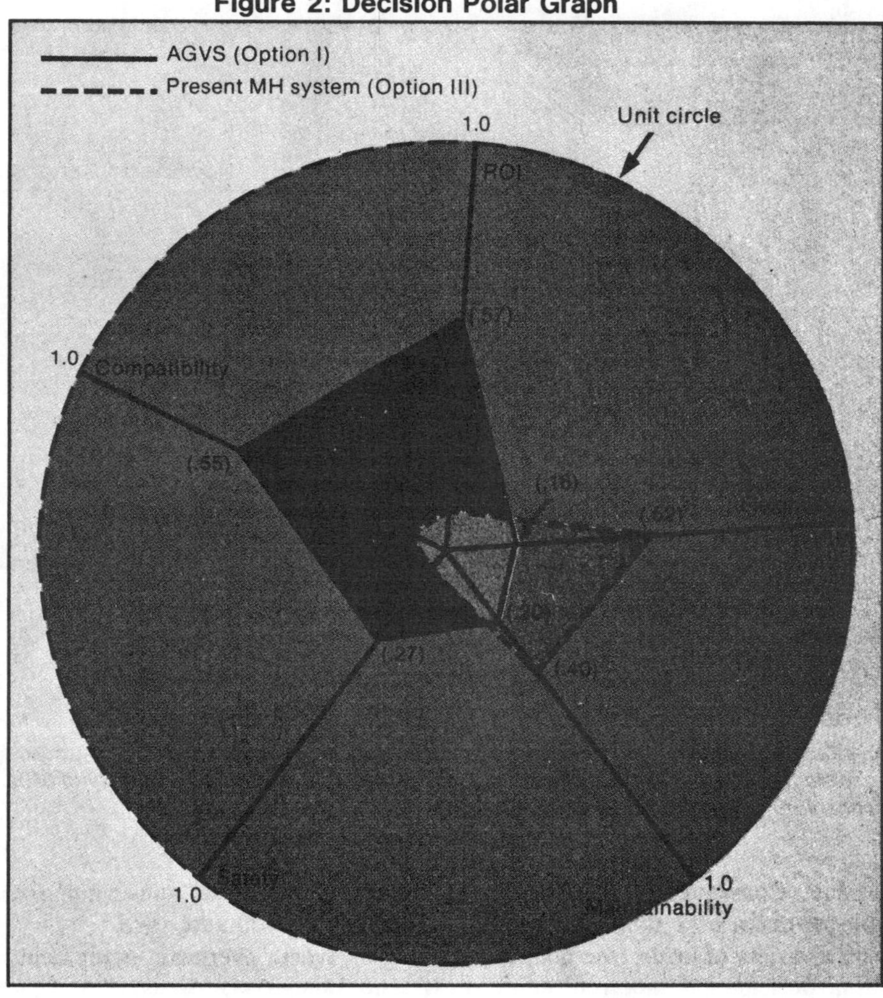

Figure 2: Decision Polar Graph

ences, and soliciting the advice of external consultants will help ensure the consideration of a broad range of alternatives.

Brainstorming sessions, the nominal Group Technique and other participative group solution-generating techniques should also be used to tap the resources of the affected and knowledgeable parties. This involvement will also bring forth the commitment of those parties to the system that is eventually implemented.

4.) *Perform the necessary calculations.* Regardless of the technique, the computer should be utilized. Computers are fast calculators, but they do not understand MH system evaluation criteria such as flexibility and safety (a situation expert systems are trying to remedy). Several multi-criteria evaluation computer programs designed specifically to facilitate multi-criteria evaluation are available. One such program was described in the Mini-Micro Computer column of the December, 1984, issue of *Industrial Engineering.*

5.) *Identify the preferred alternative.* This is easy if there is a large differential between the alternative with the highest weighted evaluation and the other alternatives. However, when the weighted evaluations are close, it is important to review the evaluation process for inconsistencies exhibited by the decision maker and possible errors in the calculations. If there is no change, the final decision can be turned over to management. Another decisive criterion is then usually added to the evaluation problem: politics.

Weighted evaluation technique

The weighted evaluation technique is a conventional and highly useful multi-criteria evaluation/decision-making tool. This technique very closely follows the evaluation procedure guidelines described previously. Criterion weights are assigned on a zero-to-100 scale. The criterion perceived as most important is assigned a weight of 100; all other criterion weights are assigned relative to that.

The weights are then normalized by dividing each criterion weight by the sum of all the weights. Each alternative is evaluated with respect to each criterion on a zero-to-100 scale, 100 representing perfect performance with respect to the given criterion.

Each evaluation is then weighted by multiplying the evaluation by its corresponding normalized criterion weight. Each alternative's overall score is then determined by summing its weighted evaluations with respect to all the criteria.

The alternative with the highest

overall weighted score is the preferred alternative. Table 1 illustrates the technique by evaluating two hypothetical MH systems (A and B) with respect to flexibility, ROI and ease of implementation.

Analytical hierarchy process

A significant drawback of simple weighted evaluation techniques is that they neglect the issue of inconsistency on the part of the decision maker or evaluator. Inconsistency arises when the evaluation results do not confirm the evaluator's preferences.

Inconsistency is a major bias in human judgment that accounts for a large portion of human forecasting errors and deficiencies in planning and evaluating. It is the result of the evaluator's lack of understanding of the problem, attitude variation and circumstance variation. The more alternatives and criteria in the evaluation problem, the more significant the inconsistency issue becomes.

An increasingly popular multi-criteria evaluation procedure which incorporates inconsistency, very closely mimics the human decision making process and facilitates acquiring information from the evaluator is the analytical hierarchy process (AHP). The AHP has been used to evaluate microcomputer systems, develop corporate strategic plans and evaluate facility plans. The process is described below.

The AHP decomposes the evaluation/decision into hierarchical levels. The hierarchy for the MH system evaluation problem is shown in Figure 1. Pairwise comparisons of elements on the same level are made with respect to the elements in the level directly above them. Degrees of preference or intensity of the evaluator in the choice for each pairwise comparison are then measured, and those measurements recorded in a matrix of comparisons. Measurements are made on the following preference scale:

- ☐ If A and B are equally important (perform equally well), assign the number 1 in row of A, column of B.
- ☐ If A is weakly more important than (weakly out-performs) B, assign the number 3 in row of A, column of B.
- ☐ If A is strongly more important than (strongly out-performs) B, assign the number 5 in row of A, column of B.
- ☐ If A is very strongly more important than (very strongly out-performs) B, assign the number 7 in row of A, column of B.
- ☐ If A is absolutely more important than (absolutely out-performs) B, assign the number 9 in row of A, column of B.
- ☐ Use even numbers (2,4,6,8) to represent compromise between comparisons.
- ☐ For inverse comparisons, such as B to A, use the reciprocal of the ranking A to B.

Table 1: Weighted Evaluation Technique

Factor	Weight	Alternatives: A Score	Alternatives: B Score
1. Flexibility	70	90	60
2. ROI	100	70	90
3. Ease of Implementation	40	100	90
Total	210		

Factor	Normalized Weight	Alternatives: A Weighted Score	Alternatives: B Weighted Score
1. Flexibility	.33	30	20
2. ROI	.48	34	43
3. Ease of Implementation	.19	19	17
Total	1.00	83	80

Table 2: Pairwise Comparisons with Respect to Flexibility

Evaluation with respect to flexibility	Alternatives: I	II	III	IV
Alternative I	1	1/2	1/4	3
Alternative II	2	1	1/3	3
Alternative III	4	3	1	4
Alternative IV	1/3	1/3	1/4	1

Table 2 shows the matrix of comparisons with respect to flexibility of four alternatives in a material handling system evaluation problem.

For example, from the matrix we can read that the evaluator feels that alternative II weakly outperforms alternative IV with respect to flexibility. A similar matrix is used to determine criterion weights, comparing flexibility, ROI, maintainability, compatibility and safety.

By normalizing the principle eigenvector of the matrix in Table 2, a vector of priorities is calculated (.16, .23, .52, .09), each value representing the evaluation/weight of the row's alternative/criterion. Normalizing requires summing the elements of each column and then dividing each element in the column by that sum.

The normalized elements are then averaged across the rows. The row average is the resulting evaluation/weight of that row's corresponding alternative/criterion.

Table 3: Calculation of the Priority Vector

	Normalized Matrix from Figure 3				
	I	II	III	IV	
I	.14	.10	.14	.27	(.14 + .10 + .14 + .27) ÷ 4 = .16
II	.27	.21	.18	.27	(.27 + .21 + .18 + .27) ÷ 4 = .23
III	.55	.62	.55	.36	(.55 + .62 + .55 + .36) ÷ 4 = .52
IV	.05	.07	.14	.09	(.05 + .07 + .14 + .09) ÷ 4 = .09

Table 4: Final, Overall Evaluation of Alternatives

Options	(ROI	Comp.	Criterion Flex.	Maint.	Safe.)	×	Criterion (Weight)
I	.57	.55	.16	.20	.27		.48
II	.28	.25	.23	.20	.14		.04
III	.10	.08	.52	.40	.06	×	.21
IV	.05	.12	.09	.20	.52		.19
							.09

	Option	Weighted Evaluation
	I	(.39)
=	II	.24
	III	.24
	IV	.13

Table 5: Criterion Importance: Pairwise Comparisons

	Criteria: ROI	Compatibility	Flexibility	Maintainability	Safety
ROI	1	7	5	4	5
Compatibility	1/7	1	1/4	1/7	1/5
Flexibility	1/5	4	1	3	4
Maintainability	1/4	7	1/3	1	5
Safety	1/5	5	1/4	1/5	1

In the example, alternative I is given an evaluation of .16 (0-1.0 scale) with respect to flexibility. Table 3 demonstrates this procedure for a normalized matrix.

Once each alternative has been evaluated with respect to each criterion and the criterion weights have been determined, an overall evaluation of each alternative material handling system can be performed. This is accomplished by multiplying the matrix of alternative evaluations by the matrix of criterion weights. Table 4 illustrates the procedure for a four-alternative, five-criterion example. The alternative with the highest weighted evaluation is preferred.

As discussed, the incorporation of the evaluator's inconsistency is a very attractive feature of the AHP. This inconsistency can be calculated through a series of additional matrix manipulations.

An inconsistency value can be calculated at each step in the evaluation process as well as for the evaluation process as a whole. Standards for inconsistency inform the evaluator of the degree of confidence that can be placed in the evaluation. (The detailed calculations of the AHP are described in *The Analytical Hierarchy Process,* by Thomas Saaty—see "For further reading.")

A case study

In 1982, a large electronics manufacturer in the Northeast contemplated a total redesign of its material handling system. The criteria upon which any system was to be evaluated were passed down from upper level management as follows:

1.) *Flexibility*—A merger with another firm was imminent; consequently the MH system had to be able to respond to an unknown variety and volume of parts.

2.) *Maintainability* — Downtime of the existing MH equipment was excessive and was limiting production capacity.

3.) *Safety*—Several serious injuries attributable to fork truck operation had occurred in the last year of operation.

4.) *Compatibility*—A strong desire was expressed for the MH system to be compatible and integrated with the AS/RS and highly automated manufacturing equipment.

5.) *ROI*—Several major projects were competing with the MH system redesign for funding.

A series of management interviews resulted in the AHP criterion importance matrix shown in Table 5.

From calculations described earlier, the criteria were assigned weights as follows:

Criterion	*Weight*
ROI	.48
Flexibility	.21
Maintainability	.19
Safety	.09
Compatibility	.03

These criteria prompted the consideration of four alternative material handling systems. The alternatives and their attributes are displayed in the box on page 48.

A team comprised of the firm's manufacturing management and material handling consultants provided input for the evaluation. The unweighted normalized AHP evaluations of alternatives are shown in Table 4.

Weighting these evaluations by the matrix of weights determined

Alternatives and Their Attributes

I. Automated Guided Vehicle System (AGVS)

—ROI: 30.7%, highest among alternatives.
—Safety: controlled travel speed, radar.
—Flexibility: add/subtract vehicles, guide path easily changed.
—Compatibility: compatible with AS/RS and CNC machines.
—Maintainability: major drawback of highly sophisticated systems.

II. Combination of AGVS and Fork Trucks

—ROI: 24.0%, 2nd highest among alternatives.
—Safety: introduction of fork trucks reduces safety.
—Flexibility: introduction of fork trucks improves flexibility.
—Compatibility: some lost with fork trucks.
—Maintainability: improved over totally automated system.

III. Fork Trucks and Manual (the existing system)

—ROI: 0%.
—Safety: human intervention leads to reduced safety.
—Flexibility: people are the most flexible material handlers.
—Compatibility: POOR.
—Maintainability: highest among the alternatives.

IV. Monorail and AGVS

—ROI: 8.0%, third highest among alternatives.
—Safety: overhead equipment improves.
—Flexibility: fixed path of monorail impairs flexibility.
—Compatibility: above average.
—Maintainability: below average.

earlier (see Table 4) yielded the following preferential ordering of alternatives:

Alternative	*Weighted evaluation*
I	.39
II	.24
III	.24
IV	.13

Alternative I, the AGVS, is clearly preferred.

The consistency of these results was "good" as opposed to "excellent." Had alternative I not been such a clear choice, it would have been necessary to reexecute those steps in the evaluation process which exhibited high inconsistency values.

Visual aids

In presenting these results to management, the value of visual aids cannot be underestimated. One very effective visual aid is a decision polar graph (see Figure 2). Each spoke represents a criterion. The end of each spoke represents a desired or maximum value for the corresponding criterion. Normalized, weighted or absolute values can be displayed. The evaluation scores of two alternatives with respect to each criterion are connected to form an area. The alternative representing the largest area is preferred.

Figure 2 compares alternatives I and III from the case study using the unweighted evaluation data from Table 4.

Conclusion

Material handling systems cannot be evaluated solely on the basis of ROI. Multi-criteria evaluation procedures are required to structure, focus and improve the evaluation process. The weighted evaluation technique and the AHP are two excellent evaluation procedures.

Inconsistency in the evaluation must also be a consideration. Inconsistency is a major flaw in human judgment. The AHP helps ensure consistent and logical evaluations.

Material handling is the least understood facet of factory operations. Thorough evaluation of material handling systems and equipment will increase our and management's understanding of this function. Increased understanding is the necessary first step in gaining acceptance for material handling in its deserved role as the factory integrator.

Acknowledgment

The author would like to express his appreciation to John Spain, vicepresident of Tompkins Associates Inc., for providing pairwise comparison trade-off information.

For further reading:

Cox, Daniel K., "Menu System Aids Decisions Made on Qualitative Data," *Industrial Engineering,* December 1984.

Goldher, Joel D., "What Flexible Automation Means to Your Business." *Modern Materials Handling,* September 7, 1984.

Klahorst, H.T., "Flexible Manufacturing Systems: Combining Elements to Lower Costs, Add Flexibility," *Industrial Engineering,* November 1981.

Makridakis, Spyros. "If We Cannot Forecast How Can We Plan?" *Long Range Planning,* 14(3); 10-20; 1981.

Saaty, Thomas L., *Decision Making For Leaders,* California: Lifetime Learning Publications, 1982.

Saaty, Thomas L., *The Analytical Hierarchy Process,* New York: McGraw Hill, 1980.

IE

Ed Frazelle works as a consultant to industry and government specializing in facilities planning and design. In addition, he is the managing editor of *IIE Transactions.* A senior member of IIE, he holds office at the region and chapter levels. He is a recipient of the Gilbreth Memorial Fellowship, among other Institute honors. Frazelle is active in several international professional and honor societies. He earned his BSIE and MSIE from North Carolina State University.

DSS Series Editor's Note

The *MIS Quarterly* has shown a special interest in the question of assessing the costs and benefits of information systems. In this second article in the series on Decision Support Systems, Peter Keen deals with this issue for DSS which, by their very nature, make traditional cost benefit analysis nearly useless. He suggests and outlines value analysis as an alternative approach, based on the incremental development of the system, and staged cost-value assessments.

Ralph H. Sprague, Jr.

Value Analysis: Justifying Decision Support Systems

By: Peter G. W. Keen

Abstract

Managers face a dilemma in assessing DSS proposals. The issue of qualitative benefits is central, but they must find some way of deciding if the cost is justified. A general weakness of the cost-benefit approach is that it requires knowledge, accuracy, and confidence about issues which for innovations are unknown, ill-defined, and uncertain. The benefit of a DSS is the incentive for going ahead. The complex calculations of cost-benefit analysis are replaced in value analysis by rather simple questions about its usefulness.

Keywords: Decision support systems, value analysis, innovation

ACM Categories: 3.5, 3.64

Introduction

Decision Support Systems (DSS) are designed to help improve the effectiveness and productivity of managers and professionals. They are interactive systems frequently used by individuals with little experience in computers and analytic methods. They support, rather than replace, judgment in that they do not automate the decision process nor impose a sequence of analysis on the user. A DSS is in effect a staff assistant to whom the manager delegates activities involving retrieval, computation, and reporting. The manager evaluates the results and selects the next step in the process. Table 1 lists typical DSS applications.[1]

Traditional cost-benefit analysis is not well-suited to DSS. The benefits they provide are often qualitative; examples cited by users of DSS include the ability to examine more alternatives, stimulation of new ideas, and improved communication of analysis. It is extraordinarily difficult to place a value on these. In addition, most DSS evolve. There is no "final" system; an initial version is built and new facilities are added in

---

[1]Detailed descriptions of each DSS shown in Table 1 can be found in the following references:

GADS:	Keen and Scott Morton [12], Carlson and Sutton [6]
PMS:	Keen and Scott Morton [12], Andreoli and Steadman [2]
IRIS:	Berger and Edelman [3]
PROJECTOR:	Keen and Scott Morton [12]
IFPS:	Wagner [16]
ISSPA:	Keen and Gambino [11]
BRANDAID:	Keen and Scott Morton [12], Little [14]
IMS:	Alter [1]

Other DSS referred to in this article are:

AAIMS:	Klaas [13], Alter [1]
CAUSE:	Alter [1]
GPLAN:	Haseman [9]

Table 1. Examples of DSS Applications

DSS	Applications	Benefits
GADS Geodata Analysis Display System	Geographical resource allocation and analysis; applications include sales force territories, police beat redesign, designating school boundaries	Ability to look at more alternatives, improved teamwork, can use the screen to get ideas across, improved confidence in the decision
PMS Portfolio Management System	Portfolio investment management	Better customer relations, ability to convey logic of a decision, value of graphics for identifying problem areas
IRIS Industrial Relations Information System	*Ad hoc* access to employee data for analysis of productivity and resource allocation	*Ad hoc* analysis, better use of "neglected and wasted" existing data resource, ability to handle unexpected short term problems
PROJECTOR	Strategic financial planning	Insight into the dynamics of the business, broader understanding of key variables
IFPS Interactive Financial Planning System	Financial modelling, including mergers and acquisitions, new product analysis, facilities planning and pricing analysis	Better and faster decisions, saving analysts' time, better understanding of business factors, leveraging managing skills
ISSPA - Interactive Support System for Policy Analysts	Policy analysis in state government; simulations, reporting, and *ad hoc* modelling	*Ad hoc* analysis, broader scope, communication to/with legislators, fast reaction to new situations
BRANDAID	Marketing planning, setting prices, and budgets for advertising, sales force, promotion, *etc.*	Answering "what if?" questions, fine-tuning plans, problem-finding
IMS Interactive Marketing System	Media analysis of large consumer database, plan strategies for advertising	Helps build and explain to clients the rationale for media campaigns, *ad hoc* and easy access to information

response to the users' experience and learning. Because of this, the costs of the DSS are not easy to identify.

The decision to build a DSS seems to be based on value, rather than cost. The system represents an investment for future effectiveness. A useful analogue is management education. A company will sponsor a five day course on strategic planning, organizational development, or management control systems on the basis of perceived need or long term value. There is no attempt to look at payback period or ROI, nor does management expect a direct improvement in earnings per share.

This article examines how DSS are justified and recommends Value Analysis (VA), an overall methodology for planning and evaluating DSS proposals. The next section illustrates applications of DSS. Key points are:

1. a reliance on prototypes,
2. the absence of cost-benefit analysis,
3. the evolutionary nature of DSS development, and
4. the nature of the perceived benefits.

The section on the Dynamics of Innovation relates DSS to other types of innovation. It seems clear that innovation in general is driven by "demand-pull" — response to visible, concrete needs — and not "technology push."

The Methodologies for Evaluating Proposals section briefly examines alternative approaches to evaluation: cost-benefit analysis, scoring techniques, and feasibility studies. They all require fairly precise estimates of, and tradeoffs between costs and benefits and often do not handle the qualitative issues central to DSS development and innovation in general. The final part of the article defines Value Analysis.

The overall issue this article addresses is a managerial one:

1. What does one need to know to decide if it is worthwhile to build a DSS?
2. How can an executive encourage innovation while making sure money is well spent?
3. How can one put some sort of figure on the value of effectiveness, learning, or creativity?

It would be foolish to sell a strategic planning course for executives on the basis of cost displacement and ROI. Similarly, any effort to exploit the substantial opportunity DSS provide to help managers do a better job must be couched in terms meaningful to them. This requires a focus on value and a recognition that qualitative benefits are of central relevance. At the same time, systematic assessment is essential. The initial expense of a DSS may be only in the $10,000 range, but this still represents a significant commitment of funds and scarce programming resources. The methodology proposed here is based on a detailed analysis of the implementation of over twenty DSS. It is consistent with the less formal approaches most managers seem to use in assessing technical innovations. Value analysis involves a two stage process:

1. *Version 0:* This is an initial, small scale system which is complete in itself, but may include limited functional capability. The decision to build Version 0 is based on:
   a. An assessment of benefits, not necessarily quantified;
   b. A cost threshold — is it worth risking this amount of money to get these benefits?

   In general, only a few benefits will be assessed. The cost threshold must be kept low, so that this decision can be viewed as a low risk research and development venture, and not a capital investment.
2. *Base System:* This is the full system, which will be assessed if the trial Version 0 has successfully established the value of the proposed concept. The decision to develop it is based on:
   a. Cost analysis: What are the costs of building this larger system?
   b. Value threshold: What level of benefits is needed to justify the cost? What is the likelihood of this level being attained?

A major practical advantage of this two stage strategy is that it reduces the risks involved in development. More importantly, it simplifies the tradeoff between costs and benefits, without making the analysis simplistic. It is also a more natural approach than traditional cost-benefit analysis; until value is established, *any* cost is disproportionate.

Decision Support Systems

The DSS applications shown in Table 1 cover a range of functional areas and types of task. They have many features in common:

1. They are *non-routine* and involve frequent *ad hoc* analysis, fast access to data, and generation of non-standard reports.
2. They often address "what-if?" questions; for example, "What if the interest rate is X%?" or "What if sales are 10% below the forecast?"
3. They have no obvious correct answers; the manager has to make

qualitative tradeoffs and take into account situational factors.

The following examples illustrate the above points:

1. **GADS.** In designing school boundaries, parents and school officials worked together to resolve a highly charged political problem. A proposal might be rejected because it meant closing a particular school, having children cross a busy highway, or breaking up neighborhood groups. In a previous effort involving redistricting, only one solution has been generated, as opposed to six with GADS over a four day period. The interactive problem solving brought out a large number of previously unrecognized constraints such as transportation patterns and walking times, and parent's feelings.

2. **BRANDAID.** A brand manager heard a rumor that his advertising budget would be cut in half. By 5:00 p.m. he had a complete analysis of what he felt the effect would be on this year's and next year's sales.

3. **IFPS.** A model had been built to assess a potential aquisition. A decision was needed by 9:00 a.m. The results of the model suggested the acquisition be made. The senior executive involved felt uneasy. Within one hour, the model had been modified and "what if" issues assessed that led to rejection of the proposal.

4. **ISSPA AND IRIS.** Data which had always been available, but not accessible, were used to answer *ad hoc*, simple questions. Previously, no one bothered to ask them.

These characteristics of problems for which DSS are best suited impose design criteria. The system must be:

1. *Flexible* to handle varied situations.

2. *Easy to use* so it can be meshed into the manager's decision process simply and quickly.

3. *Responsive* because it must not impose a structure on the user and must give speedy service.

4. *Communicative* because the quality of the user-DSS dialogue and of the system outputs are key determinants of effective uses especially in tasks involving communication or negotiation. Managers will use computer systems that mesh with their natural mode of operation. The analogy of the DSS as a staff assistant is a useful one.

Many DSS rely on prototypes. Since the task the system supports is by definition non-routine, it is hard for the user to articulate the criteria for the DSS and for the designer to build functional specifications. An increasingly popular strategy is thus to use a flexible DSS "tool" such as APL, or a DSS "Generator" [15]. These allow an initial version of a "Specific DSS" to be delivered quickly and cheaply. It provides a concrete example that the user can react to and learn from. It can be easily expanded or modified. The initial system, Version 0, clarifies the design criteria and specifications for the full DSS. Examples of this two phase strategy include:

1. **ISSPA** — built in APL. Version 0 took seventy hours to build and contained nineteen commands. The design process began by sketching out the user-system dialogue. New user commands were added as APL functions. Ten of the forty-eight commands were requested by users, and several of the most complex ones were entirely defined by users.

2. **AAIMS** — an APL-based "personal information system" for analysis of 150,000 time series. The development was not based on a survey or user requirements, nor on any formal plan. New routines are tested and "proven" by a small user group.

3. **IRIS** — a prototype was built in five months and evolved over a one year period. An "Executive language" interface was defined as the base for the DSS and a philosophy was adopted of "build and evaluate as you go."

4. **CAUSE** — There were four evolutionary versions. A phased development was used to build credibility. The number of routines was expanded from 26 to 200.

There have been several detailed studies of the time and the cost needed to build a DSS in APL. A usable prototype takes about three weeks to deliver. A full system requires another twelve to sixteen weeks.[2]

End-user languages similarly allow fast development. One such DSS "generator" is Execucom's IFPS (Interactive Financial Planning System), a simple, English-like language for building strategic planning models. The discussion below is based on a survey of 300 IFPS applications in 42 companies.[3] The models included long range planning, budgeting, project analysis, evolution of mergers, and acquisitions.

The average IFPS model took five days to build and contained 360 lines (the median was 200). Documented specifications were developed for only 16%. In 66% of the cases, an analyst simply responded to a manager's request and got something up and running quickly. Cost-benefit analysis was done for 13%, and only 30% have any objective evidence of "hard" benefits. 74% of the applications replace manual procedures.

[2]See Grajew and Tolovi [8] for a substantiation of these figures. They built a number of DSS in a manufacturing firm to test the "evolutive approach" to development.

[3]IFPS is a proprietary product of Execucom, Inc., in Austin, Texas. The survey of IFPS users is described in Wagner [19].

Given that most of the responding companies are in the Fortune 100, this indicates the limited degree to which managers in the planning functions make direct use of computers.

Most DSS are built outside data processing, generally by individuals who are knowledgeable about the application area. Table 2 gives figures on where requests for IFPS applications came from and how they are built.

The IFPS users were asked to identify the features of the language that contributed most to the success of the DSS. In order of importance, these are:

1. speed of response,
2. ease of use,
3. package features (curve-fitting, risk analysis, what-if?),
4. sensitivity analysis, and
5. time savings.

The evolutionary nature of DSS development follows from the reliance on prototypes and fast development. There is no "final" system. In most instances, the system evolves in response to user learning. A major difficulty in designing DSS is that many of the most effective uses are unanticipated and even unpredictable. Examples are:

1. **PMS** — the intended use was to facilitate a portfolio based rather than security based approach to invest-

Table 2. IFPS Development Process

	Data Processing	Staff Analyst	Middle Management	Top Management
Who requested the application	0	4	30	66
Who built it	3	53	22	22
Who uses the terminal	0	70	21	9
Who uses the output	0	6	42	52

ment. This did not occur, but the DSS was invaluable in communicating with customers.

2. **GPLAN** — the DSS forced the users (engineers) to change their roles from analysts to decision makers.

3. **PROJECTOR** — the intended use was to analyze financial data in order to answer preplanned questions and the actual use was as an educational vehicle to alert managers to new issues.

Usage is also very personalized, since the managers differ in their modes of analysis and the DSS is under their own control. For example, six users of PMS studied over a six month period differed strongly in their choice of operators (see Table 3).[4]

The benefits of DSS vary; this is to be expected given the complex situational nature of the tasks they support and their personalized uses. The following list shows those frequently cited in DSS case studies, together with representative examples.[5] Table 4 summarizes the list.

[4]See Andreoli and Steadman [2] for a detailed analysis of PMS usage.

[5]This list is taken verbatim from Keen, "Decision Support Systems and Managerial Productivity Analysis" [8].

1. **Increase in the Number of Alternatives Examined**
   - Sensitivity analysis takes 10% of the time needed previously.
   - Eight detailed solutions generated versus one in previous study.
   - Previously took weeks to evaluate a plan; now takes minutes, so much broader analysis.
   - Users could imagine solutions and use DSS to test out hypotheses.
   - "No one had bothered to try price/profit options before."

2. **Better Understanding of the Business**
   - President made major changes in company's overall plan, after using DSS to analyze single acquisition proposal.
   - DSS alerted managers that an apparently successful marketing venture would be in trouble in six month's time.
   - DSS is used to train managers; gives them a clear overall picture.
   - "Now able to see relationships among variables."

Table 3. Relative Use of DSS Operators (PMS)

	percentage of use by each manager						
Operator	A	B	C	D	E	F	**percentage of use by all users**
Table	22	22	38	22	76	57	47
Summary	40	10	30	8	0	38	17
Scan	0	26	5	24	0	0	4
Graph	14	4	13	30	5	0	8
Directory	2	0	0	0	1	4	1
Others	22	38	14	16	18	1	23

Table 4. DSS Benefits

	Easy to measure?	benefit can be quantified in a "bottom line" figure?
1. Increase in number of alternatives examined	Y	N
2. Better understanding of the business	N	N
3. Fast response to unexpected situations	Y	N
4. Ability to carry out *ad hoc* analysis	Y	N
5. New insights and learning	N	N
6. Improved communication	N	N
7. Control	N	N
8. Cost savings	Y	Y
9. Better decisions	N	N
10. More effective teamwork	N	N
11. Time savings	Y	Y
12. Making better use of data resource	Y	N

3. **Fast Response to Unexpected Situations**

- A marketing manager faced with unexpected budget cut used the DSS to show that this would have a severe impact later.

- Helped develop legal case to remove tariff on petroleum in New England states.

- Model revised in twenty minutes, adding risk analysis; led to reversal of major decision made one hour earlier.

4. **Ability to Carry Out *Ad Hoc* Analysis**

- 50% increase in planning group's throughput in three years.

- The governor's bill was published at noon "and by 5 pm I had it fully costed out."

- "I can now do QAD's — quick-and-dirties."

- System successfully used to challenge legislator's statements within a few hours.

5. **New Insights and Learning**

- Quickened management's awareness of branch bank problems.

- Gives a much better sense of true costs.

- Identified underutilized resources already at analysts' disposal.

- Allows a more elegant breakdown of data into categories heretofore impractical.
- Stimulated new approaches to evaluating investment proposals.

6. **Improved Communication**
   - Used in "switch presentations" by advertising agencies to reveal shortcomings in customer's present agency.
   - Can explain rationale for decision to investment clients.
   - Improved customer relations.
   - "Analysis was easier to understand and explain. Management had confidence in the results."
   - "It makes it a lot easier to sell (customers) on an idea."
7. **Control**
   - Permits better tracking of cases.
   - Plans are more consistent and management can spot discrepancies.
   - Can "get a fix on the overall expense picture."
   - Standardized calculation procedures.
   - Improved frequency and quality of annual account reviews.
   - Better monitoring of trends in airline's fuel consumption.
8. **Cost Savings**
   - Reduced clerical work.
   - Eliminated overtime.
   - Stay of patients shortened.
   - Reduced turnover of underwriters.
9. **Better Decisions**
   - "He was forced to think about issues he would not have considered otherwise."
   - Analysis of personnel data allowed management to identify for the first time where productivity gains could be obtained by investing in office automation.
   - Increased depth and sophistication of analysis.
   - Analysts became decision makers instead of form preparers.
10. **More Effective Team Work**
    - Allowed parents and school administrators to work together exploring ideas.
    - Reduced conflict — managers could quickly look at proposal without prior argument.
11. **Time Savings**
    - Planning cycle reduced from six man-days spread over twenty elapsed days to one half day spread over two days.
    - "Substantial reduction in manhours" for planning studies.
    - "(My) time-effectiveness improved by a factor of 20."
12. **Making Better Use of Data Resource**
    - Experimental engineers more ready to collect data since they knew it would be entered into a usable system.
    - "More cost-effective than any other system (we) implemented in capitalizing on the neglected and wasted resource of data."
    - Allows quick browsing.
    - "Puts a tremendous amount of data at manager's disposal in form and combinations never possible at this speed."

Table 4 adds up to a definition of managerial productivity. All the benefits are valuable but few of them are quantifiable in ROI or payback terms.

In few of the DSS case studies is there any evidence of formal cost-benefit analysis. In most instances, the system was built in response to a concern about timeliness or scope of analysis, the need to upgrade management skills, or the potential opportunity a computer data resource or modelling capability provides. Since there is little *a priori* definition of costs and benefits, there is little *a posteriori* assessment of gains. A number of DSS failed in their aims, but where they are successful, there is rarely any formal analysis of the returns. Many of the benefits are not proven. In managerial tasks there is rarely a clear link between decisions and outcomes, and a DSS can be expected to *contribute* to better financial performance, but not directly cause it. In general, managers describe a successful DSS as "indispensable" without trying to place an economic value on it.

The Dynamics of Innovation

DSS are a form of innovation. They represent:

1. a relatively new concept of the role of computers in the decision process,
2. an explicit effort to make computers helpful to managers who on the whole have not found them relevant to their own job, even if they are useful to the organization as a whole.
3. a decentralization of systems development and operation, and often a bypassing of the data processing department, and
4. the use of computers for "value added" applications rather than cost displacement.

There is much literature on the dynamics of technical innovations in organizations.[6] Its conclusions are fairly uniform and heavily backed by empirical data.

Surveys of the use of computer planning models support these conclusions. In nine cases studied[7] the decision to adopt planning models was based on:

1. comparison with an ongoing system which involves examining either a manual or partially computerized system and deciding that some change is desirable,
2. comparison with a related system, such as a successful planning model in another functional area,
3. initiation of a low cost project, and
4. comparison with competitors' behavior resulting in the use of a "reference model" which reduces the need to estimate the impact of a model not yet constructed on improved decisions and performance.

Even in traditional data processing applications, the emphasis on value rather than cost is common. A survey of all the proposals for new systems accepted for development in a large multinational company found that even though cost-benefit analysis was formally required, it was used infrequently.[8] The two main reasons for implementing systems were:

1. mandated requirements, such as regulatory reports, and
2. identification of one or two benefits, rarely quantified.

Traditional cost-benefit analysis is effective for many computer-based systems. It seems clear, however, that it is not used in innovation. This may partly be because innovations involve R&D; they cannot be predefined and clear specifications provided. There is some evidence that there is a conflict in organizations between groups concerned with performance and those focused on cost. In several DSS case studies, the initiators of the system stress to their superiors that the project is an investment in R&D, not in a predefined product.

Surveys of product innovations consistently find that they come from customers and users rather than centralized technical or research staff. Well over three-quarters of new products are initiated

[6]See Tornatzky, *et al.* [16].

[7]See Planning, "How Managers Decide to Use Planning Models," [3] *Long Range Planning*, Volume 13, April 1980.

[8]See Ginzberg [8].

by someone with a clear problem looking for a solution.[9] Industrial salesmen play a key role as "gatekeepers" bringing these needs to the attention of technical specialists., Even in the microprocessor industry, the majority of products are stimulated in this way by "demand-pull," not by "technology-push."[10]

Case studies indicate that DSS development reflects the same dynamics of innovation as in other technical fields. Table 5 states the same dynamics of innovation as in other technical fields.

Methodologies for Evaluating Proposals

There are three basic techniques used to evaluate proposals for computer systems in most organizations:

1. cost-benefit analysis and related ROI approaches — this views the decision as a *capital investment*,
2. scoring evaluation — this views it in terms of *weighted scores*, and
3. feasibility study — this views it as *engineering*.

Each of these is well-suited to situations that involve hard costs and benefits, and that permit clear performance criteria. They do not seem to be useful — or at least used — for evaluating innovations or DSS.

Cost-benefit analysis is highly sensitive to assumptions such as discount rates and residual value. It needs artificial and often arbitrary modifications to handle qualitative factors such as the value of improved communication and improved job satisfaction. Managers seem to be more comfortable thinking in terms of perceived value and then asking if the cost is reasonable. For example, expensive investments on training are made with no effort at quantification. The major benefits of DSS listed in Table 4 are mainly qualitative and uncertain. It is difficult to see how cost-benefit analysis of them can be reliable and convincing in this context.

Scoring methods are a popular technique for evaluating large-scale technical projects, such as the choice of a telecommunications package, especially when there are multiple proposals with

[9]See Utterback [14].

[10]See von Hippel [15].

Table 5. Dynamics of DSS Innovation

Innovations are value-driven	Main motivation for DSS is "better" planning, timely information, *ad hoc* capability, *etc.*
Early adopters differ from late adopters	DSS are often initiated by line managers in their own budgets; once the system is proven other departments may pick it up.
Informal processes are central	DSS development usually involves a small team; key role of intermediaries knowledgeable about the users and the technology for the DSS; data processing rarely involved; frequently DSS are "bootleg" projects.
Cost is a secondary issue	Costs are rarely tracked in detail; DSS budget is often based on staff rather than dollars; little charge out of systems (this may reflect item below).
Uncertainty reduced by trialability, ease of understanding, clear performance value	Use of prototypes, emphasis on ease of use.

varying prices and capabilities. Scoring techniques focus on a list of desired performance characteristics. Weights are assigned to them and each alternative rated. For example:

characteristic	*weight*	*alternative*	*weighted score*
response time	.30	15	4.5
ease of use	.20	20	4.0
user manual	.10	17	1.7

Composite scores may be generated in several ways: mean rating, pass-fail, or elimination of any alternative that does not meet a mandatory performance requirement. Cost is considered only after all alternatives are scored. There is no obvious way of deciding if alternative A, with a cost of $80,000 and a composite score of 67, is better than B, with a cost of $95,000 and a score of 79.

Feasibility studies involve an investment to identify likely costs and benefits. They tend to be expensive and to focus on defining specifications for a complete system. They rarely give much insight into *how* to build it, and assume that the details of the system can be laid out in advance. DSS prototypes are a form of feasibility study in themselves. They are a first cut at a system. Some designers of DSS point out that Version "0" can be literally thrown away. Its major value is to clarify design criteria and establish feasibility, usefulness, and usability. The differences between a prototype and a feasibility study are important:

1. The prototype moves the project forward, in that a basic system is available for use and the logic and structure of the DSS already implemented.

2. The prototype is often cheaper, if the application is suited to APL or an end-user language.

3. The feasibility study is an abstraction and the prototype is concrete. Since DSS uses are often personalized and unanticipated, direct use of the DSS may be essential to establishing design criteria.

There is no evidence that any of these methods are used in evaluating DSS, except occasionally as a rationale or a ritual. More importantly, almost every survey of the dynamics of innovation indicates that they do not facilitate innovation and often impede it.

Value Analysis

The dilemma managers face in assessing DSS proposals is that the issue of qualitative benefits is central, but they must find some way of deciding if the cost is justified. What is needed is a systematic methodology that focuses on:

1. value first, cost second,
2. simplicity and robustness. Decision makers cannot, and should not have to, provide precise estimates of uncertain, qualitative future variables,
3. reducing uncertainty and risk, and
4. innovation, rather than routinization.

The methodology recommended here addresses all these issues. It relies on prototyping which:

1. factors risk, by reducing the initial investment, delay between approval of the project, and delivery of a tangible product; and
2. separates cost and benefit, by keeping the initial investment within a relatively small, predictable range.

If an innovation involves a large investment, the risk is high. Since estimates of costs and benefits are at best approximate, the decision maker has no way of making a sensible judgment. Risk is factored by reducing scope. An initial system is built at a cost below the capital investment level; the project is then an R&D effort. It can be written off if it fails. By using the DSS one identifies benefits and establishes value. The designer is also likely to learn something new about how to design the full system. The prototype accomplishes the same things as a feasibility study, but goes further in that a real system is built.

The benefit of a DSS is the incentive for going ahead. The complex calculations of cost-benefit analysis are replaced in value analysis by simple questions that most managers naturally ask and handle with ease:

1. What exactly will I get from the system?
   - it solves a business problem,
   - it can help improve planning, communication, and control, and
   - it saves time.
2. If the prototype costs $X, do I feel that the cost is acceptable?

Obviously the manager can try out several alternatives — "If the prototype only accomplishes two of my three operational objectives, at a lower cost of $Y, would I prefer that?" The key point is that value and cost are kept separate and not equated. This is sensible only if the cost is kept fairly low. From case studies of DSS, it appears that the cost must be below $20,000 in most organizations for value analysis to be applicable.

This first stage of value analysis is similar to the way in which effective decisions to adopt innovations are made. It corresponds to most managers' implicit strategy. The second stage is a recommendation; there is no evidence in the literature that it is widely used, but it seems a robust and simple extension of Version "0". Once the nature and value of the concept has been established the next step is to build the full DSS. The assessment of cost and value now needs to be reversed:

1. How much will the full system cost?
2. What threshold of values must be obtained to justify the cost? What is the likelihood they will occur?

If the expected values exceed the threshold, no further quantification is required. If they do not, then there must either be a scaling down of the system and a reduction in cost, or a more detailed exploration of benefits.

Value analysis follows a general principle of effective decision making — simplify the problem to make it manageable. A general weakness of the cost-benefit approach is that it requires knowledge, accuracy, and confidence about issues which for innovations are unkown, ill-defined, and uncertain. It therefore is more feasible to:

1. Establish value first, then test if the expected cost is acceptable.
2. For the full system, establish cost first, then test if the expected benefits are acceptable.

Instead of comparing benefits against cost, value analysis merely identifies relevant benefits and tests them against what is in effect a market price: "Would I be willing to pay $X to get this capability?" It is essential that the benefits be accurately identified and made operational. The key question is how would one know that better planning has occurred? The prototype is in effect an experiment in identifying and assessing it.

Figure 1 illustrates the logic and sequence of value analysis. The specific details of the method are less important than the overall assumptions, which have important implications for anyone trying to justify a DSS whether as a designer or user. Marketing a DSS requires building a convincing case. Figure 1 can be restated in these terms:

1. Establish value — the selling point for a DSS is the specific benefits it provides for busy managers in complex jobs.

2. Establish cost threshold — "trialability" is possible only if the DSS is relatively cheap and installed quickly. If it costs, say, $200,000, it is a capital investment, and must be evaluated as such. This removes the project from the realm of R&D and benefits as the focus of attention to ROI and tangible costs and inhibits innovation.

3. Build Version "0" — from a marketing perspective this is equivalent to "strike while the iron is hot." Doing so is possible only with tools that allow speedy development, modification, and extension.

4. Assess the prototype — for the marketer this means working closely with the user and providing responsive service.

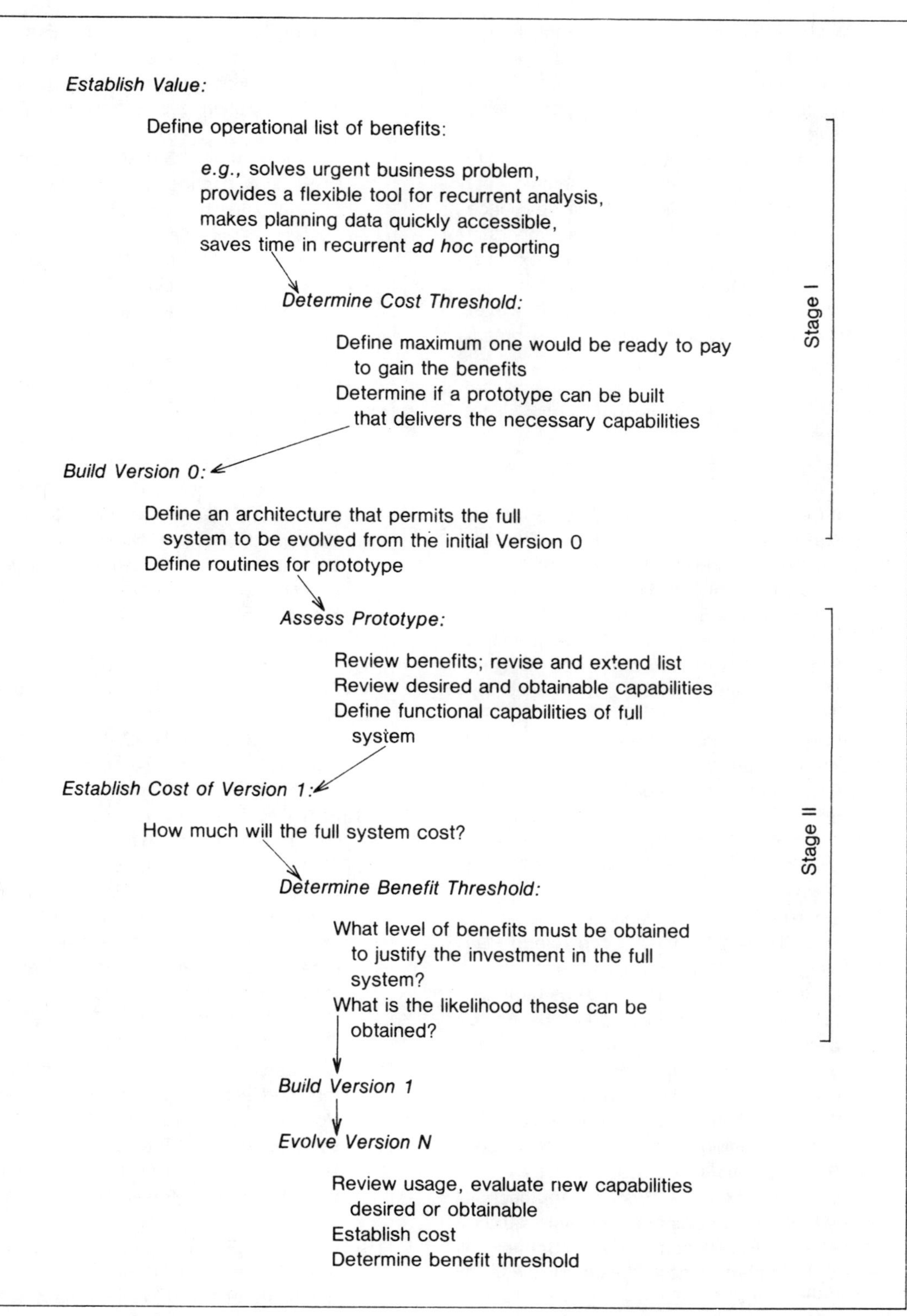

Figure 1. Value Analysis

Two analogies for DSS have been mentioned in this article: the staff assistant and management education. The strategy used to justify DSS depends upon the extent to which one views such systems as service innovations and investments in future effectiveness as opposed to products, routinization, and investment in cost displacement and efficiency. The evidence seems clear — DSS are a potentially important innovation. Value is the issue, and any exploitation of the DSS approach rests on a systematic strategy for identifying benefits, however qualitative, and encouraging R&D and experimentation.

References

[1] Alter, S. *Decision Support Systems: Current Practice and Continuing Challenges*, Addision-Wesley, 1980.

[2] Andreoli, P. and Steadman, J. "Management Decision Support Systems: Impact on the Decision Process," Master's thesis, Sloan School of Management, Massachusetts Institute of Technology, 1975.

[3] Berger, P. and Edelman, F. "IRIS: A Transaction Based DSS for Human Resources Management," in Carlson, E.D., ed., "Proceedings of a Conference on Decision Support Systems," *Data Base*, Volume 8, Number 3, Winter 1977, pp. 22-29.

[4] Blanning, R. "How Managers Decide to Use Planning Models," *Long Range Planning*, Volume 13, April 1980.

[5] Carlson, E.D., ed. "Proceedings of a Conference on Decision Support Systems," *Data Base*, Volume 8, Number 3, Winter 1977.

[6] Carlson, E.D. and Sutton, J.A. "A Case Study of Non-programmer Interactive Problem Solving," IBM Research Report, RJ13H2, San Jose, California, 1974.

[7] Ginzberg, M.J. "A Process Approach to Management Science Implementation," Ph.D. dissertation, Sloan School of Management, Massachusetts Institute of Technology, 1975.

[8] Grajew, J. and Tolovi, J., Jr. "Conception et Mise en Oeuvre des Systems Interactifs d'aide a la Decision: l'approche Evolutive," Doctorial dissertation, Universite des Sciences Sociales de Grenoble, Institut d'Administration des Entreprises, France, 1978.

[9] Haseman, W.D. "GPLAN: An Operational DSS," in Carlson, E.D., ed., "Proceedings of a Conference on Decision Support Systems," *Data Base*, Volume 8, Number 3, Winter 1977, pp. 73-78.

[10] Keen, P.G.W. "Decision Support Systems and Managerial Productivity Analysis," paper presented at the American Productivity Council Conference on Productivity Research, Houston, Texas, April 1980.

[11] Keen, P.G.W. and Gambino, T. "The Mythical Man-Month Revisited: Building A Decision Support System in APL," paper presented at the APL Users Meeting, Toronto, Canada, September 1980.

[12] Keen, P.G.W. and Scott Morton, M.S. *Decision Support Systems: An Organizational Prespective,* Addison-Wesley, 1978.

[13] Klaas, R.L. "A DSS for Airline Management," in Carlson, E.D., ed., "Proceedings of a Conference on Decision Support Systems," *Data Base*, Volume 8, Number 3, Winter 1977, pp. 3-8.

[14] Little, J.D.C. "BRANDAID," *Operations Research*, Volume 23, Number 4, May 1975, pp. 628-673.

[15] Sprague, R.H., Jr. "A Framework for the Development of Decision Support Systems," *MIS Quarterly,* Volume 4, Number 4, December 1980, pp. 1-26.

[16] Tornatzky, L.G., *et al.* "Innovation Processes and Their Management: A Conceptual, Empirical, and Policy Review of Innovation Process Research," National Science Foundation Working Draft, October 19, 1979.

[17] Utterback, J.M. "Innovation in Industry and the Diffusion of Technology," *Science*, Volume 183, February 1974.

[18] von Hippel, E. "The Dominant Role of Users in the Scientific Instrument Innovation Process," *Research Policy,* July 1976.

[19] Wagner, G.R. "Realizing DSS Benefits with the IFPS Planning Language," paper presented at the Hawaii International Conference on System Sciences, Honolulu, Hawaii, January 1980.

About the Author

Peter G. W. Keen, *Assistant Professor at the Sloan School of Management at M.I.T., received his DBA from Harvard University and has been on the faculties of the Harvard Business School, Stanford University, and the Wharton School at the University of Pennsylvania.*

His publications include the first book in the Decision Support series, Decision Support Systems: An Organizational Perspective with M.S. Scott Morton, *and articles on cognitive style, implementation, design and development of DSS, and political aspects of data and software engineering. His current research includes a study of support systems for policy analysts in educational finance, funded by the Ford Foundation. He is a member of the editorial policy board and coordinator for the DSS area in a new journal,* Human Systems Management.

Reprinted from **Industrial Engineering,** November 1982.

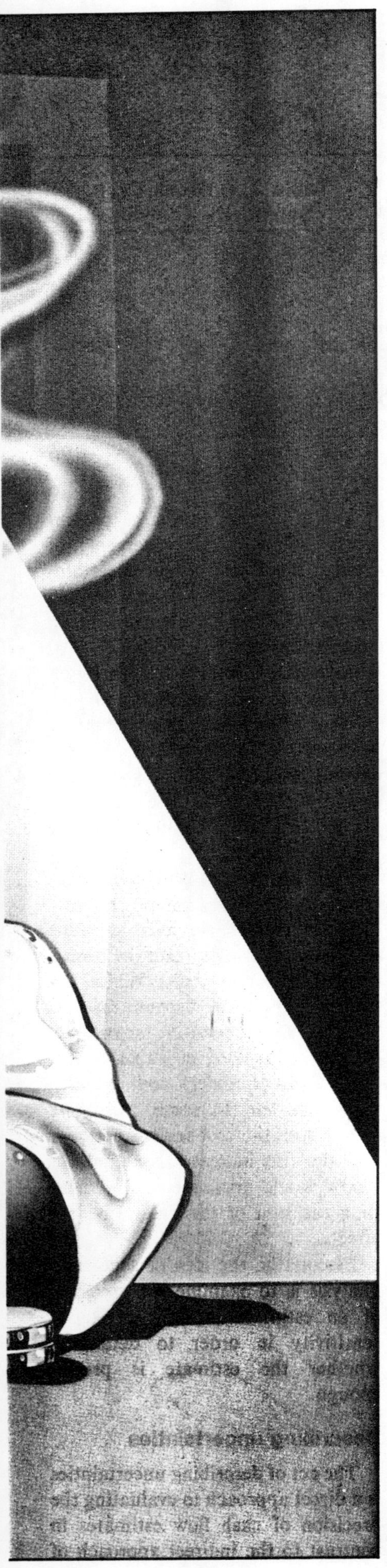

Risk Analysis Method Can Help Make Firms' Investments Less Of A Gamble

By James R. Buck
The University of Iowa

The transition from textbooks to engineering practice produces some surprises. One of these surprises is the realization that the nice, crisp numerical examples in textbooks translate into fuzzy and uncertain images in the real world. Costs and income estimates are imprecise regarding both amount and timing.

Although such estimates may be educated best-guesses, they are not perfect predictors of the future. As a result, some projects may cost more money than they save or produce other risks.

The question, then, is, "How should consideration be given to the economic risk associated with these uncertainties in practical situations?" That question is the focus of this article.

Environmental uncertainties

Uncertainties are those elements of the future which cannot be perfectly predicted. The timing of a machine failure that requires an overhaul is an example. Some others are equipment utilization percentages, prices for raw materials, payments for goods sold, weather, accidents, fraction-lot defectives and sales volumes for various product lines.

Each of these examples shows a change in the environment which cannot be controlled or infallibly predicted. Also, any one of these changes can affect the amount or timing of cash flows of a project and alter the project economics. Moreover, some of these elements are more unpredictable than others and therefore offer more uncertainty.

Most of these environmental uncertainties can be studied for greater predictability. Work sampling can be performed to improve an estimate of equipment utilization. Scrap or quality control records can be investigated for better knowledge of fraction-lot defectives. National weather-service studies have made weather prediction more accurate.

Although such studies can help reduce uncertainties, there will always be some uncertainty left no matter how much study is devoted to the subject. In other words, there is no way to make economic decisions without some uncertainties.

Economic risk

Uncertainties are unknowns and

Table 1: Present Worth Changes Due to Factor Changes of ±20%

Factor	Actual Factor Conditions			
	Decrease to 0.8	Δ%	Increase to 1.2	Δ%
Investment (8,000 to 12,000)	4,642.41	+76	642.41	−76
Savings Per Year (3,200 to 4,800)	113.93	−96	5,170.89	+96
Life (4 yr or 6 yr)	1,013.42	−62	3,976.12	+50
Discount Rate (.16 or .24)	3,766.78	+43	1,646.76	−38

Table 2: A Break-Even Sensitivity Analysis

Factor	Δ% In Actual Amount
Investment	+26
Savings per Year	−21
Life	−31
Discount Rate	+60

unpredictables in the factors which affect project cash flows. The effects of these unknowns and unpredictables on economic criteria are the risks.

Contrary to popular belief, more uncertainty in the uncontrolled factors does not necessarily mean more risk. That depends mostly on our knowledge of the relationships between the factors we cannot control and their economic implications. These are the reasons it is important to *manage* risks rather than just let them happen.

The relationship between uncertainties and that which is perceived as risk is critical. Some uncertainties may have more risk effect than others, and some may counterbalance others.

Risk is typically thought of as an exposure to undesired consequences. The worse the consequences, or the greater the exposure, or both, the greater the risk. In project economics, risky consequences are higher costs or lower cost savings. More specifically, lower net present worths and higher probabilities of those consequences denote more risk. Therefore, projects with higher expected net present worths provide greater economic safety.

A typical example

Suppose that your firm is considering a project that requires a $10,000 investment and promises to yield $4,000 cost savings per year over the next five years. Normally one would say that this project's 2.5-year payback period would give a rate of return on the project, before taxes, of about 1/2.5 = 0.40, or 40%. That is an appealing project at first glance. The net present worth of this project, using a minimum attractive rate of return of 20% continuous, is:

$$P = -10{,}000 + 4{,}000\,[1-e^{-0.2(5)}]/0.2 = \$2{,}642$$

and this surplus in net present worth is comforting. Then the inevitable question arises: "But how good are those numbers?"

Sensitivity analysis

Many people use "sensitivity analysis" as a backwards approach to answering the question about the accuracy of the cost and time estimates. Essentially, this form of analysis indicates the amount of change in an economic criterion (e.g., present worth) that will result in a certain percentage change in each individual part of all component cash flows. These individual parts of the cash flows are the sources of uncertainty, and so sensitivity analysis tells how much the present worth changes with a change of perhaps 20% in the amount of cost savings per year or in the duration of cost savings.

Table 1 shows results of present worth change (and percent of change) when each of four uncertainty factors is decreased or increased by 20%. Greater changes mean greater sensitivity.

Another approach to sensitivity analysis involves finding the change in the uncertainty factor that is required to bring about a zero present worth. Table 2 shows the results of this second procedure, with which smaller changes denote greater sensitivity.

Either of these sensitivity analyses shows that the amount of cost savings per year is the most sensitive factor of uncertainty, and the investment amount is next. While the project life and the discount rate are successively less sensitive factors, it is interesting to note, in Table 1, that the effects of under- and overestimates are *not* the same. Also note that if both the cost savings per year and the life increased, net present worth would increase 156%, more than the sum of the two individual effects.

Essentially, the idea of sensitivity analysis is to examine the precision of an estimate in light of factor sensitivity in order to determine whether the estimate is precise enough.

Describing uncertainties

The act of describing uncertainties is a direct approach to evaluating the precision of cash flow estimates in contrast to the indirect approach of

Table 3: A Gallery of Probability Functions for Describing Uncertainties

Random Variable	Form and Range	Probability p(x)	Statistics	Remarks
Normal	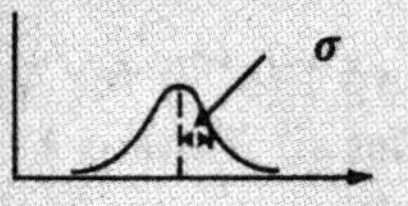 $-\infty$ to $+\infty$; μ, σ	$\frac{1}{\sigma\sqrt{2\pi}} e^{-(x-u)^2/2\sigma^2}$	$E(x)=\mu$ $V(x)=\sigma^2$ $M(x)=\mu$	p(x) is symmetrical about μ
Gamma	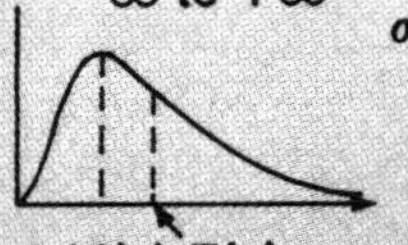 0 M(x) E(x); 0 to $+\infty$; a, b	$\frac{x^{(a-1)}e^{-x/b}}{(a-1)!b^a}$	$E(x)=ab$ $V(x)=ab^2$ $M(x)=(a-1)b$ if $a>1$, otherwise $M(x)=0$	Positively Skewed
Negative Exponential	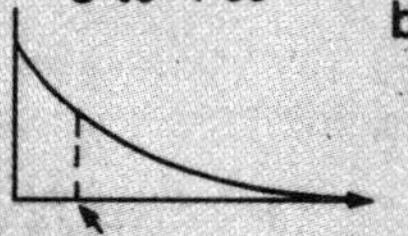 0 E(x); 0 to $+\infty$; a, b	$\frac{1}{b}e^{-x/b}$	$E(x)=b$ $V(x)=b^2$ $M(x)=0$	Same as gamma when $a=1$
Rectangular	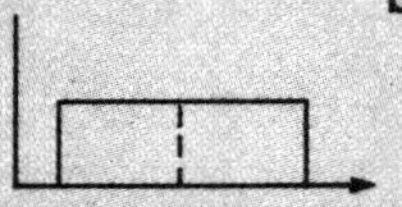 a E(x) b; a to b; a, b	$\frac{1}{b-a}$	$E(x)=(a+b)/2$ $V(x)=(b-a)^2/12$ No mode	p(x) is symmetrical about E(x) (strictly finite)
Triangular	 a b E(x) c; a to c; a, b, c	$\frac{2(x-a)}{(c-a)(b-a)}\ a\leq x\leq b$ $\frac{2(c-x)}{(c-a)(c-b)}\ b\leq x\leq c$	$E(x)=(a+b+c)/3$ $V(x)=[a(a-b)+c(c-a)+b(b-c)]/18$ $M(x)=b$	p(x) is positively skewed if $c-b>b-a$ and negatively skewed if the reverse (strictly finite)

sensitivity analysis. This act bears many similarities to the estimation of cash flows and timing. A great deal of research has shown that people can do a remarkably good job of making subjective estimates of uncertainties (for example, see Sheridan and Ferrell, "For further reading").

One well known way of describing uncertainties is by recognizing that each factor of a cash flow estimate is really a random variable. With that realization, the remaining job is to describe the random variable as accurately as possible.

Perhaps this description can be obtained by first identifying the general dispersional form or shape of the random variable as a probability function and then fitting the selected function to the situation. This fitting involves finding the parameter values of the selected function.

Table 3 is a gallery of typical probability functions with comments about the functional form. Most of these functional forms are distinct from each other in range or functional shape, so that the estimators can probably rule out all but one.

For example, the normal form is symmetrical with high central probability and low tail probability, whereas the rectangular form, also symmetrical, has equal probability over the range. Through reasoning such as that, most of the functional forms can be eliminated and a reasonable form can be chosen.

As will be shown later, many of the measures of economic risk are not highly sensitive to the chosen shape of the probability function as long as it is a reasonable description.

After the selected form is chosen for each factor, a specific uncertainty estimate can be obtained by specifying one or two statistics of the form. Estimates can be made subjectively

Table 4: Summary of Uncertainty Descriptions

Cost Savings Stream:	
Amount/Year	Normal μ = 4,000; σ = 500
Duration	Rectangular a = 4 b = 6
Start Delay	Negative Exponential b = ½
Overhaul:	
Amount	Arbitrary E(F) = \$1,000 V(F) = \$200
Timing	Rectangular a = 2 b = 4
Investment:	Normal μ = 10,000 and σ = 31.62 E(P) = \$10,000; V(P) = \$1,000

or by analogy. The procedure is illustrated below by an example.

In a cost savings stream, both the amount saved per year and the duration of these savings are uncertain. Suppose that the estimator thought about the amount of annual savings and felt that the normal form of random variable best described the anticipated uncertainty of that amount. If the expected annual flow is estimated at \$4,000 and the expected variation is \$500, $\mu = 4{,}000$ and $\sigma = 500$ and this completes the uncertainty description for the annual cash flow amount.

Next the *duration* of these savings is considered. Here the estimator may feel that the rectangular form best describes this uncertainty and that the least and greatest duration are respectively four and six years. Those two duration estimates correspond to the two parameters, a and b, in the rectangular uncertainty case. Each of the other cash flow factors can be considered in a similar fashion, and the estimator could arrive at the uncertainty descriptions shown in Table 4.

Table 5: Formula for Computing $E(e^{-jt})$ for Various Uncertainty Descriptions of Timing t

Uncertainty Description of t	$E(e^{-jt})$
Uniform from a to b	$[e^{-jt} - e^{-jb}]/[j(b-a)]$
Normal with mean μ and std. deviation σ	$e^{-j\mu} + (j^2\sigma^2/2)$
Exponential with b	$(1+jb)^{-1}$
Gamma with parameters a and b	$(1+jb)^{-a}$
Triangular with parameters A, B, and C	$\dfrac{2}{c-a}\,\dfrac{e^{-ja}-e^{-jb}}{j^2\,(b-a)} + \dfrac{e^{-jc}-e^{-jb}}{j^2\,(c-b)}$
Arbitrary with mean μ, standard deviation σ, and symmetrical	$(1+\sigma^2 j^2/2)e^{-j\mu}$

Source: Young, D. and Contreras, L., "Expected Present Worths of Cash Flows under Uncertain Timing," *The Engineering Economist*, Vol. 20, No. 4, 1975.

Expected present worth

To find out how these uncertainties affect the project risk, the analyst must either assume statistical independence or describe how the individual uncertainties relate to each other. If they are all independent of each other, the outcome of any one factor does not affect the outcome of any other. When that assumption appears reasonable, closed-form analysis, as shown below, can be used. Otherwise a computer simulation analysis (see Sullivan and Orr, page 42 of this issue) is recommended.

Once the uncertainty is described for a cash flow factor, the expected present worth for the cash flow can be computed directly. To illustrate this computation, suppose that the cash flow under consideration is an overhaul cost. This cost is a single future cash flow F that occurs at some time t during the project life.

Two uncertainties exist, however, the amount of the project cost and the timing of the cash outflow. The expected present worth is equal to the expected cost $E(F)$ times the factor $E(e^{-jt})$ which accounts for the flow timing. Depending upon the timing uncertainty, the factor $E(e^{-jt})$ can be computed directly from the formula in Table 5 and the uncertainty description. When the overhaul timing for this project seems equally likely at any time between two and four years, the uncertainty is described by the rectangular form with $a = 2$ and $b = 4$. At a minimum attractive rate-of-return before taxes set at $j = 20\%$, the expected timing factor from the Table 5 formula yields:

$$E(e^{-jt}) = [e^{-.2(2)} - e^{-.2(4)}]/[0.2(4\text{-}2)] = 0.55248$$

Table 6: The Project's Expected Present Worth as the Sum of the Signed Component Present Worths*

Cash Flow Component	Expected Present Worth
Investment:	\$−10,000
Cost saving per year:	+11,448
Overhaul:	− 552
Sum:	\$ 896

*+ denotes inflows and − denotes outflows.

For an expected overhaul cost of \$1,000 when it occurs, the expected present worth is:

$$E(P) = 1{,}000(0.55248) = \$552.48.$$

That expected present worth is a few dollars more than the present worth would be if the timing were known to be exactly three years.

Table 5 also shows an uncertainty

Table 7: Present Worth Variance Formula for Single Future Cash Flows F With Uncertain Timing or Uniform Annual Flows A With Uncertain Durations After the Start

Probability Density Function on Timing	Variance of Present Worth Due to Uncertain Timing of a Single Future Flow for Duration of a Uniform Flow A
Rectangular probability from a to b	$F\left[E(F)^2 + V(F)\right]\left\{\frac{e^{-2ja} - e^{-2jb}}{2j(b-a)} - \frac{[e^{-ja} - e^{-jb}]^2}{2j^2(b-a)}\right\} + V(F)\left\{\frac{e^{-ja} - e^{-jb}}{j(b-a)}\right\}$
	$A\left\{\frac{E(A)^2 + V(A)}{j^2}\right\}\left\{\frac{e^{-2ja} - e^{-2jb}}{2j(b-a)} - \left[\frac{e^{-ja} - e^{-jb}}{j(b-a)}\right]^2\right\} + \frac{V(A)}{j}\left\{1 - \frac{e^{-ja} - e^{-jb}}{j(b-a)}\right\}$
Gamma probability timing parameters a and b	$F\left\{E(F)^2 + V(F)\right\}\left\{\left(\frac{1}{1+2jb}\right)^a - \left(\frac{1}{1-jb}\right)^{2a}\right\} + V(F)\left(\frac{1}{1-jb}\right)^a$
Negative exponential a = 1	$A\frac{E(A)^2 + V(A)}{j^2}\left\{\left(\frac{1}{1+2jb}\right)^a - \left(\frac{1}{1-jb}\right)^{2a}\right\} + \frac{V(A)}{j}\left\{1 - \left(\frac{1}{1-jb}\right)^a\right\}$
Normal with timing parameter u and o	$F\left\{E(F)^2 + V(F)\right\}\left\{e^{-2ju + 2j^2o^2} - e^{-2ju + j^2o^2}\right\} + V(F)e^{-2jutj^2o^2/2}$
	$A\frac{E(A)^2 + V(A)}{j^2}\left\{e^{-2ju + 2j^2o^2} - e^{-2ju + j^2o^2}\right\} + \frac{V(A)}{j}\left\{1 - e^{-jutj^2o^2/2}\right\}$

Source: Rosenthal, R. E., "The Variance of Present Worth of Cash Flows under Uncertain Timing," *The Engineering Economist*, Vol. 23, No. 3, 1978.

description that is called an arbitrary form. When the uncertainty description is fuzzy but symmetrical, this formula may be used.

In the case of the overhaul timing, with a three-year expectation and a timing variance of one-third of a year (standard deviation = 0.58 years), which is the same as with the uniform uncertainty form, the timing present worth factor with an arbitrary form is:

$$E(e^{-jt}) = [1 + (1/3)\,(0.2)^2/2]e^{-.2(3)} = 0.5525$$

in which any difference is negligible. Differences in the selected form of the uncertainty usually have only a minor effect on the calculated expected present worth.

When the cash flow is a uniform cash flow stream for an uncertain time *duration,* the expected present worth is:

$$E(P) = E(A)\ [1 - E(e^{-jt})]/j$$

where *E(A)* is the expected cash flow amount per year and $E(e^{-jt})$ is found from the Table 5 formula and the duration uncertainty description. For an expected cost savings amount per year of $4,000 but with a standard deviation of $500 and a duration uncertainty that is rectangular from four to six years, the expected present worth is:

$$E(P) = 4{,}000\ [1 - (0.37034)]/.2 = \$12{,}593$$

which is slightly less present worth than would result from a known duration of five years.

Another source of uncertainty can be the start of a cost savings cash flow. Sometimes an essential part for a project is delayed, and that delay holds back the start of the uniform cost savings. If the delay uncertainty in starting the cost savings cash flow is a negative exponential form with an expected delay of one-half year, the delay present worth factor from Table 5 is:

$$E(e^{-jt}) = [1 + 0.2(0.5)]^{-1} = 0.90909$$

and the expected present worth of the same cost savings stream *with the delay* is:

$$E(P) = 0.90909(12{,}593) = \$11{,}448$$

In this way the described uncertainties of a cash flow can be combined to find the expected present worth.

It also follows that the expected present worth of the entire project consists of the algebraic sum of the component present worths, with negatively signed costs and positively signed cash inflows. Consequently, the cash flow uncertainties of the project with the uncertainty descriptions shown in Table 4 would create the expected present worths and the total expected present worth shown in Table 6.

Present worth variances

Although expected present worths reflect the environmental uncertainties as an average economic value, that measure does not directly show the economic risk. What is also needed is the present worth variability. The variance, or the square root of this variance (the standard deviation), describes the expected amount of variation in present worth about that expected. When the expected net present worth is divided by the standard deviation of present worth, the smaller the ratio, the greater the risk.

With the cash flow factors of uncertainty as described, present worth variances can be calculated

Table 8: Selection Ratios and Probabilities

Ratio	0.25	0.5	0.75	1.0	1.5	1.65	2.0	2.33
Probability	40%	31%	23%	16%	7%	5%	2%	1%

Table 9: The Project's Present Worth Variance as the Sum of the Component Variances Without Regard to Sign

Component Cash Flow		Variance of Present Worth V(P)	$\sqrt{V(P)}$
Investment:		$1,000,000	1,000
Cost Savings/Year A:		6,507,640	2,551
Overhaul:		26,230	162
	Sum:	+7,533,870	2,745

Table 10: Evaluating Economic Risk of Projects

Expected Range of Project Present Worth:

From $E(P) - \sqrt{V(P)} = 896 - 2{,}745 = \$-1{,}849$

To $E(P) + \sqrt{V(P)} = 896 + 2{,}745 = \$+3{,}641$

Probability of a Positive Present Worth:

$$\frac{E(P)}{\sqrt{V(P)}} = \frac{896}{2{,}745} = 0.326 = Z$$

P(Z) = Probability of a positive present worth = 0.626

1 − P(Z) = Probability of a negative present worth = 0.374

much as in the expected present worth case. However, the calculations become a bit more lengthy, and the timing and cash amount variances need to be computationally merged. Table 7 gives these formulas.

In the case of the overhaul cost in the project for which the expected futuıe cost is $1,000 and the standard deviation of this cost is $200, the variance of present worth can be found directly from Table 7 using the rectangular uncertainty case. The resulting present worth variance is $26,320, and its square root of $162 denotes a variation of present worth around the expected $552. That is, the expected range of present worth for this overhaul is $552 ± $162, or from $390 to $714.

Once these present worth variances are computed for each cash flow stream, the project variance is simply the *sum* of the variances, regardless of whether a cash flow is a receipt or an expenditure.

Table 8 shows a summary of the calculations for this project according to the uncertainty descriptions shown in Table 4. It is interesting to note that all of the needed calculations flow directly from those uncertainty descriptions, which means that a programmable calculator or a small computer could greatly aid in the analysis.

Considering economic risk

Once the expected present worth and its variance have been determined, the remaining question is, "How can this information be used meaningfully to successfully manage economic risk?"

One suggestion is to rank prospective projects by their riskiness. An approximation of the riskiness of a project is the probability that the project will *not* yield a positive net present worth. If the expected present worth is negative, there is less than a fifty-fifty chance of an economic success. With a positive expected net present worth, simply divide the standard deviation of worth into that expectation, and the resulting ratio is approximately equal to a standardized normal deviate.

Tables of these standard deviates denote corresponding probabilities of negative net present worths. A few selection ratios and the associated probabilities are shown in Table 8.

If a maximum acceptable risk can be set, all projects above a maximum probability (or below a minimum ratio) can be eliminated, or at least postponed for later consideration. Alternatively, the expected range of net present worth of a risky project is the expected present worth plus or minus one standard deviation.

To illustrate the evaluation of economic risk with the project example used in Tables 4, 6 and 9, consider the expected range of present worth. This range should be the expected present worth minus and plus the present worth standard deviation as shown in Table 10. Another method is to find the probability that a positive expected worth will occur. Table 10 shows this calculation where the chances favoring a successful project are greater than 6 out of 10.

Final project selection from among those projects remaining depends upon management's perception of present worth and risk. When the ranking by expected net present worth is the same as the ranking using the above ratio, projects with higher expected present worths also have a lower risk and there is no conflict. Otherwise a tradeoff must be made between the projects with higher expected present worth and risk and those which are lower in both attributes.

If a tradeoff rate R can be viewed as a dollar less in present worth which is acceptable to 1% more risk, a risk-adjusted measure P' can be made for prospective projects as:

$$P = E(P) - Rp(P < O)$$

where $p(P < O)$ is the probability that the resulting present worth is *not* positive. Projects can then be ranked on the P' criterion for economic acceptability in an uncertain environment.

Figure 1: Importance of Considering Net Present Worth Variance

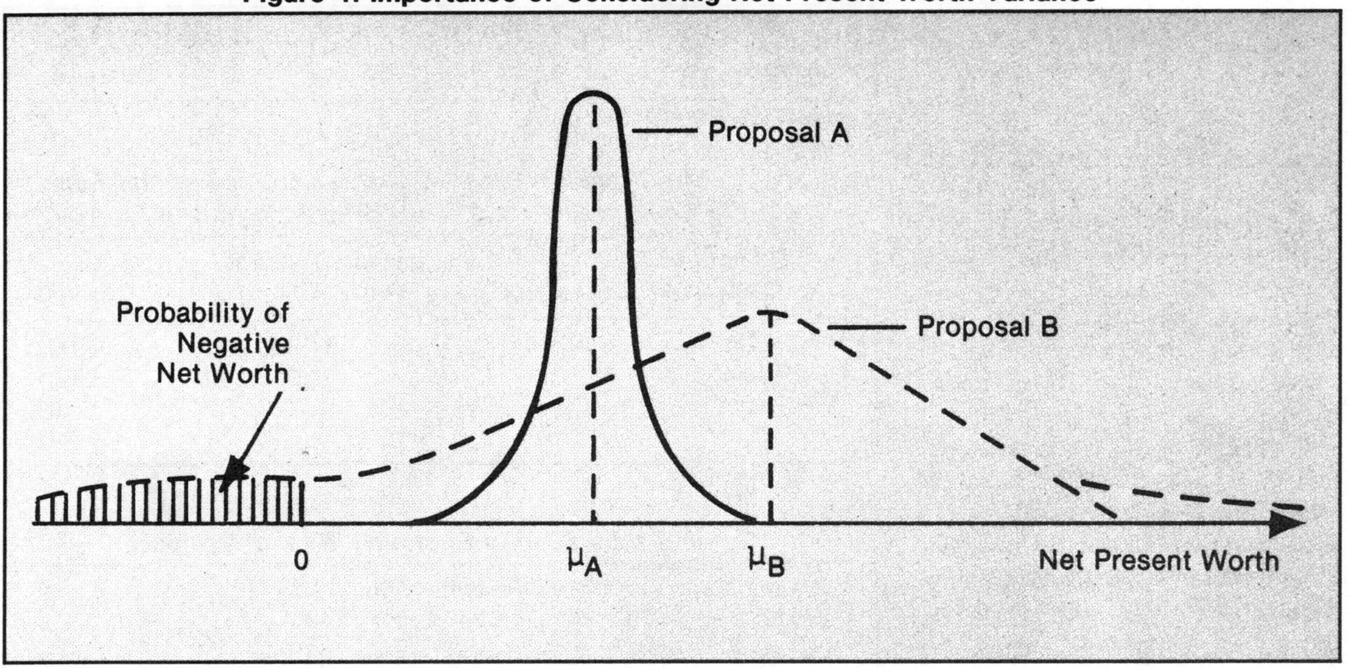

Figure 1 illustrates the importance of considering net present worth variance in addition to expected present worths. Proposal A has less expected present worth than proposal B, but the latter proposal involves far more variance. Part of proposal B's variance is beneficial, carrying improved present worth, and part is *risky* where there is negative present worth. The probability of negative present worth is a measure of the proposal's exposure to poor economic consequences. In the final choice between these proposals the decision maker must weigh the prospective benefits against that risk.

Summary

Economic risk analysis was very limited in the past. However, a lot has happened in the past 10 to 20 years, and some of that improvement is reflected in the above.

Almost any company can use the ideas shown here. Computations can be done by hand, but a programmable calculator or a microcomputer will help. The major message in this article is that risk analysis is within easy grasp, and economic risk is also.

Because of space limitations, only the case of continuous cash flow models is shown. However, the same forms of analysis can be used with serial flows. This series case is shown in Tanchoco and Buck and in Tanchoco, Buck and Egbelu along with other functional forms of cash flows beyond single future flows or uniform flows.

These functional forms with continuous cash flows are extended in Zinn, Lesso and Motazed and in Tanchoco, Buck and Leung. Another type of extension of risk evaluation, known as loss integrals, is given by Buck.

For further reading:

Buck, J.R., "Truncated Linear Gaussian Loss Integrals," *American Institute of Industrial Engineers Transactions,* Vol. 9, No. 2, 1977.

Rosenthal, R.E., "The Variance of Present Worth of Cash Flows under Uncertain Timing," *The Engineering Economist,* Vol. 23, No. 3, 1978.

Sheridan, T.B. and Ferrell, W.R., *Man-Machine Systems Information Control and Decision Models of Human Performance,* MIT Press, Cambridge, MA, 1979.

Sullivan, W.G. and Orr, R.G., "Monte Carlo Simulation Analyzes Alternatives in Uncertain Economy," *Industrial Engineering,* November, 1982.

Tanchoco, J.M.A. and Buck, J.R., "A Closed-Form Methodology for Computing Present Worth Statistics of Risky Discrete Cash Flows," *American Institute of Industrial Engineers Transactions,* Vol. 9, No. 3, 1977.

Tanchoco, J.M.A., Buck, J.R., and Leung, L.C., "Modeling and Discounting of Continuous Cash Flows Under Risk," *Engineering Costs and Production Economics,* Vol. 5, 1981.

Tanchoco, J.M.A., Buck, J.R. and Egbelu, P.J., "Analysis of Capital Expenditure Projects with Uncertain Discrete Cash Flows," *Omega,* Vol. 10, No. 3, 1982.

Young, D. and Contreras, L., "Expected Present Worths of Cash Flows under Uncertain Timing," *The Engineering Economist,* Vol. 20, No. 4, 1975.

Zinn, C.D., Lesso, W.G., and Motazed, B., "A Probabilistic Approach to Risk Analysis in Capital Investment Projects," *The Engineering Economist,* Vol. 22, No. 4, 1977.

James R. Buck is chairman of the systems engineering division and the program of industrial and management engineering at the University of Iowa. He received his bachelor's degree in civil engineering and his master's in structural engineering at Michigan Technological University and a PhD in industrial engineering at the University of Michigan. His research has concentrated primarily on the area of decision making. Buck is a member of the Institute of Industrial Engineers, the Institute of Management Science and the Human Factors Society.

Reprinted from ***Industrial Engineering,*** *November 1982.*

Monte Carlo Simulation Analyzes Alternatives In Uncertain Economy

By William G. Sullivan, P.E.
and R. Gordon Orr
The University of Tennessee, Knoxville

Uncertainty is present in virtually all engineering economy studies, because these studies customarily concern the dollars and "sense" of what is expected to happen in the future. Our best estimates of future conditions are likely not to be accurate, and we know it.

The analysis centers, therefore, on answering the basic question: "In an uncertain environment, will the best alternative available now really make (or save) money for the company?" The answers to such questions are of critical importance to the continued financial success of your company.

There are numerous methods for taking uncertainty in engineering economy studies into account. The following six methods are quite popular in industry and government:

☐ *Break-even analysis* is commonly utilized when the selection among alternatives is heavily dependent on a single factor—such as capacity utilization—that is uncertain. A break-even point for the factor at which two alternatives are equally desirable from an economic standpoint is found. It is then possible to choose between the alternatives by estimating the most likely value of the uncertain factor and comparing this estimate to the break-even value.

☐ *Sensitivity analysis* is often employed when one or more factors are subject to uncertainty. The basic questions that sensitivity analysis attempts to resolve are: (1) What is the behavior of the measure of merit (e.g., net present worth) to a particular percent of change—positive or negative—in each individual factor? (2) What is the amount of change in a particular factor that will cause a reversal in preference for an alternative?

☐ *Optimistic—pessimistic estimation* of factors included in an engineering economy study has been used to establish a range of extreme values for the economic measure of merit. If the range falls *entirely* in the acceptable region (e.g., net present worth is zero or greater), there is little doubt that the alternative is desirable.

Often, though, all factors estimated under conservative (pessimistic) conditions result in an unacceptable alternative.

However, this method directs attention to the best and worst outcomes of going ahead with an alternative and requires managerial judgment to make the final determination to accept or reject it.

☐ *Risk-adjusted minimum attractive rates of return* are sometimes utilized to deal with future uncertainties. This method involves the use of higher interest rates for alternatives that are classified as "highly uncertain" and lower ones for projects for which there appear to be fewer uncertainties.

☐ *Reduction of the useful life* of an alternative is another means for attempting to include explicitly the effects of uncertainty. Here the estimated project life is reduced by a fixed percentage, for instance 50%, and each alternative is evaluated regarding its acceptability over only this reduced lifespan.

☐ *Probability functions* for uncertain elements can be estimated and directly incorporated into the analysis of alternatives. This approach involves various statistical concepts and makes use of various descriptive measures (e.g., expected value and variance) for summarizing uncertainty in the before- or after-tax analysis of one or more alternatives.

Two procedures based on formal

Table 1: Summary of Data Inputs to the Computer Program

Deterministic Analysis:

- ☐ Cash Flows:
  - —Single amounts
  - —Annuities
  - —Linear gradients
  - —Geometric gradients
- ☐ Project Life
- ☐ Interest Rate

Probabilistic Analysis:

Types of Factors	Types of Distributions			
	Discrete	Uniform	Normal	Triangular
☐ Cash Flows Independent of Project Life:	X	X	X	X
—Single amounts	X	X	X	X
—Annuities	X	X	X	X
☐ Project Life	X	X	X	X
☐ Cash Flows Dependent on Project Life:				
—Single amounts	X			

Deterministic and Probabilistic Analysis:

☐ (Combinations of all types of factors and distributions shown under each)

probabilistic concepts are in widespread use: (1) Closed-form analysis of the measure of merit and (2) Monte Carlo simulation of the measure of merit.

This article illustrates the Monte Carlo simulation method of dealing with uncertainty, which is generally more versatile than the other methods mentioned above. A companion article by James Buck (see page 34) explains the essential ingredients of closed-form analysis.

Monte Carlo analysis

Ready access to electronic computers has resulted in increased use of Monte Carlo simulation as an important tool for analyzing project uncertainties. Monte Carlo simulation generates a large number of random outcomes for probabilistic factors so as to imitate the randomness inherent in the engineering economy problem. In this manner a solution to a rather complex problem can be inferred from the behavior of these random outcomes.

The first step in a simulation analysis is the construction of an analytical model that represents the actual investment opportunity. This may be as straightforward as developing an equation for the net present worth of a proposed industrial robot on an assembly line or as complex as examining the effects of various U.S.-imposed trade barriers on your firm's sales of a proposed new product in international markets.

The second step is the development of a probability distribution (either discrete or continuous) for each uncertain factor in the model from subjective or historical data. (Procedures for estimating subjective probabilities are given in Chapter 3 of Hull and in chapter 6 of Sullivan and Claycombe—see "For further reading").

Sample outcomes are randomly generated using the probability distribution for each uncertain factor and are then utilized to determine a trial outcome for the model. By repeating this sampling process a large number of times with the aid of a computer, a frequency distribution of trial outcomes for a desired measure of merit, such as net present worth, is created.

Table 2: Subjective Probability Distributions

Investment:	Normally distributed with a mean of −\$90,000 and a standard deviation of \$2,000.
Useful Life:	Uniformly distributed with a minimum life of 10 years and a maximum of 14 years.
Annual Revenue:	\$50,000 with a probability of 0.4. \$55,000 with a probability of 0.5. \$60,000 with a probability of 0.1.
Annual Expense:	Normally distributed with a mean of −\$38,000 and a standard deviation of \$2,700.

The resulting frequency distribution can then be used to answer questions about the measure of merit. One such question would be, "What is the probability that the investment opportunity will have a net present worth greater than zero?"

Proper implementation of this procedure, coupled with a representative analytical model, results in an approximation of a particular measure of profitability. But how many simulation trials are necessary for an accurate approximation? In general, the greater the number of trials, the more accurate the approximation will be.

One method of determining whether enough trials have been conducted is keeping a running average of results. At first, this average will vary a great deal from trial to trial. The amount of change between successive averages should decrease as the number of simulation trials increases. Eventually this running average should level off and permit

Table 3: After-Tax Cash Flows for Renting versus Purchasing a Computer System (most likely estimates shown)

ABC (rent):

Year	Before-tax Cash Flow*	After-tax Cash Flow
1-5	−$21,600	−$11,664

*Rental fees are payable at the end of each year.

XYZ (purchase):

Year	Before-Tax Cash Flow	ACRS Depreciation**	Taxable Income	Income Taxes	After-tax Cash Flow
0	−$60,000	—	—	—	−$60,000
1	− 4,800***	−$12,000	−$16,800	+$ 7,728	+ 2,928
2	− 4,800	− 19,200	− 24,000	+ 11,040	+ 6,240
3	− 4,800	− 14,400	− 19,200	+ 8,832	+ 4,032
4	− 4,800	− 9,600	− 14,400	+ 6,624	+ 1,824
5	− 4,800	− 4,800	− 9,600	+ 4,416	+ 384

**1986 ACRS percentages are used in this example (year 1—20%; year 2—32%; year 3—24%; year 4—16%; year 5—8%).
***This is the sum of −$2,400/yr for taxes and insurance and −$2,400/yr for maintenance.

Table 5: Simulation Results for Example 3 (1,000 trials)

	Inflation Rate (%/year) 5%		7%		10%	
	NPW	Relative Frequency	NPW	Relative Frequency	NPW	Relative Frequency
	−$37,598	0.102	−$30,249	0.097	−$18,227	0.109
	− 23,375	0.194	−13,449	0.196	3,521	0.194
	− 12,243	0.498	297	0.504	22,680	0.496
	− 3,895	0.206	11,370	0.203	38,215	0.203
Expected value of NPW:	−$15,458		−$3,169		$17,383	

an accurate approximation to be made.

A computer program

To permit a wide variety of engineering economy problems subject to uncertainty to be investigated, a Fortran program that accepts many types of probabilistic and deterministic data and operates on an interactive ("conversational") basis was written. The program allows typical data to be entered easily by means of a series of questions that permit the user to "talk" with the computer.

The data required in before- and after-tax economy studies include numerous types of cash flows and information on project life and interest rates. A summary of the capabilities of the computer program to deal with many forms of data is shown in Table 1. Additional features of the simulation code are illustrated in the three examples that follow.

The first example demonstrates the before-tax analysis of a single project in which all factors except the interest rate are probabilistic. (The interest rate is not generally regarded as a probabilistic quantity, but it can be allowed to vary through multiple computer runs.)

Example 2 gives an after-tax analysis of two options for obtaining computer services (renting versus buying) and includes numerous uncertain quantities.

Example 3 illustrates a before-tax analysis of a situation in which several key factors are thought to be deterministic while others are probabilistic. In this example, variable inflation rates are incorporated into geometric gradients of cash flows.

In each example the factors are assumed to be *independent,* whether they are certain or uncertain, to simplify the presentation. It should be noted, however, that dependencies among factors can be dealt with using Monte Carlo simulation (see, for example, Kryzanowski, Lusztig and Schwab).

Example 1

This example illustrates how Monte Carlo simulation can be used to analyze a capital investment opportunity being considered by a large manufacturer of air conditioning equipment. The subjective probability distributions have been estimated for the four uncertain factors as shown in Table 2.

The management of this company wishes to determine the probability that the investment will be a profitable one using an interest rate of 10%. To answer this question, the net present worth (NPW) of the venture will be simulated.

To demonstrate the use of the computer simulation program men-

Table 4: Summary of Key Statistical Measures

Statistical Measure	Alternative: Rent	Buy
Expected NPW	−$43,300	−$48,600
Standard deviation of NPW	5,280	980

Figure 1: Data Input for Computer Simulation Program

```
THE FOLLOWING PROGRAM USES MONTE CARLO SIMULATION
TECHNIQUES AS APPLIED TO RISK ANALYSIS PROBLEMS OF
ENGINEERING ECONOMY.
WILL YOU BE USING A REMOTE PRINTER FOR OUTPUT? (Y OR N) [Y]
INPUT A RANDOM NUMBER BETWEEN 1 AND 1000. [284]
HOW MANY ITERATIONS DO YOU WISH TO RUN? [10000]
WHAT INTEREST RATE (PERCENT) IS TO BE USED? [10]

THE DATA FOR EACH RANDOM VARIABLE INVOLVED MAY BE FORMULATED AS FOL-
LOWS:

  1. SINGLE VALUE OR ANNUITY
  2. SINGLE VALUE WITH ARITHMETIC GRADIENT
  3. SINGLE VALUE WITH GEOMETRIC GRADIENT
  4. DISCRETE DISTRIBUTION
  5. UNIFORM DISTRIBUTION
  6. NORMAL DISTRIBUTION
  7. A SERIES OF YEARLY CASH FLOWS
  8. SALVAGE VALUE DEPENDENT ON PROJECT LIFE
  9. TRIANGULAR DISTRIBUTION

INFORMATION FOR INITIAL CASH FLOW:
 DISTRIBUTION IDENTIFICATION NUMBER = [6]
 MEAN VALUE = [-90000]
 STANDARD DEVIATION = [2000]

INFORMATION FOR YEARLY CASH FLOW:
  THIS CASH FLOW MAY CONSIST OF A NUMBER OF DIFFERENT
  ELEMENTS WHICH MAY FOLLOW DIFFERENT DISTRIBUTIONS.
  PLEASE INPUT THE DATA ONE ELEMENT AT A TIME AND YOU WILL
  BE PROMPTED FOR ADDITIONAL INFORMATION.
  DISTRIBUTION IDENTIFICATION NUMBER = [4]
  NUMBER OF VALUES = [3]
  INPUT VALUES IN ASCENDING ORDER.
  VALUE 1 = [50000]
                    WITH PROBABILITY = [0.4]
  VALUE 2 = [55000]
                    WITH PROBABILITY = [0.5]
  VALUE 3 = [60000]
                    WITH PROBABILITY = [0.1]
  IS THERE ADDITIONAL ANNUAL CASH FLOW DATA? (Y OR N) [Y]

  DISTRIBUTION IDENTIFICATION NUMBER = [6]
  MEAN VALUE = [-38000]
  STANDARD DEVIATION = [2700]
  IS THERE ADDITIONAL ANNUAL CASH FLOW DATA? (Y OR N) [N]

INFORMATION FOR SALVAGE VALUE:
  DISTRIBUTION IDENTIFICATION NUMBER = [1]
  CASH VALUE = [0]

INFORMATION FOR PROJECT LIFE:
  DISTRIBUTION IDENTIFICATION NUMBER = [5]
  MINIMUM VALUE = [10]
  MAXIMUM VALUE = [14]
STATISTICAL SUMMARY:
  EXPECTED VALUE OF PRESENT WORTH =                           15381.00
  VARIANCE OF PRESENT WORTH =                             868537050.00
  STANDARD DEVIATION OF PRESENT WORTH =                       29470.95
  PROBABILITY THAT PRESENT WORTH IS GREATER THAN ZERO =          0.688

  EXPECTED VALUE OF ANNUAL WORTH =                             2228.67
  VARIANCE OF ANNUAL WORTH =                               18565643.00
  STANDARD DEVIATION OF ANNUAL WORTH =                         4308.79
  PROBABILITY THAT ANNUAL WORTH IS GREATER THAN ZERO =           0.688
```

tioned above, the computer queries and responses of the user (shown in boxes) for Example 1 are shown in Figure 1. The questions are followed by simulation statistics for 1,870 trials. (This many trials were needed for the cumulative average present worth to stabilize to a variation of ±0.5%.) A histogram of results is shown in Figure 2.

The histogram indicates that the median net present worth of this investment is $15,234 and that the dispersion of present worth trial outcomes is considerable. The standard deviation of simulated trial outcomes is one measure of this dispersion.

Based on Figure 2, 68.8% of all simulation outcomes show a net present worth of zero or more. Consequently, this project may be too risky for the small company to undertake because the "down-side" risk of failing to realize at least a 10% return on the investment is about 3 chances out of 10. Perhaps another investment should be considered.

Example 2

A large, profitable company is investigating the possible installation of an electronic computer that will automate most of its accounting, billing and inventory control functions. It will also supply management with some additional services not now available.

Two types of equipment are being considered. The ABC computer system can be rented on a five-year basis for $1,800 per month. Due to various conditions in the rental agreement, it is believed that annual rent will be normally distributed, with a mean cost of $21,600 and a standard deviation of $2,000. This includes a maintenance contract.

In addition, the period of needed service is uncertain. It is thought to be four, five or six years with probabilities of 0.1, 0.6 and 0.3, respectively.

Figure 2: Results of Computer Simulation Program

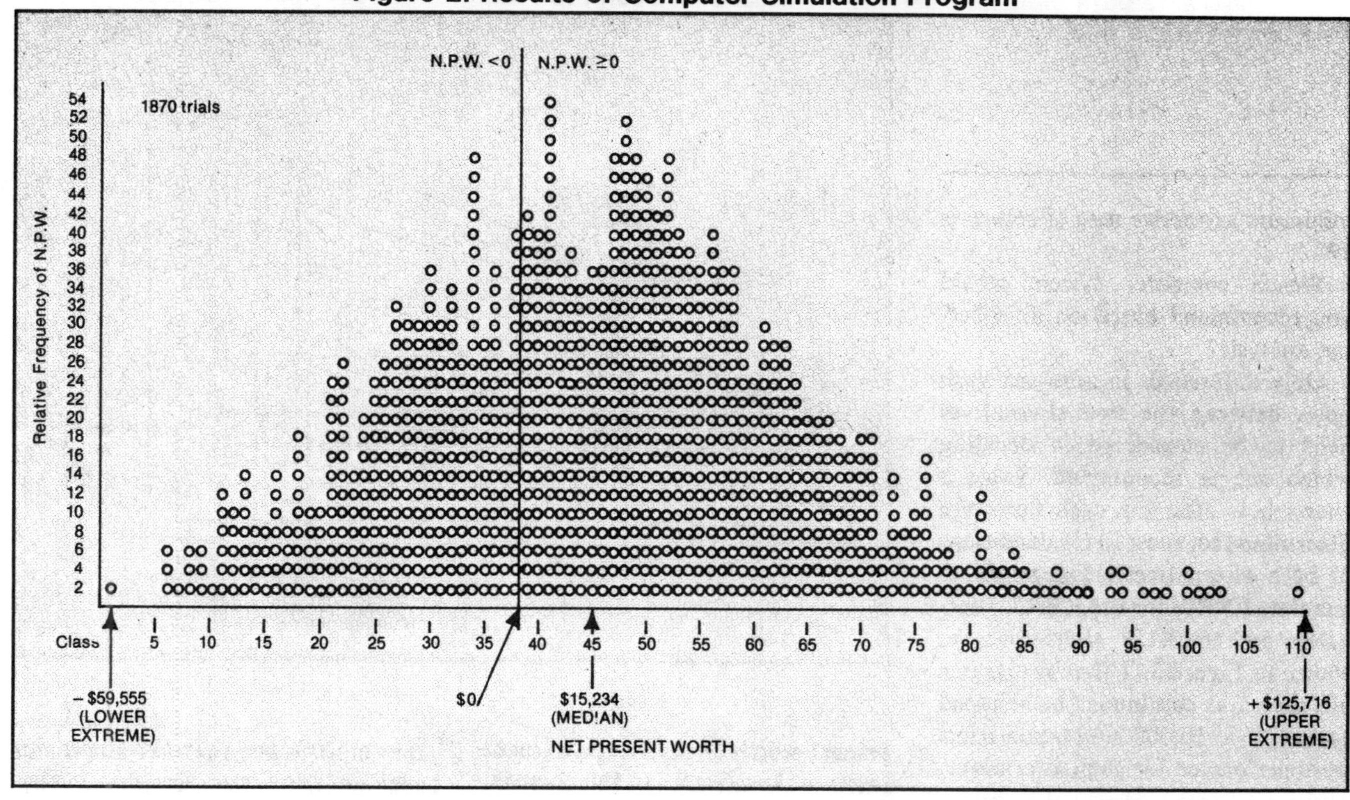

Figure 3: Expected Net Worths of Rent and Buy Alternatives

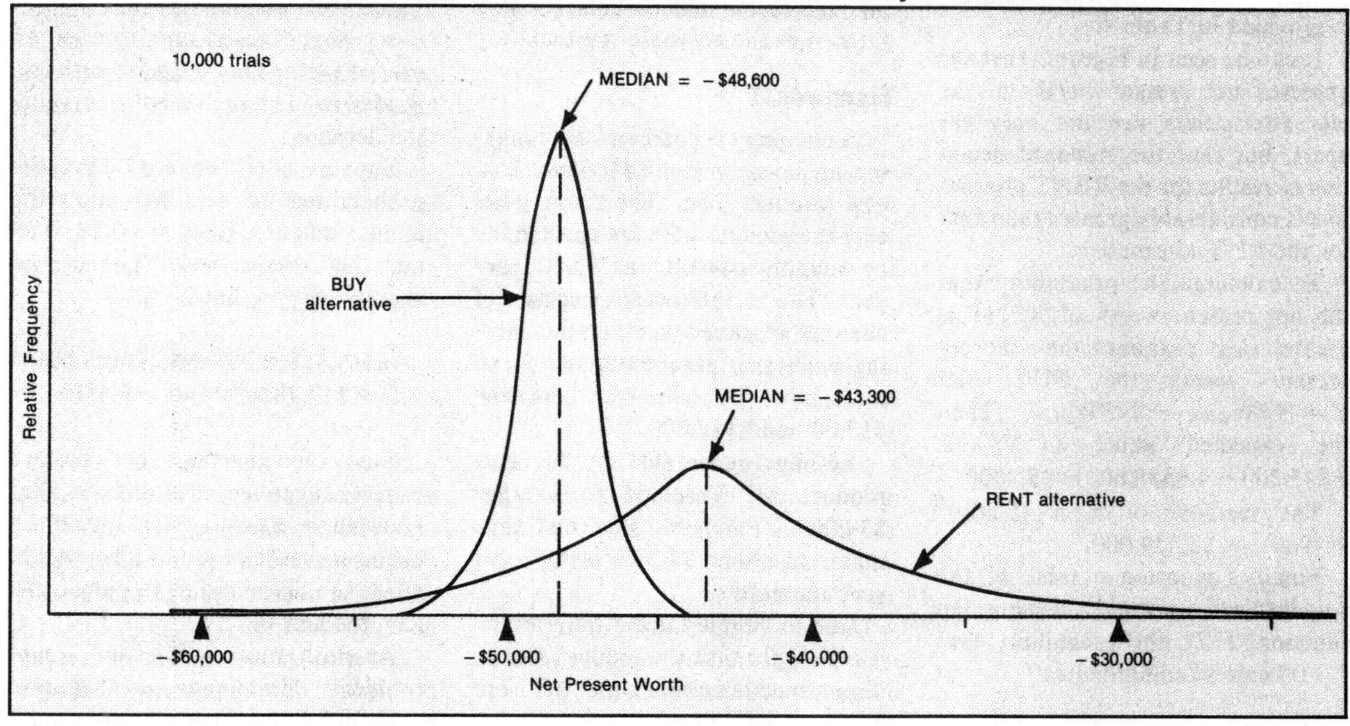

The XYZ computer system would have to be purchased at an expected cost of $60,000. This initial investment, which is subject to uncertainties in the bidding process, is believed to be normally distributed with a standard deviation of $1,000. A maintenance contract can be obtained at a cost of $2,400 per year. Taxes and insurance on owned equipment are about 4% of the first cost per year.

If either computer system is installed, one employee whose annual salary is now $18,000 will be upgraded, with an accompanying salary increase of $1,200 per year plus associated fringe benefits.

Because of the rapid technological development in the field, the company feels that if the XYZ computer system is purchased, it will be fully written off using accelerated cost recovery system (ACRS) depreciation over a recovery period of five years. The firm's incremental income tax rate is 46% and its after-tax minimum attractive rate of return is 12%.

Which computer system would you recommend based on an after-tax analysis?

Only differences in after-tax cash flows between the two alternatives need to be considered in deciding which one to recommend. Table 3 shows how after-tax cash flows are determined for most likely conditions of both alternatives. Histograms of simulated NPW for the RENT alternative and the BUY alternative are shown in Figure 3. These histograms are shown as continuous bell-shaped curves since 10,000 simulation trials were performed for each alternative. (A normal curve is expected because of the Central Limit Theorem.) A summary of key statistical measures is provided in Table 4.

It can be seen in Figure 3 that the expected net present worths of the two alternatives are not very far apart, but that the standard deviation of results for the RENT alternative is considerably greater than that for the BUY alternative.

To calculate the probability that the net present worth of RENT is greater (less negative) than the net present worth of BUY, let $Y = NPW_{RENT} - NPW_{BUY}$. Then the expected value of Y is $-\$43,300 - (-\$48,600) = \$5,300$.

The variance of Y is $(5,280)^2 + (980)^2 = 28,839,000$.

Finally, by using a table of the standardized normal distribution function, $F(Z)$, the probability that $Y \geq 0$ can be computed as:

$$\Pr(Y \geq 0) = \Pr\left[Z \geq \frac{0 - 5300}{\sqrt{28,839,000}}\right] = \Pr[Z \geq -0.987] \simeq 0.84.$$

There is a high probability, therefore, that the RENT alternative will be less expensive after taxes than the BUY alternative despite the greater dispersion in the simulated net present worth of the former. Uncertainty in key factors in the comparison has been explicitly investigated through Monte Carlo simulation, and the recommended choice would be to rent the computer system.

Figure 4: Cash Flow Diagram of Proposed Project

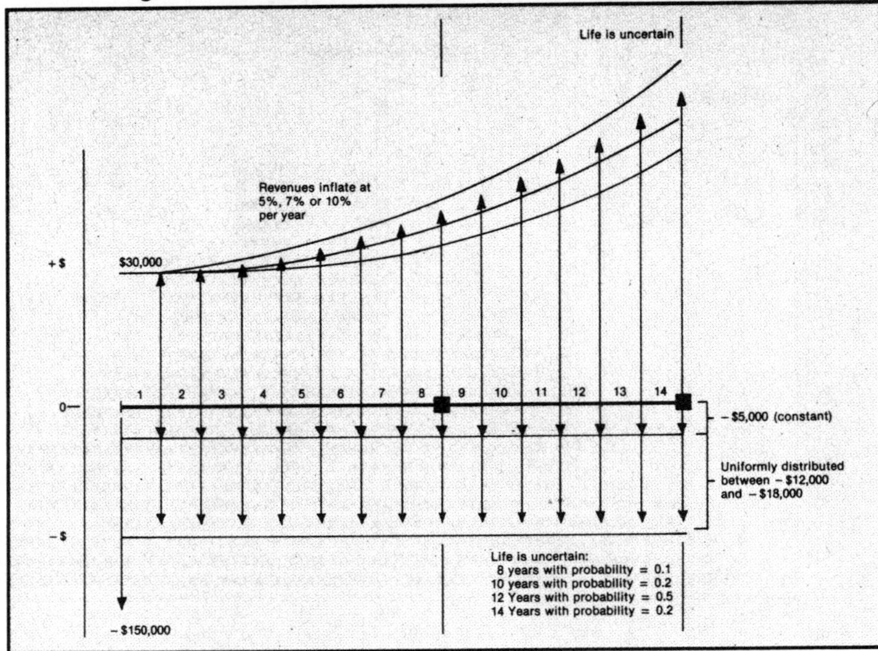

Example 3

A company is contemplating making an investment of $150,000 in a new product line. The fixed costs over the product's life are expected to be roughly constant at $5,000 per year. The variable costs, because of changes in capacity utilization during each year, are estimated to be uniformly distributed between $12,000 and $18,000.

Revenues generated by the new product are expected to average $30,000 in the first year and then inflate at either 5%, 7% or 10% per year thereafter.

Due to a highly uncertain market, it is thought that the product will be discontinued in either eight, 10, 12 or 14 years, with probabilities of 0.1, 0.2, 0.5 and 0.2, respectively.

The firm's minimum attractive rate of return is 20%. Should this venture be undertaken?

To help visualize the uncertainty inherent in this proposed project, its cash flow diagram is presented in Figure 4.

The simulated NPWs for this investment opportunity at an interest rate of 20% per year are shown for each inflation rate scenario in Table 5.

If the inflation rate scenarios in Table 3 are weighted by their subjective probabilities of coming true, an overall average net present worth can be determined and used in making the decision.

Suppose it is believed that the probabilities of 5%, 7% and 10% annual inflation rates are 0.20, 0.60 and 0.20, respectively. The overall average net present worth is:

$$-\$15,458(0.20) - \$3,169(0.60) + \$17,383(0.20) = -\$1,516$$

Since the average net present worth is negative, it is unlikely that knowledge regarding the variability of this measure of profitability would affect the decision not to pursue the new product line.

As with most studies of actual problems, developing an accurate model of the decision situation and estimating the data required for input into the computer are the major challenges to the skill of the analyst.

The computer program that was developed here can be applied to numerous types of problems. The resultant development of frequency histograms for a simulated measure of merit, such as net present worth, can be utilized as a visual aid to assist with decision making in simple or complex studies. In simple cases the closed-form analysis of a problem will lead to an exact solution, whereas Monte Carlo simulation provides an approximate solution.

For further reading:

Canada, J. R. and White, J. A., *Capital Investment Decision Analysis for Management and Engineering,* Prentice-Hall Inc., Englewood Cliffs, NJ, 1980.

Hull, J. C., *The Evaluation of Risk in Business Investment,* Pergamon Press, New York, 1980.

Kryzanowski, L.; Lusztig, P. and Schwab, B., "Monte Carlo Simulation of Capital Expenditure Decisions—A Case Study," *The Engineering Economist,* Vol. 18, No. 1, Fall 1972.

Sullivan, W. G. and Claycombe, W. W., *Fundamentals of Forecasting,* Reston Publishing Co., Reston, VA, 1977. IE

William G. Sullivan, P.E., is professor of industrial engineering at the University of Tennessee-Knoxville. He received his PhD from Georgia Institute of Technology and is a registered professional engineer. Sullivan is director of the engineering economy division of IIE and past chairman of the engineering economy division of ASEE.

R. Gordon Orr is an associate industrial engineer with the Federal Systems Division of IBM Corp. He is currently working in the area of hardware cost engineering in Manassas, VA. An associate member of IIE and AACE, Orr holds both BSIE and MSIE degrees from the University of Tennessee, where his primary areas of interest were capital investment analysis and information systems design.

Justifying Multimachine Systems: An Integrated Strategic Approach

Nallan C. Suresh, Jack R. Meredith, University of Cincinnati, Cincinnati, Ohio

Abstract

Current problems in justifying multimachine systems are analyzed and an integrated framework is developed in which both strategic and tactical considerations are included. At the strategic level the tools for project selection are reviewed, followed by the economic analysis normally required for these systems. Special tools, notably simulation, are illustrated through an example situation.

***Keywords**: Capital Equipment Justification, Economic Justification, Multimachine Systems, Flexible Manufacturing Systems, Group Technology.*

Introduction

In recent years the problems relating to the economic justification of new manufacturing technologies have received increasing attention. Particularly, multimachine systems like group technology (GT) cells, flexible manufacturing systems, robotic multimachine systems, and also machining centers have exposed some of the limitations of traditional methods of equipment justification. Before we begin to analyze these problems, it is necessary to define the terms involved, and also briefly trace the evolution of these technologies.

Almost all of these technologies with perhaps the exception of GT, can be explained on the basis of continuing developments in automation. Going by chronological order, we consider GT first. GT was developed as a formal discipline by Mitrafanov and others in the USSR in the period following World War II. GT involves, primarily, the rearrangement of machines in the typical job shop on the basis of component specialization instead of process specialization. Thus, a GT factory consists of cells of machines devoted to part families instead of cells based on functional specialization. Classification and coding methods, and later noncodification methods like the production flow analysis (PFA), were developed to identify part families.

It is a misconception to treat GT as being synonymous with classification and coding methods. As per this larger view, part family identification methods, computer aided process planning (CAPP), cellular manufacturing, period batch control (an MRP-type system with no lot sizing and operating like the JIT system) are all subsets of the GT philosophy. Here we are primarily concerned with the cellular manufacturing aspects. The first textbook on GT authored by Mitrafanov appeared at

Editor's Note: This paper was awarded the 1985 Alfred V. Bodine/SME Award for studies in machine tool economics. This award is presented annually to a business or engineering graduate student and administered by the SME Engineering Education Foundation.

approximately the same time as numerical control was developed at the Massachusetts Institute of Technology in the late 1950s.

To provide definitions for numerical control (NC), computer numerical control (CNC), direct numerical control (DNC), machining centers, and FMSs, we adopt the approach taken by Cook[1] of M.I.T. We consider a typical general purpose machine of the pre-NC era. The operations involved are to (1) move the workpiece to the machine, (2) load the workpiece onto the machine and rigidly and accurately affix it to the machine, (3) select the proper tool and insert it into the machine, (4) establish and set machine operating conditions, (5) control machine motion enabling the tool to execute the desired function, (6) sequence different tools, conditions and motions until all operations possible on the machine are complete, and (7) unload the part from the machine. In the pre-NC general purpose machine, all seven functions are done manually.

With the advent of NC, step (5) in the above sequence was automated, and the control of the machine was taken away from the operator and made automatic by information stored on a punched tape. Thus, complex operations requiring high levels of skill were automated and a new profession, that of the part programmer, emerged. The next major development was automatic tool changing (ATC) systems, whereby step (3) was automated. This was done under the control of punched tape information and by providing a bank of tools for performing a variety of operations. In these 'machining centers', several operations could now be combined on the same machine.

The next step was the development of the CNC, which essentially replaced the punched tape medium with core memory, magnetic disks, and so on. Soon the computer was made to control the operations of several machines, which gave rise to the DNC. In this system, material transfers are not computer controlled and automated. The next logical development was that of the FMS, in which the material handling is also integrated into the system. In the FMS at the machine level, all seven of the above steps are automated.

We also include robotic multimachine systems in addition to the above. These systems are collections of machine tools, conveyors, gauging equipment, and robots, in which the robot services the machines. Typically in these systems, apart from performing material and tool handling, the robot serves as a source of communication among the machines and controls the operation cycles.

It must also be mentioned that our primary focus of interest is the typical batch manufacturing situation. The majority of manufacturing situations fall under this category of small batch—large variety. The characteristics of this context are well known. For example, as much as 95% of the manufacturing cycle times consist of queuing and transporting times, the operations generally follow the functional or job shop mode, in nearly 75% of the cases the lot sizes are less than 50 units, in the U.S. nearly 95% of the machines used belong to the pre-NC technology, and only 31% of the machines are less than 10 years of age, compared to a figure of 61% in Japan.

The concept of grouping machines into cells devoted to part families is still in the incipient stage, even with GT which is a pre-NC development. And the implementation of new manufacturing technologies has been disconcertingly slow, as seen from comparative statistics of robot, CNC, and FMS populations in the U.S. with respect to other countries.

In the following sections, we first summarize the current problems in justification, followed by a literature review. A framework for justification, in which both strategic and tactical considerations are included, is proposed next. At the strategic level, the basic tools for project selection are then described, which is followed by the economic analysis normally required for multimachine systems. Special tools used for justification, notably simulation, are illustrated through an example situation.

Justification Problems

Based on the voluminous literature available concerning the current problems in justification, these problems can be broadly summarized under the following categories.

High Capital Costs and Risk. High capital costs and the risk involved seem to be major deterrents to investing in new manufacturing technologies. Almost all of the new technologies involve integration. For example, machining centers combine several operations previously done on separate machine tools. The FMS brings to a single cell the operations done at various work centers in the conventional factory, in which even material handling is automated and computer controlled. Thus, the capital costs turn out to be much higher and the payback period longer than conventional machine tools. The new technologies also pose higher risks due to as yet unfamiliar technology, the control of which is mainly through software, and the fact that they have system wide implications. Thus, coupled with the inherent bias toward short-term results and risk aversion, the high capital costs, risk levels and sophistication of technology have contributed to the slow adoption rates.

Myopic Approaches to Equipment Justification. Limited approaches to equipment justification are perhaps the most often cited reason concerning justification problems. It has been pointed out on numerous occasions that American industry primarily relies on the payback method in considering investment proposals. The drawbacks of the payback method are well known—ignoring the returns after the payback period, its emphasis on liquidity, etc. With respect to new technologies, the use of this method practically guarantees rejection. The most recent survey of machine tool justification practices confirms the use of this method as the primary justification procedure in many firms, with an average payback period hurdle used being a very myopic 2.91 years.[2] The main attraction of this method is its simplicity, and importantly, the fact that it is a joint measure of risk and return. The payback method avoids the complications of the present value methods.

Inappropriate Capital Budgeting Procedures. Gold[3] points out that capital budgeting procedures for the new technologies are mere extensions of the procedures used for conventional technologies. These procedures tend to (1) be bottom-up processes for generating new equipment proposals, (2) be subject to narrow levels of analysis as for conventional machines, (3) depend on traditional capital budgeting procedures for their evaluation, and (4) virtually guarantee that the companies will either defer their purchase or, once purchased, apply them too narrowly. The emphasis on quick returns and short-term oriented reward systems preclude the consideration of long, uncertain projects.

The Difficulty of Quantifying Indirect Benefits. In recent years, it has been increasingly pointed out that the bulk of the benefits arising from the new technologies are not in direct labor or materials, but in the so-called indirect benefit category. For instance, in the case of numerical control, it was estimated by Warner and Swasey[4] that indirect savings amounted to 58.9% of the direct savings. With higher levels of integration, as in the FMS, indirect savings could amount to several times that of direct savings. The capital budgeting procedures used in the industry are still based on a first order analysis as opposed to a second order analysis. The major benefits of GT, FMSs etc., are the throughput time reductions, work-in-process (WIP) inventory reductions, and so on. Typically with inadequate tracking systems, many major savings areas have been excluded in the justification exercise, which also contribute to showing longer payback periods for the new technologies. Additionally, there is considerable scope for conjecture when it comes to evaluating such savings items as 'strategic value of quality increases'. Existing capital budgeting procedures have also been seen to be bottom-up, too quantitatively oriented, and tend to emphasize single-valued criteria such as ROI, and do not favorably consider conjectures concerning strategic advantages.

Prediction of Benefits over an Extended Time Period. The complexities of new technologies not only pose problems for near-term operating patterns, but also for predicting long-term performance and estimation of cash flows. These are augmented by the longer investment planning horizons required. Also, the cash flows are affected not only by market determined factors, but operating policies as well. For instance, the scheduling policies for an FMS have a strong bearing on the realization of returns. This seems to be another distinguishing characteristic of the new technologies—the investment decisions are influenced by the scheduling policies.

Technological Uncertainties. The rapid evolution of new technologies has created a sizable knowledge gap among both manufacturing and financial professionals. This has reinforced a 'wait and see' attitude, favoring rejection of proposals. Even stand alone robot projects are viewed as radical investment alternatives in many firms and vague fears about labor resistance and other such matters prevail.

Inadequacies in Costing Methods. Conventional cost accounting has its origins in mass production and classical job shop situations. In the job shop situation, a major focus of control is direct labor through elaborate time standards and work measurements. As the material proceeds from work center to work center, overhead allocations are typically made on the basis of direct labor. But in cellular manufacturing, as in GT cells or FMSs, the cell becomes the cost center, obviating the need for elaborate measurements within the cell. Changes required for cost accounting have been felt in the U.S. in FMS situations[5] and in the U.K. in GT contexts.[6] From a justification standpoint, the difficulty of evaluating indirect benefits with the present inadequate tracking systems poses the main problem.

Differing Nature of Operations. The new technologies introduce different methods of operation and control. In cellular manufacturing, as in GT and FMSs, the part family is the focus of interest as opposed to the operation in the classical job shop. The operations are geared to taking advantage of component similarities and different operating philosophies based on economies of scope. Software is the primary means of control and simulation is a daily necessity. All these have contributed to the justification problem.

Analysis Still Based on Subsystems and Suboptimization. Mathematical models based on operations research have been characterized by single-valued criteria and tend to focus on narrow problem definitions.[7] Also, engineering economy continues to be formula oriented and emphasizes such factors as cost of capital, tax rates, inflation rates and the corresponding sensitivity analyses. These are by and large applicable for the traditional, stand alone technologies at the tactical level of decision making, but new tools are required for integrated systems, emphasizing multiple-criteria decision making approaches, and capital budgeting procedures geared to system wide analyses. Also the prediction of cash flows arising from systems is an aspect which has not been given sufficient attention.

A Brief Literature Review

In considering the literature devoted to the justification of new manufacturing technologies, we start with that devoted to robotics, even though most of these relate to the stand alone robotic justification problem.

References 8, 9, 10, 11, 12, 13, 14, 15, 16, and 17 are fairly representative of the voluminous literature devoted to robot justification. Fleischer[12] categorizes these into three classifications—those utilizing accounting methods, payback methods and the discounted cash flow method. He also provides an exhaustive listing of the economic consequences under the broad categories of Plant and Equipment, Operation and Maintenance, and Product-related aspects. Fleischer further addresses the choice of parameters for the cost of capital, inflation and tax rates. Sullivan[15] distinguishes between the tactical nature of many of the conventional machine tool justification procedures and the more strategic nature of robotic investment.

The major features emerging from the literature devoted to robotic justification are that almost all of them: (1) address only the issues of stand alone robots, (2) identify all the direct and hidden costs and benefits, and (3) utilize conventional financial analysis methods for capital budgeting. Many of these articles also deal with the issues of cost of capital, inflation, tax rates, etc., and the corresponding sensitivity analysis. The major fact which emerges from all these is that, generally, 'first order effects' dominate robotic applications, and these are usually sufficient to justify these systems. The payback periods are also seen to be shorter for these systems, ranging from six months to five years and are very strongly related to the nature of the application. For instance, loading/unloading applications can be justified largely within a two year time frame.[14]

In the case of machining centers, where several operations are combined, 'second order effects' become more predominant. Generally, this increases with the level of integration of the system. Perhaps the most comprehensive work on machining centers and NC technology has been done by Fullerton of Warner and Swasey.[4,18,19] The major focus in these methods is to track down and quantify all the indirect savings, through graphs and models which depict the value added as a function of the cycle time. An example of this is seen in *Figure 1*.

For instance, it is shown in *Figure 1* that WIP reductions arise not only due to reduced throughput time, but also due to reduced unit costs and better scheduling. Meyer[20] uses these procedures for the case of multimachine systems with very little modification. It is easy to show that the Warner and Swasey methods are applicable to these situations as well. Klahorst[21] identifies, exhaustively, 14 critical

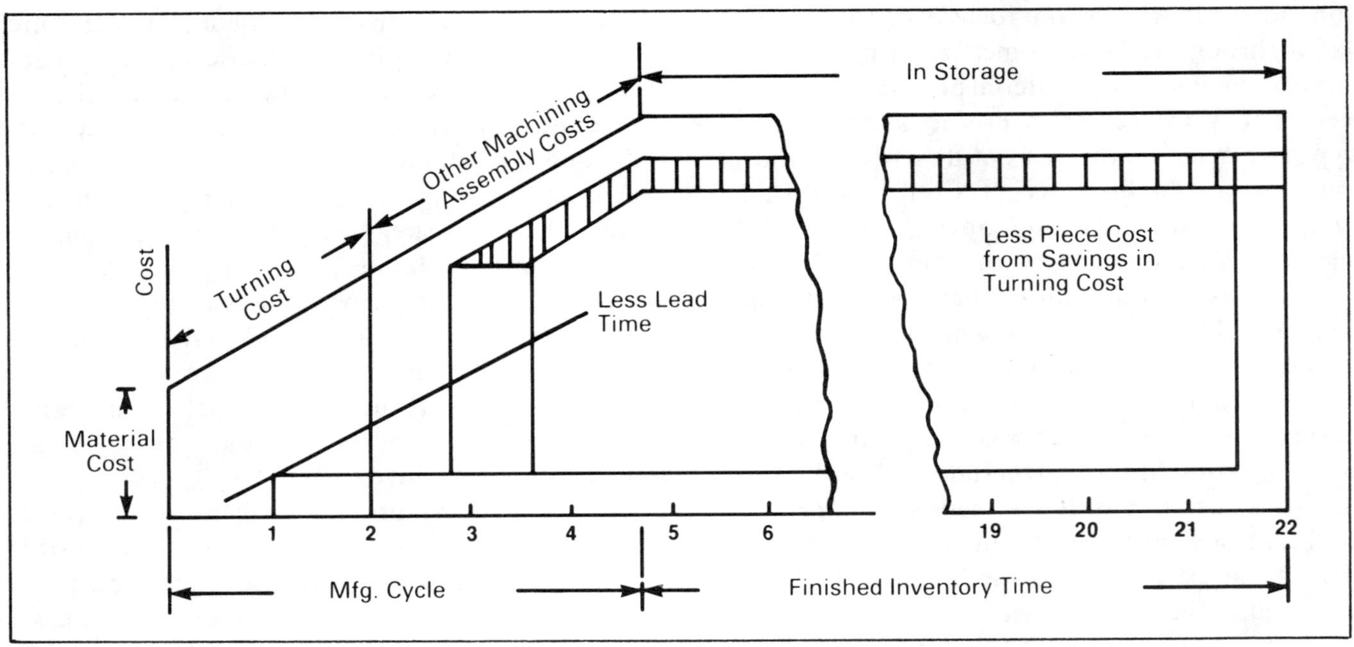

Figure 1
**Warner & Swasey Procedures:
An Example Chart**

cost factors to be considered in the justification of these systems.

Recently, the FMS justification problem has begun to receive attention. Hutchinson[22,23] was one of the first to deal with the economics of flexible automation. While comparing transfer lines and hard automation, with respect to flexible automation, he points out that the ability to incrementally acquire capacity is a significant economic advantage. In a later paper,[23] Hutchinson evaluates the economic advantage of flexibility for a specific situation, comparing transfer lines and flexible systems through a simulation approach. Gold,[3] Goldhar[24] and others have pointed out the basic economic characteristics of these systems. Kulatilaka[25] analyzes the impact of flexible automation on the systematic risk parameter beta (a financial measure of risk) and concludes that the overall effect is ambiguous. Suresh and Meredith[26] highlight the generic characteristics of new manufacturing technologies and approach the FMS justification problem from a cellular manufacturing perspective. Kester[27] points out in a recent article that the opportunities created by flexible automation are similar to stock options and takes an options pricing approach. Also, Shewchuck[28] emphasizes the need for an explicit recognition of 'soft correlations', e.g., the so-called strategic benefits in the justification exercise.

A recent survey of justification methods used for robotics and FMSs in the aerospace industry reveals some interesting findings.[29] Until now, the robotics and FMS investment decisions have rarely been driven by economic factors, although economic analysis is used as one element of the decision process. No stochastic methodologies were used for investment decision making. A broader approach to investment analysis, with economic factors as one element, has been advocated.

The Justification Framework

The technologies coming under the umbrella term of computer integrated manufacturing (CIM) can be grouped hierarchically, based on the level of system integration. For instance, stand alone technologies, machining centers, manufacturing cells, and FMSs can be represented as shown in *Figure 2*. Many managerial characteristics begin to unfold from this perspective, relating to financial justification, implementation of technology, production control, cost accounting and other aspects. Here the primary focus of interest is the implications of this hierarchical view for the justification of these technologies.

As we move from stand alone installations at the lower tactical level, there is a progressive

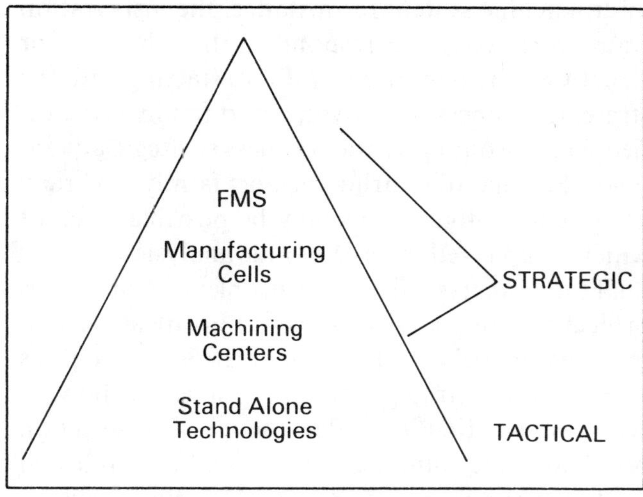

Figure 2
A Hierarchical View

integration of systems culminating in complete automation. It can be seen that while the technology has been integrating upward, capital budgeting procedures have remained at the tactical level, emphasizing economic factors and single-valued criteria like the ROI. There is a definite gap in the upper areas with respect to justification procedures and risk analysis methods.

Traditionally, since the technology related to stand alone machine tools, justification was primarily conducted using bottom-up procedures. (In fact, the very term 'justification' has some bottom-up connotations.) Higher level technologies are sought to be justified using 'first order oriented' procedures which have worked well until now. The new technologies however, require a revamping of justification procedures and savings calculation approaches to focus mainly on the so-called indirect benefits.

The upper levels are also characterized by a less formal discipline and a lack of generally accepted procedures. Kester[27] points out that in justification exercises, there is frequently contention between two groups, i.e., the strategists and the quantitative analysts. Invariably, even though the strategists are fairly realistic about the costs and benefits, their arguments fail due to a lack of structured methodology and discipline. What is currently required is an explicit recognition of inherently nonquantifiable factors in this arena, but at the same time developing structured methodology for strategic capital budgeting. The tangibility of costs and benefits reduce as we go upward in the hierarchy and the quantification

Table 1
Justification Characteristics: Stand Alone *vs* Integrated Systems

	Stand Alone	Integrated Systems
Level in the hierarchy	Tactical	Higher than tactical
Capital cost	*Generally* lower	*Generally* higher
Payback period	*Generally* shorter	*Generally* longer
Operation analysis	Easier	More complex
System impact	Localized	More system wide
Major justification components	First order effects	Second order effects
Product cost	Estimation easier	Estimation more difficult
Justification criteria	Single-valued, financial	Multiple criteria, and qualitative factors
Justification procedures	More structured	Less structured (at present)
Justification procedures	Bottom-up	Top down (should be)
Implementation	*Generally* easier	*Generally* more difficult
Project risk	*Generally* lower	*Generally* higher
System output	Completed operation(s)	Completed part
Unit of production planning and control	Completed operation(s)	Completed parts and part family

becomes progressively more difficult. At present, a dilemma exists as to whether one should pursue the quantification exercise with greater zeal or whether there should be an explicit recognition of non-quantifiable factors. We feel that both are needed, but programmed decision making procedures using single-valued criteria such as ROI should be avoided. This is essentially a world of multiple-criteria as in any strategic decision (for example a plant location decision).

The impact of the technologies is localized at the lower levels and more system wide at the higher levels. First order effects dominate the justification at lower levels, but second order effects must be identified and used for justification at the higher levels. Again, it is easy to analyze the operations at the lower levels, but it becomes increasingly analytically intractable at the higher levels. This is precisely the reason why simulation has become an invaluable tool for the analysis and control of higher level technologies.

The implications of the hierarchical model for other managerial concerns like implementation of technology, production control, and cost accounting are dealt with at length elsewhere,[30] and the justification implications are summarized in *Table 1*.

Thus, the overall justification framework has to be a top-down exercise, as many have proposed, and the justification or the economic analysis forms one segment of the overall exercise. In the case of a

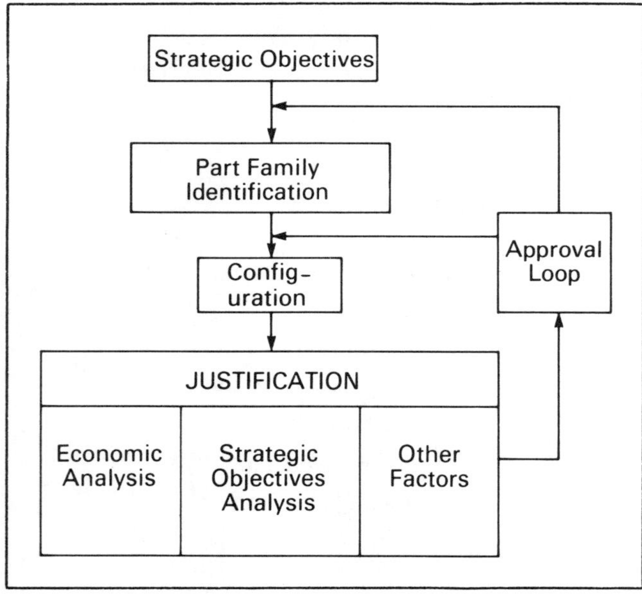

Figure 3
The Justification Framework

multimachine system for instance, the justification framework would correspond to the schematic or structure shown in *Figure 3*. First, starting with the strategic business objectives, a part family is identified which would give the business strategic advantages. For manufacturing the part family, a variety of system configurations may be possible, each of which is subjected to an economic analysis as well as qualitative analysis. Thus, the problem encompasses project selection as well as project justification, and includes financial and economic criteria as well as the so-called strategic factors. In the following sections, we first consider the project selection problem using multiple criteria methods followed by the economic analysis required for multimachine systems.

Strategic Aspects of Justification

At the strategic level, the problem is one of project selection based on multiple criteria which include economic factors as well as other factors. Capital rationing is involved when several projects are contending for limited resources. Thus, a typical batch manufacturer may be faced with many possible automation as well as preautomation projects like constructing a GT classification and coding system, or a computer aided process planning (CAPP) system. Project selection is then based on a variety of factors. These might include, for instance, the flexibility offered by a system, technological leadership in a particular area, ease of implementation, customer image, quality improvements, and so forth. . .

Many project selection methods have been developed and the choice of a project selection model itself depends on the following criteria: realism, capability of the selection model, flexibility, ease of use, and cost (the justification exercise itself should be justifiable). This section is based largely on project selection methodology as expounded by Meredith and Mantel.[31] Broadly, the selection models can be classified as either numeric or non-numeric.

Non-Numeric Models. Non-numeric models include the Sacred Cow, Operating Necessity, Competitive Necessity, Product Line Extension, and the Comparative Benefit Model. These models are briefly explained below.

The Sacred Cow. The initiation of a project purely on the suggestion of a senior and powerful official in the organization. This is the only rationale for selecting the project.

The Operating Necessity. The initiation of a project purely to keep an important system operating. The basic issue is to examine if the system is worth maintaining at the estimated cost of the project.

The Competitive Necessity. We hear this reasoning very often in the context of CIM—to invest purely to keep up with the global competition. Like Operating Necessity, Competitive Necessity also involves bypassing detailed economic analyses.

The Product Line Extension. Using this type of model, a project to develop new products would be judged on the degree to which it fits the firm's existing product line, fills a gap or strengthens a weak link. Detailed profitability analyses are not warranted.

The Comparative Benefit Model. This model involves the ranking of several projects on the basis of 'relative merit'. The Q-Sort method involves dividing the projects into three groups—good, fair, and poor—based on relative merit. When each group has eight or more members, the projects are sorted within each group. The process may be carried out by an individual or a committee.

Numeric Models. Numeric models include Financial (or Economic) Models and Scoring Models.

Financial Models. Traditional justification methods utilize these models as the sole measure of acceptability. These include the Payback Period, Average Rate of Return, Discounted Cash Flow, Internal Rate of Return, Profitability Index and several other profitability models like Pacifico's method, Dean's Profitability method, and so on. Most of these models are described extensively elsewhere and will not be discussed here.

Scoring Models. These models include the Unweighted 0-1 Factor Model, Unweighted Factor Scoring Model, the Weighted Factor Scoring Model, the Constrained Weighted Factor Scoring Model, Integer Programming Approach (as in the Dean and Nishry Model), and the Goal Programming Approach. There are other important methods like the Analytic Hierarchy Process,[32] and multicriteria methods utilizing utility functions. Of these, the Weighted Scoring Model is the most popular and is a very simple technique to use. Basically, the Weighted Scoring Model involves evaluating a project with respect to various criteria which have been assigned weights and computing the total worth of a project. In the Analytic Hierarchy Process, a series of pair wise comparisons are made, which reduces the inconsistencies which could be present in the Weighted Scoring Model.

The main advantages of the scoring models are the multiple criteria capability, simplicity and ease of use, consistency with management policies, and flexibility. On the other hand, the simplicity of these models creates a temptation to include a large number of factors, or to add sophistication. However, most of the extra factors carry small weights and only have a marginal bearing on the selection of a project.

The use of a multicriteria decision making approach such as scoring models is vital for strategic level decisions and has to form an integral part of the justification exercise. This has not been the case until now. Traditional procedures used in the industry have only emphasized economic analysis based purely on single-valued, financial criteria. In the next section, we address the economic analysis for multimachine systems.

The Economic Analysis

The savings arising from multimachine systems can be traced throughout the manufacturing cycle: in the preproduction stage, the production stage and the postproduction stage. The major benefits are listed in *Table 2*. It is important to identify all the benefits and include them in the justification. Many of these are interrelated. For instance, the configuration of a multimachine system requires the identification of a part family, which in turn creates benefits in the upstream areas by way of design nonproliferation in the future, cost avoidances, and so on. Typically, the exclusion of these important benefits has tended to lengthen the payback periods computed for these systems.

It is necessary to understand the basic economic characteristics of these systems. First, by dedicating a group of machines to a part family, most or all of the operations for the parts are carried out in the cell. This reduces material handling, and the bulk of the queuing and transporting times which account for as much as 95% in the regular job shop is eliminated. This also reduces the WIP inventories.

Setup times are reduced by sequencing based on part similarities as in the GT cell. In the new technologies, the setup merely involves downloading

Table 2
An Overview of the Benefits

Preproduction Stage:	a) Increased productivity in design and process planning.
	b) Reduced lead time for preproduction activities.
	c) Integration of design, process planning and manufacturing.
	d) Standardization benefits, design nonproliferation.
	e) Better quality, reliability, maintainability.
	f) Savings in tool design and investments.
	g) Increased accuracy and better data management.
Production Stage:	h) Reduction in manufacturing lead times.
	i) Reduction in work in process inventories.
	j) Predictability of operations.
	k) Flexibility and sensitivity to market.
	l) Reduced scrap rates.
	m) Reduction in material handling.
	n) Increased capacity due to setup time reductions, economical small lot production.
	o) Better control over operations.
	p) Savings in floor space.
	q) Synergism of the above.
Postproduction Stage:	r) Better competitive position.
	s) Better and faster response to variety with smaller lots.
	t) Better delivery promises, reduced and more predictable lead times.
	u) Reduced finished parts inventory.
	v) Increased capacity leading to ability to handle increased sales.
	w) Better reliability, maintainability, etc.

the appropriate part programs and tool changeovers. Operation overlapping can be easily accomplished due to the proximity of the next-operation machine which brings down the cycle time even more. Quality improvements, reduced floor-to-floor times, and savings in floor space are other advantages. Production control is easier since there is no operation-wise division of responsibility as in the conventional job shop. All these are very synergistic. With reduced lot sizes, manufacturing can be done just prior to the date of need as per the period batch control (or JIT) philosophy. Operations become more predictable due to the shorter cycle times and the predictability arising from automation. Flexibility, which has many dimensions, is also a major advantage.

From the above, it can be seen that indirect savings form the major savings element and every effort should be made to exhaustively compute these savings. The use of special methods like the Warner and Swasey procedures are required for this exercise. Klahorst[21] mentions 14 critical cost factors to be considered in the economic evaluation of these systems. These factors include:

1. Direct Labor Costs.
2. Machine Setup Costs.
3. Tooling Costs.
4. Materials Handling Costs.
5. Part Inspection Costs.
6. Equipment Maintenance Costs.
7. Shop Supervisor Costs.
8. Production Control Costs.
9. Manufacturing Engineering Costs.
10. Plant Facilities Costs.
11. Inventory Costs.
12. Fixturing Costs.
13. Prototype and New-part Cost.
14. Rework and Scrap Costs.

The use of a modified burden rate is also suggested, which involves separately examining the key components of the burden and subtracting their effects from the overall shop burden. We could also include the savings in floor space in the above list.

Reductions in throughput times and WIP inventory can also be easily estimated. The throughput time reductions in switching over to cellular manufacturing can be estimated as follows. Assuming a perfect cell in which all the operations are completed for a part family, for a given part i, let S_{1i} be the total setup time in the present arrangement. In the cell, since the throughput time is purely a function of the setup time, the operation time, and the batch size, the new throughput time T_i', would equal at its upper limit, $S_{1i} + T_{mi}*Q$, where T_{mi} is the total machining time per unit and Q is the batch size. This is assuming no overlapped operation. Even with this assumption, T_i' represents the elimination of a large part of the queuing and transporting times. However, with cellular manufacturing, the setup time is lower, says S_{2i}.

Thus, in *Figure 4*, line A represents the throughput time as a function of the batch quantity, with the assumption of no operation overlap and no setup time reductions. Line B represents a more realistic case with overlaps and setup time reductions. The actual time taken will depend on the specifics of a situation, i.e., the scheduling rule, the current job mix, etc., and will fall in between the two lines if the scheduling proceeds as is normally done in GT cells.

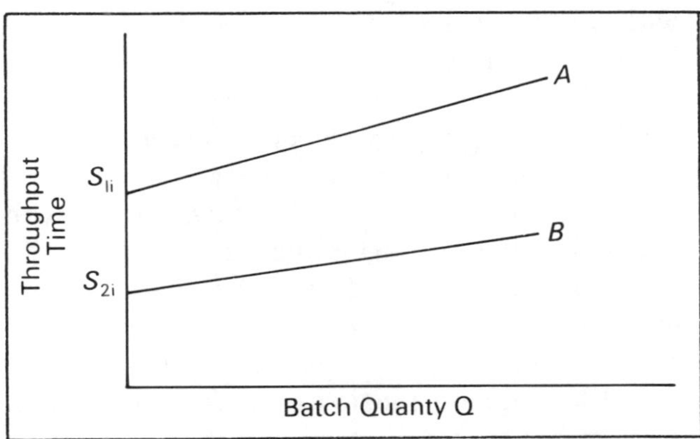

Figure 4
Throughput Time *vs* Batch Quantity for a Cell

From a justification standpoint, Line A itself represents a substantial reduction and can be used as a conservative estimate. It is also useful to remember that when installing a cell in the midst of a regular job shop, the latter is the critical path with respect to overall cycle times. Inadequate realizations of throughput time reduction and inventory increases elsewhere are frequent occurrences in practice.

The WIP inventory savings can be calculated next. For part i, in the conventional system, the inventory value $I_i = (C_{mi} + C_{vi}/2)*T_i*D_i$, where C_{mi} is the unit material costs, C_{vi} is the value added per unit, T_i is the throughput time, and D_i is the annual demand. The new inventory value I_i' now equals $(C_{mi}' + C_{vi}'/2)*T_i'*D_i$ as per the new values. These savings are then aggregated for all items in the part family.

However, the throughput time reductions and WIP inventory reductions are dependent on the job mix and the scheduling policies. These can be evaluated through a simulation approach. We illustrate this in an example situation below.

For the economic analysis, stochastic methods have hitherto been rarely used. This has become a major necessity for the new manufacturing systems. Even though simulation is widely used for analyzing the impact of various scheduling rules on GT cells or FMSs, and also for configuring such systems, it has not yet formed an integral part of the justification exercise. We illustrate the capabilities of a simulation-based economic analysis in the next section. There is currently considerable scope for utilizing risk analysis methods and concepts such as stochastic dominance as part of an integrated approach to the justification problem.

An Illustrative Example

A computer program was written in SIMSCRIPT II.5 and run on the Amdahl 470 computer at the University of Cincinnati for a hypothetical FMS configuration whose economic viability was to be investigated. The choice of SIMSCRIPT as the simulation language was essentially due to its modularity and high-level characteristics, especially its English-like statements.

The program involves simulating the configuration over an investment planning horizon of 10 years. The system is designed to produce a part family in which there are currently 10 parts. It is expected that the number of parts is likely to grow and the sales volume of the individual parts will probably shrink. This has been observed with new manufacturing technologies, i.e., customers require more and more custom-designed parts when the ability to cater to such variety exists.[33] Each part is designed to have a life cycle as shown in *Figure 5*. The maximum sales, life of the part, number of operations per part, month of introduction, and its routing are randomly generated. The values created for new parts are as follows:

NEW.PARTS—

The number of new parts introduced in the year. This is a uniform random variable between the values of LOW.LIMIT and UP.LIMIT which are given by:

LOW.LIMIT = YEAR.NUMBER + 3

UP.LIMIT = YEAR.NUMBER*2 + 6

NO.OF.OPNS.—

The number of operations for the part are randomly generated between 5 to 8 operations.

SALES—

The maximum sales for the part is a uniformly varying random variable between USALES and LSALES as given by:

LSALES =

(Investment Horizon - Current year)*25 + 100

USALES =

(Investment Horizon - Current year)*60 + 400

CREATE.ROUTING—

This is a routine which creates a routing consisting of various operations whose machines and times are randomly generated. The operation time is described as a 2-Erlang random variable with a mean equal to the mean operation time in the selected machine.

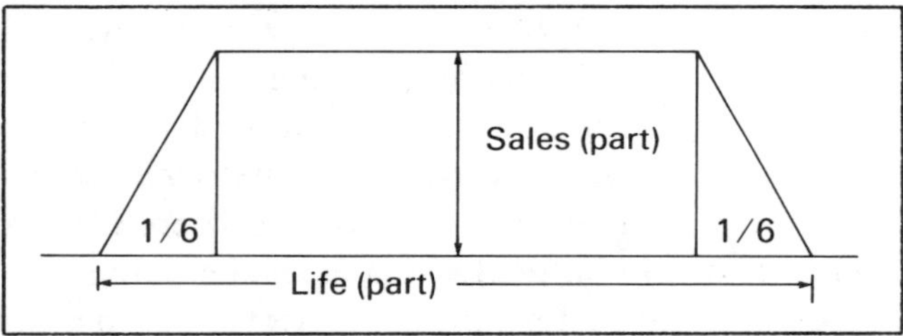

Figure 5
Part Life Cycle Assumed

Every part created has a set called ROUTING which consists of elements called OPERATION whose attributes include the machine and the operation time.

The starting configuration involves eight types of machines with one machine of each type. The distance matrix is read in which gives the distances between the various machines and load/unload station which is station 9. The configuration can be easily altered by changing any of these in the data portion of the program. The system also consists of transporters, one unit to start with, with a travel speed which is read in. There are buffer spaces in front of each machine which are initially unlimited.

The simulation proceeds as follows. Every replication involves running the simulation for 10 years on a month by month basis, i.e., for 120 months. For every replication, the product mix consisting of current parts as well as newly introduced parts is first created for the entire 10 years. This makes it simple to calculate the month by month loads imposed on the system by each part. Within each replication, for every month the parts in the set PRODUCT.MIX are examined to check if any demand exists for the part in the month. If yes, a job is created for the part and filed into a set called POOL. The set POOL is created on the basis of increasing total processing times of jobs. This is to ensure the application of the Shortest Processing Time (SPT) rule for scheduling. For every job, TASKS exist for each operation to be performed. The set POOL is created as a set for the SPT rule by defining it as a SET RANKED BY LOW JB.PR.TIME in the PREAMBLE (see *Appendix 1*). By changing this one statement, experiments with other priority rules can be tried. The inputs required are shown in *Appendix 2*.

The jobs in the set POOL are loaded onto the FMS. Initially, the job waits for the transporter to be available and is then taken to the first operation machine. The job waits in the buffer space before the machine if the machine happens to be busy. The simulation then proceeds as in any multimachine, multijob situation. This is accomplished in the PROCESS JOB.LOAD portion of the program.

For every job, the flow time is monitored apart from waiting times in queue. For the machines, the queue statistics and utilization levels are time averaged. A sample output is shown in *Appendix 3*. The output can be made more concise if required. In the job summary part of the output, the jobs are listed as per the SPT rule, i.e., in the order of loading. For each job, the total processing time and the cycle time are given. The machine summary then gives the utilization and the queue statistics.

From the sample output, it can be seen that in month 21 for replication 1, the overall utilization for machines happens to be the highest. Also the total flow time works out to 1788 hours compared to available hours of 352 in a given month, assuming 2 shifts and 22 working days. Increasing the number of machine types 4 and 5 to 2 units each brings the flow time to 1456 hours as seen in *Appendix 4*. The utilization of the transporter increases very marginally. Increasing the number of type 1 machines to 2 units further reduces the flow time to 1426 hours.

Thus, it can be seen that throughput times can be arrived at through simulation instead of being calculated on an item by item basis as done earlier, thereby allowing the system to be configured to optimize the flow times and the capital costs. Reducing the flow times increases the capital costs, but decreases inventory costs and results in other tactical and strategic advantages.

Using such a generalized simulation program, a variety of questions can be answered from the justification point of view:

1. A worst case scenario with respect to product proliferation and part volume/variety characteristics can be accomplished by changing the four or five statements which determine the part generation during the simulation run.
2. Given the load generated during the investment planning period, the capacity requirements can be arrived at. As Hutchinson[22] points out, one of the real advantages of these systems is the capability to incrementally acquire capacity and postpone the capital commitments to a later period. This leads to lower capital costs as well as risk reduction.
3. The configuration of the system can be easily varied by changing the number of machines and changing the distance matrix values to suit different layouts. Also, the number and travel speeds of the transport mechanism can be varied.
4. The model can be reprogrammed to test other types of multimachine systems, such as a group technology cell or robotic multimachine system, to manufacture the same part family. The capital costs in the case of GT will be lower, but higher labor costs and greater operation time variabilities are involved.
5. Randomized equipment breakdowns and their impact on the overall system performance can also be investigated with this model.
6. Additional modules for performing cash flow analysis and net present value analysis can be included in the above program. The effects of tax rates, inflation, cost of capital and other such factors can be determined. Alternatively, these costs and savings could be computed using a spread sheet analysis. The simulation provides the performance measures of the system like the manufacturing cycle time which has a bearing on WIP inventory and other costs. Various scheduling rules, lot size factors, overlapped operations and so on all produce different cash flows and their impact can be easily analyzed by the simulation/spread sheet approach.

In the use of simulation, comparisons between alternative systems are usually made on the basis of the expected value of such performance measures as the throughput time. However, due to the stochastic nature of the process, using expected value comparisons often results in erroneous conclusions. In such cases, it is preferable to consider the entire distribution of the performance measures based on several replications, and subject them to a stochastic dominance type of analysis (*Appendix 5*). There are three degrees of stochastic dominance and efficient algorithms exist for performing comparisons among several alternatives. The experimental design, the statistical methodology behind simulation, and all the other aspects of simulation are well developed areas.

Appendix 1

```
PREAMBLE
  RESOURCES
    EVERY MACHINE HAS A MC.ID AND A NO.OF.UNITS AND A MEAN.OP.TIME AND A
      MC.Q.DELAY AND A NO.WKG.MC
    EVERY TRANSPORTER HAS A TR.LOCN AND A TR.SPEED
      DEFINE MC.ID, NO.OF.UNITS, NO.WKG.MC, TR.LOCN AS INTEGER VARIABLES
      DEFINE MEAN.OP.TIME, MC.Q.DELAY, TR.SPEED AS REAL VARIABLES
    PROCESSES
      EVERY LOAD HAS A MONTH
      EVERY JOB.LOAD HAS A JOB.NUMBER
  TEMPORARY ENTITIES
    EVERY JOB HAS A JB.PR.TIME AND A JB.PART AND A JB.QTY AND A JB.FL.TIME
      AND A JB.MC.Q.DELAY AND A JB.TR.Q.DELAY
      AND BELONGS TO A POOL
      AND OWNS A SEQUENCE
      DEFINE JB.FL.TIME, JB.PR.TIME, JB.MC.Q.DELAY, JB.TR.Q.DELAY
      AS REAL VARIABLES
      DEFINE JB.PART, JB.QTY AS INTEGER VARIABLES
      THE SYSTEM OWNS THE POOL
    EVERY TASK HAS A TK.JOB AND A TK.MC AND A TK.TIME
      AND BELONGS TO A SEQUENCE
      DEFINE TK.JOB, TK.MC AS INTEGER VARIABLES
      DEFINE TK.TIME AS A REAL VARIABLE
    EVERY PART HAS A SALES AND A LIFE AND A ST.MONTH AND A NO.OF.OPNS
      AND OWNS A ROUTING
      AND BELONGS TO A PRODUCT.MIX
    EVERY OPERATION HAS AN OPN.PART AND AN OPN.MC AND AN OPN.TIME
      AND BELONGS TO A ROUTING
      THE SYSTEM OWNS THE PRODUCT.MIX
      DEFINE SALES, LIFE, ST.MONTH, NO.OF.OPNS, OPN.PART, OPN.MC
        AS INTEGER VARIABLES
      DEFINE OPN.TIME AS A REAL VARIABLE
      DEFINE OP.ID AS AN INTEGER, 1-DIMENSIONAL ARRAY
      DEFINE DISTANCE  AS AN INTEGER, 2-DIMENSIONAL ARRAY
      DEFINE POOL AS A SET RANKED BY LOW JB.PR.TIME
      DEFINE PRODUCT.MIX AS A FIFO SET
      DEFINE ROUTING AS A FIFO SET
      DEFINE SEQUENCE AS A FIFO SET
      DEFINE HORIZON, NO.CURR.PARTS, NO.SHIFTS, NO.DAYS, NO.PARTS,
        REPLICATIONS AS REAL VARIABLES
      DEFINE ST.TIME AS A REAL VARIABLE
      TALLY MC.MEAN.DELAY AS THE MEAN OF JB.MC.Q.DELAY
      TALLY TR.MEAN.DELAY AS THE MEAN OF JB.TR.Q.DELAY
      TALLY MEAN.MC.Q.DELAY AS THE MEAN OF MC.Q.DELAY
      ACCUMULATE MC.UTILN AS THE AVERAGE OF N.X.MACHINE
      ACCUMULATE MAX.Q AS THE MAXIMUM AND AVE.Q AS THE AVERAGE OF
        N.Q. MACHINE
      ACCUMULATE TR.UTILN AS THE AVERAGE OF N.X.TRANSPORTER
      ACCUMULATE MEAN.NO.WKG.MC AS THE AVERAGE OF NO.WKG.MC
      DEFINE HOURS TO MEAN UNITS
END
```

Appendix 2

Input Data

Data	Description
10 10 1	No. of current parts, planning horizon, lot factor
8	No. of types of machines
0.03 0.03 0.045 0.05 0.05 0.02 0.02 0.05	Mean operation times
420 30 5 -10	Current part data: sales (max), life, no. of operations, start month
300 24 6 -5	
450 50 7 -12	
800 36 6 -20	
500 24 8 -12	
550 36 5 -12	
300 30 6 0	
650 48 7 -2	
700 24 5 0	
800 36 6 -5	
22 2 30	No. of days, shifts, replications
1 50 5	No. of machines of each type, MTBF, MTTR
1 50 5	
1 50 5	
1 50 5	
2 50 5	
1 50 5	
1 50 5	
1 50 5	
2.5	Transporter speed
0 5 10 15 20 25 30 35 5	Distance matrix
5 0 5 10 15 20 25 30 10	
10 5 0 5 10 15 20 25 15	
10 10 5 0 5 10 15 20 20	
20 15 10 5 0 5 10 15 25	
25 20 15 10 5 0 5 10 30	
30 25 20 15 10 5 0 40	
5 10 15 20 25 30 35 40 0	

Appendix 3

Part Life Cycle and Routings (A Sample)

PRODUCT MIX REPLICATION 1

PART	SALES	LIFE	# OPNS	ST.MONTH	*** ROUTINGS *** OPERATION	M/C	OP.TIME
1229388	420	30	5	-10	1221652	8	.0247
					1221684	5	.7145
					1221716	2	.0441
					1221748	7	.0692
					1221780	1	.1533
1229436	300	24	6	-5	1221812	7	.0840
					1221844	3	.2048
					1221876	8	.0577
					1221908	1	.1154
					1221940	2	.1324
					1221972	6	.0655

1229484	450	50	7	-12	1222004	1	.0913
					1222036	5	.1447
					1222068	2	.0547
					1222100	3	.1494
					1222156	7	.0271
					1222188	8	.1817
					1222220	4	.0990
1229532	800	36	6	-20	1222252	6	.0420
					1222284	1	.0291
					1222316	7	.0268
					1222348	5	.3061
					1222380	3	.0161
					1222412	2	.1176
1229580	500	24	8	-12	1222444	6	.0801
					1222476	2	.1626
					1222508	5	.1252
					1222540	8	.1252
					1222572	4	.1143
					1222604	1	.1404
					1222660	7	.0992

JOB SUMMARY FOR MONTH 21

JOB	PART	QTY	PROCESSING TIME	FLOW TIME
1351236	1312820	260	79.0926	430.0820
1346132	1295164	275	126.3857	844.9810
1351412	1312916	215	137.1680	936.5229
1351500	1312964	319	162.5517	534.4189
1346220	1295212	300	183.8988	561.2610
1349484	1289636	276	187.5044	1658.9053
1351324	1312868	496	213.7942	548.1333
1351148	1312772	347	246.2121	906.9133
1346396	1295308	525	294.7224	705.5500
1349132	1289444	504	301.0259	755.2615
1347972	1289396	370	307.7344	652.9556
1347708	1289252	438	319.4146	1252.8584
1346308	1295260	650	325.7151	1788.8723
1346044	1295020	450	336.5852	1599.1689
1347796	1289300	439	382.0789	1134.7461
1347884	1289348	864	397.1780	1017.6018
1349220	1289492	594	405.7874	1001.0781
1347520	1289204	800	537.9644	1723.7307
1349396	1289588	891	755.9702	1350.6851
1349308	1289540	825	1267.1458	1780.7551

MACHINE SUMMARY FOR MONTH 21

M/C	# UNITS	MEAN.OP.TIME	MEAN.DELAY	UTILIZATION	MAX.Q	AVE.Q
1	1	0.100	148.1261	0.4943	7.0	1.325
2	1	0.100	15.8720	0.2972	2.0	0.133
3	1	0.150	27.1840	0.3974	2.0	0.167
4	1	0.220	346.7511	1.0000	5.0	3.101
5	1	0.170	257.7013	0.8415	8.0	2.593
6	1	0.050	2.5938	0.1911	1.0	0.022
7	1	0.050	10.9535	0.1687	3.0	0.098
8	1	0.100	53.3675	0.5049	4.0	0.507

TOTAL FLOW TIME = 1788.873 UTILIZATION OF TRANSPORTER = 0.0003

Appendix 4

JOB SUMMARY FOR MONTH 21

JOB	PART	QTY	PROCESSING TIME	FLOW TIME
1351236	1312820	260	79.0926	243.6401
1346132	1295164	275	126.3857	514.8147
1351412	1312916	215	137.1680	636.0969
1351500	1312964	319	162.5517	290.2539
1346220	1295212	300	183.8988	271.7498
1349484	1289636	276	187.5044	652.3000
1351324	1312868	496	213.7942	481.5095
1351148	1312772	347	246.2121	641.4011
1346396	1295308	525	294.7224	630.1748
1349132	1289444	504	301.0259	711.0325
1347972	1289396	370	307.7344	823.5808
1347708	1289252	438	319.4146	779.4045
1346308	1295260	650	325.7151	776.1719
1346044	1295020	450	336.5852	949.9465
1347796	1289300	439	382.0789	900.6560
1347884	1289348	864	397.1780	532.3943
1349220	1289492	594	405.7874	927.3601
1347520	1289204	800	537.9644	883.8982
1349396	1289588	891	755.9702	1086.6096
1349308	1289540	825	1267.1458	1456.3645

MACHINE SUMMARY FOR MONTH 21

M/C	# UNITS	MEAN.OP.TIME	MEAN.DELAY	UTILIZATION	MAX.Q	AVE.Q
1	1	0.100	201.1873	0.6072	7.0	2.210
2	1	0.100	20.5208	0.3650	2.0	0.211
3	1	0.150	71.9657	0.4882	2.0	0.544
4	1	0.220	20.9006	1.2283	5.0	0.230
5	1	0.170	40.5664	1.0337	5.0	0.501
6	1	0.050	5.7920	0.2348	1.0	0.060
7	1	0.050	18.7370	0.2072	3.0	0.206
8	1	0.100	85.3138	0.6202	4.0	0.996

TOTAL FLOW TIME = 1456.366 UTILIZATION OF TRANSPORTER = 0.0004

Appendix 5

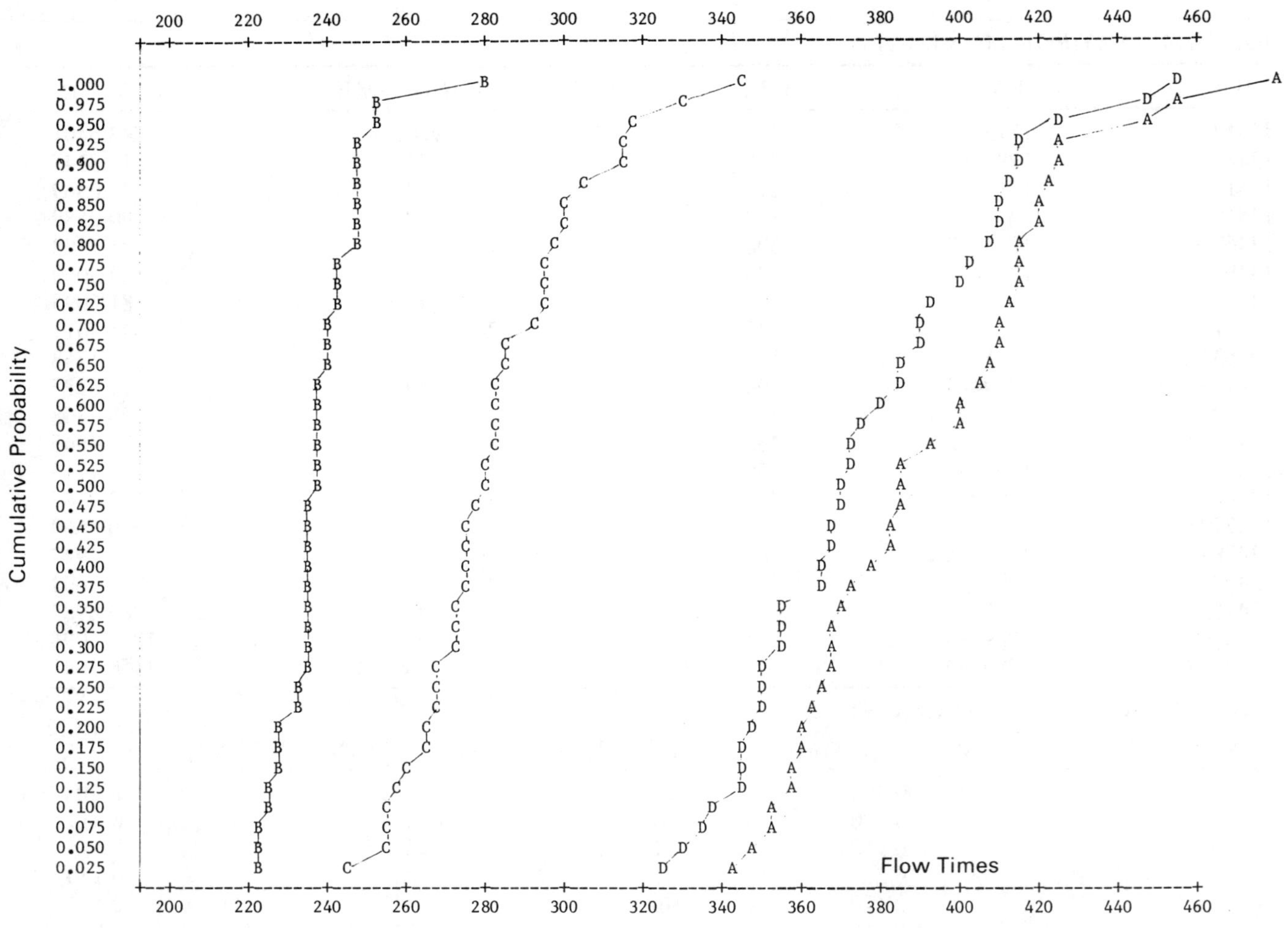

References

1. N.H. Cook. "Computer-managed Parts Manufacture," *Scientific American*, February 1975, pp. 21-29.
2. R.J. Fotsch. "Machine Tool Justification Policies: Their Effect on Productivity and Profitability," *Journal of Manufacturing Systems*, Volume 3, No. 2, pp. 169-195.
3. B. Gold. "CAM Sets New Rules for Production," *Harvard Business Review*, Volume 60:6, November/December 1982, pp. 88-94.
4. *Guidebook for Planning Machine Tool Investment*, (*American Machinist*), 1981.
5. D. Gerwin. "Control and Evaluation in the Innovation Process: The Case of Flexible Manufacturing Systems," *IEEE Transactions on Engineering and Management*, August 1981.
6. J.L. Burbidge. *Group Technology in the Engineering Industry*, Mechanical Engineering Publications Ltd., London, 1979.
7. E.S. Buffa. "Research in Operations Management," *Journal of Operations Management*, Volume 1:1, 1981.
8. J.A. Behuniak. "Economic Analysis of Robot Applications," *Proceedings of Robots IV Conference*, Society of Manufacturing Engineers, 1979.
9. T. Bublick. "The Justification of an Industrial Robot', *Proceedings of Finishing '77 Conference and Exposition*, Society of Manufacturing Engineers, 1977.
10. J. Engelberger. *Robotics in Practice*, Kogan Page Ltd., 1980.
11. J. Engelberger. "Robots Make Economic and Social Sense," *Atlanta Economic Review*, July/August, 1977.
12. G.A. Fleischer. "A Generalized Methodology for Assessing the Economic Consequences of Acquiring Robots for Repetitive Operations," *1982 Annual IIE Conference Procedings*.
13. P.Y. Huang, P. Ghandfaroush. "Procedures Given for Evaluating, Selecting Robots," *Industrial Engineering*, April 1984.
14. R.N. Stauffer. "Equipment Acquisition for the Automatic Factory," *Robotics Today*, April 1983, pp. 37-40.
15. W.G. Sullivan, "Replacement Decisions in High Technology Industries," *Proceedings of IIE Conference*, 1984, pp. 119-128.
16. W.R. Tanner. "Selling the Robot—Justification for Robot Installation," Technical Paper MS78-702, Society of Manufacturing Engineers, 1978.
17. J.P. Van Blois. "Economic Models: The Future for Robotic Justification," *Proceedings of 13th Industrial Symposium on Industrial Robots and Robots 7*, Society of Manufacturing Engineers, April 1983, pp. 4-24 to 4-31.
18. B.T. Fullerton. *Machine Replacement for the Shop Manager*, Huebner Publication, 1965.
19. Warner and Swasey Company. "Introduction to the Concept of Total Measurement of Direct and Indirect Savings," Form 71003, 1971.
20. R.J. Meyer. "A Cookbook Approach to Robotics and Automation Justification," *Robots VI Conference Proceedings*, Society of Manufacturing Engineers, March 1982.

21. H.T. Klahorst. "How to Justify Multimachine Systems," *American Machinist*, September 1983, pp. 67-70.
22. G.K. Hutchinson. "Production Capacity: CAM vs. Transfer Line," *Industrial Engineering*, September 1976, pp. 30-35.
23. G.K. Hutchinson, J.R. Holland. "The Economic Value of Flexible Automation," *Journal of Manufacturing Systems*, Volume 1, No. 2, pp. 215-228.
24. J. Goldhar, M. Jelinek. "Plan for Economies of Scope," *Harvard Business Review*, Volume 61:6, pp. 141-148.
25. N.A. Kulatilaka. "A Managerial Decision Support System to Evaluate Investments in FMSs," *Proceedings of First ORSA/TIMS Special Interest Conference on FMS*, Ann Arbor, Michigan, August 1984.
26. N.C. Suresh, J.R. Meredith. "A Generic Approach to Justifying an FMS," *Proceedings of First ORSA/TIMS Special Interest Conference on FMS*, Ann Arbor, Michigan, August 1984.
27. W.C. Kester. "Today's Options for Tomorrow's Growth," *Harvard Business Review*, March/April 1984, pp. 153-160.
28. J. Shewchuck. "Justifying Flexible Automation," *American Machinist*, October 1984, pp. 93-96.
29. J.A. Simpson. "Investment Justification of Robotic Technology in Aerospace Manufacturing," Air Force Business Management Research Center, Wright-Patterson AFB, Ohio, BRMC-83-5080.
30. N.C. Suresh. Unpublished Doctoral Dissertation (forthcoming), College of Business Administration, University of Cincinnati, Cincinnati, Ohio.
31. J.R. Meredith, S. J. Mantel Jr. *Project Management: A Managerial Approach*, John Wiley & Sons, 1985.
32. T.L. Saaty. *The Analytic Hierarchy Process*, McGraw-Hill, New York, 1980.
33. J.R. Meredith. "Peerless Saw Company," Case Study, College of Business Administration, University of Cincinnati, Cincinnati, Ohio, 1985.

Bibliography

Holmes, J.G. "Justifying a Robot Machining System in Batch Manufacturing," *Robotics Today*, Summer 1979.

Inaba, H., Sakakibara, S. "Flexible Automation—Unmanned Machining and Assembly Cells with Robots," *Proceedings of the 1st International Conference on FMS*, Brighton, United Kingdom, 1982.

Kaplan, R.S. "Measuring Manufacturing Performance: A New Challenge for Managerial Accounting Research," *Accounting Review*, Volume LVIII:4, October 1983, pp. 686-705.

Tanner, W.R. *Industrial Robots*, Society of Manufacturing Engineers, Dearborn, Michigan, 1979.

Author(s) Biography

Nallan C. Suresh is currently a Ph.D. student at the University of Cincinnati, Cincinnati, Ohio, specializing in Operations Management and Information Systems. He has an MBA degree from McMaster University, Ontario, Canada and an undergraduate degree in engineering from the Indian Institute of Technology, Madras, India. He has four years of industrial experience as a systems analyst specializing in manufacturing applications. He is certified by the American Production and Inventory Control Society and his publications have appeared or will appear in the *International Journal of Production Research, Production and Inventory Management, Journal of Operations Management, ORSA/TIMS Proceedings* and the *Robots 9 Proceedings*. Mr. Suresh is a member of RI/SME, APICS, IIE, ORSA, TIMS and AIDS. He was selected by the American Institute of Decision Sciences as an outstanding doctoral student in 1984.

Jack R. Meredith is an Associate Professor and Director of Graduate and Undergraduate Programs in Industrial and Operations Management at the University of Cincinnati. He received his undergraduate degrees in engineering and mathematics from Oregon State University. He obtained his MBA and Ph.D. degrees in business administration from the University of California, Berkeley. Dr. Meredith has held positions with Ampex, Hewlett-Packard, TRW and Douglas Aircraft Company. He is the editor of *Operations Management Review* and his articles have appeared in *Operations Research, Management Science, Computers and Industrial Engineering, Health Care Systems, American Journal of Public Health, Journal of Operations Management*, and the *International Journal of Production Research*. He authored *Fundamentals of Management Science* (with E. Turban), *The Hospital Game, The Management of Operations* (Wiley, 1984), and *Project Management* (with S.J. Mantel Jr., Wiley, 1985). Dr. Meredith is a member of CASA/SME, RI/SME, IIE, TIMS, AIDS, APICS, and OMA.

STRATEGIC CONSIDERATIONS

In this section, we deal with the larger picture of automation in the justification process, especially the abstract, qualitative, and non-monetary aspects of the technology. Strategic issues are discussed as well as the impact on employees, the form of the justification proposal, and even the politics required to sell the proposal.

The article by Grierson reviews the overall situation and reiterates the firm's original need for new technology. The strategic issues are described in relation to the cost and other tactical issues normally considered in the evaluation of automation proposals. In Kaplan's article, the intangible benefits are contrasted even more strongly with the tangible benefits. Then in the article by Primrose and Leonard, a number of new issues surface in terms of additional strategic benefits and denials of commonly accepted benefits.

Gerwin moves into a new realm and discusses the people and organizational issues, as well as reporting on how firms have decided to acquire automation. The article by Meredith picks up the theme of championing postulated in the previous article and describes how this process works for adopting new technology.

The last three articles give suggestions for presenting the justification proposal to upper management. Baker makes some suggestions based on General Electric's experience with justifying automation in a number of its plants. Weaver includes an excellent format for the justification report which should help it be viewed very positively by upper managers. And in the last article, Staples advises on personalizing the technology's benefits to appeal to each type of board member.

Reprinted from ***Production Engineering,*** *September 1984.*

Integrated Manufacturing—The Concepts

THE MANAGERS: Maestros Or Magicians?

"Manufacturing management today needs to think strategically about how to compete five and ten years in the future, and to establish a long-life corporate architecture to implement the strategy."

By Donald K. Grierson

The American factory is a victim of neglect. A typical U.S. manufacturing plant today is characterized by obsolete machine tools, with 80 to 90% of the work performed manually. Failing quality and soaring inventories are their legacy.

America's machine tools are among the oldest in use in any industrialized nation; two-thirds of them are over 10 years old. An engineer who doesn't keep up with new technology is out of date in less than ten years. So it shouldn't be surprising that much of our antiquated manufacturing establishment is incapable of competing successfully in world markets. Just as engineers must constantly upgrade their skills, factories must also be constantly upgraded if they are to maintain cost leadership in their fields.

Modernization today equates to factory automation. Yet, sad to say, automation systems in the U.S. are nearly nonexistent. Probably no more than 5% of American factories outside of the automotive industry are automated to any major degree. Why is this so?

Some say that U.S. management has tunnel vision. When business is poor, management tends to assume it will be poor forever, so there is no incentive to invest in new capital equipment. And when business is booming they see no need to "fix the roof when it isn't raining." Some companies do invest during good times, of course, but they do it primarily to increase capacity. They are often not looking five years out to when times might not be so good, and they might want to operate more efficiently or use their excess capacity to make some other product, or simply to make a better quality product that will allow them to increase market share.

I think one of the main reasons American management is having so much difficulty focusing on the automation challenge is that it has never had to think about it before. For the most part, present-day factories were in place when most of today's CEOs got their jobs. And whenever another factory was needed to meet demand, companies simply cloned an existing factory, giving little or no thought to how they might improve on what they already had. Sometimes a few machine tools were upgraded with numerical control, but for the most part the new factory was simply a shinier version of the old. And that was a very comforting, non-threatening way for manufacturing management to operate.

Now, management is being forced to rethink the whole idea of the factory, and take a long, hard look at why American factories have been losing market share to their world competitors.

Playing catchup. Unlike U.S. industry, companies in Japan do not automate because there is a three-year, or five-year payback; they take a much longer-term view of investment. Japanese companies are automating because they believe that the only way they can be competitive in the world marketplace is to drive toward the lowest possible level of cost. So this becomes a corporate goal. The mission statement for the corporation becomes, "we will be the low-cost high-quality producer in the world for the business we are in, and we will make the investments needed to accomplish this."

If those investments have a seven-year payback, that is not a deterrent. The idea is to be the cost leader, and to get the benefits of being a very profitable business over the long haul. What is impressive about what is happening in Japan is not only that they are ahead of us in many areas, but that the rate at which they are improving is much faster than in the U.S.

Here are people that are already in a leadership mode, setting goals to reduce their costs by an additional 30% per year. How long is it going to take for U.S. companies to catch up to a target accelerating that rapidly? The situation is similar to the Red Queen's Race in Lewis Carroll's *Alice in Wonderland.* U.S. industry today has to run as fast as it can simply to prevent the Japanese from extending their lead. And it must run twice as fast to catch up.

For some manufacturers who have gotten as far behind as they have in the Red Queen's Race, the only hope they really have of catching up is to begin applying new managerial approaches and advanced technology to their manufacturing processes so as to change completely the basis of competitive advantage now held by the Japanese. The old fairy tale of the plodding but dedicated tortoise overtaking the faster but undisciplined hare doesn't apply here.

Management commitment. Automation planning has to start at the top and work down. It is important that the CEO understand where he wants to take the business over the next five years and that it is possible to get there. And that he can, in the future, have a computer-integrated manufacturing system for his business.

In making automation decisions, management has to look at a different set of benefits than short-term ROI. It has to put values on soft attributes, things the factory will be able to do that it was not able to do in the past. The conventional financial approaches of "how many people are you going to eliminate" and "what kind of payback can you get" have to be modified. The appropriation process has to be rethought and redefined.

The entire management team needs to realize that there is not a single solution to making American factories more competitive. There are, instead, a number of solutions. And these solutions are not necessarily contradictory—they are often complementary.

One solution is getting better people. Japanese companies assign platoons of young engineers to work on the factory floor, and these engineers identify and remove bottlenecks. It is an approach we could profitably follow. Another solution is to apply the managerial techniques we already have available to us. And a third is the installation of computer-integrated automation systems on a planned basis, as part of a well-defined, future-friendly architecture for the company.

Integration. In the automated factory, there is no such thing as a clean cut between engineering and manufacturing. We have, for some time, been convinced that an integrated approach involving computer-aided design, computer-aided engineering, and computer-aided manufacturing is the right approach. This requires that management understand that the benefits you get from integrating manufacturing and engineering are not necessarily savings in direct labor. More often, the most significant benefits are those you get by eliminating up to one-third of the development cost of a new product.

Over the past two years, significant advances have been made in automation hardware and systems integration that have helped to close the gap between engineering and manufacturing. We know more now about the importance of simulation, and creating dynamic models of how the factory is going to operate. Individual products at the same time have been improved, with more process knowledge built into them, such as in intelligent vision systems for welding applications. We also better understand the need for the integration of production management techniques with the automation products. Indeed, knowing how to dynamically model, and therefore how to schedule the flow of batch products through a flexible manufacturing system, is more important to the future of

"Management has to put values on soft attributes, things the factory will be able to do that it was not able to do in the past."

Donald K. Grierson is a senior vice president and the group executive of General Electric Co.'s Industrial Electronics Business Group in Charlottesville, Va., and is responsible for GE's thrust in factory automation. Mr. Grierson holds a mechanical engineering degree from Ohio University and an MBA from Xavier University. He began his career with GE in 1957, and his previous assignments have included general manager of the Aircraft Engine Group's Aviation Service Dept., general manager of the Battery Business Dept. and Carboloy Systems Dept., general manager of the Metalurgical Business Div., and vice president and general manager of the Lamp Products Div.

the business than is the FMS itself.

Managerial techniques. Management approaches to improving manufacturing processes include materials resources planning (MRP), optimized production timetable (OPT), and the Japanese concept of just-in-time inventory system management, or Kanban.

MRP and OPT are fundamentally extensions of data processing capability. They assist in the scheduling functions, and help control the total function in some respects. They work reasonably well in complicated job shops, where large numbers of pieces are produced in small batches. However, as the factory gets bigger with more individual products being components of other products, we have found that MRP and OPT have their limitations.

"Just-in-time parts delivery is seen by some as a universal cure, but it really only works well in continuous-flow manufacturing operations."

Kanban, or just-in-time parts delivery, is seen by some as a universal cure, but it really works well only in continuous-flow manufacturing operations. And there it works superbly. We have found that where we can use this approach, the dollars we take out of inventory can more than pay for the hardware.

But, job shops on the one hand, and continuous-flow processes on the other, are the exceptions rather than the rule in American manufacturing plants. It is the batch and assembly operations that today represent 70% of the problem of managing manufacturing. And there really aren't a lot of management tools that work, which is why we see the need in this type of operation for widespread application of technology like CAD/CAM, group technology, distributed numerical control, and computer-integrated manufacturing.

Getting started. The best starting point is a plan based on a description of your desired business results — what the projected impact of installing a major automated system will be on the performance of your business—and an assessment of your present system. The plan should cover each business area, with a planning horizon of at least three years and preferably five years. Automation projects cannot be accomplished all at once, which is why planning must be long-range. But the sooner the planning begins, the sooner the actual automation system implementation can start.

As you look at the factory, there are some clues on where to get started. The production schedule starts with market information and customer orders, which are usually captured in a computer. So these data can often be moved directly from order entry to the factory, if no engineering activity is needed. Your business information system people are a good in-house resource for starting the planning process.

Engineering is another likely starting point. Computer-aided design and engineering analysis probably offers the biggest savings for the lowest cost, so look closely at automating drafting and design early in the program.

Software packages and engineering workstations which distribute operations management support to the end user provide a low-cost entry point to factory automation. Flexible manufacturing cells allow small diversified batches to be produced as economically as large,

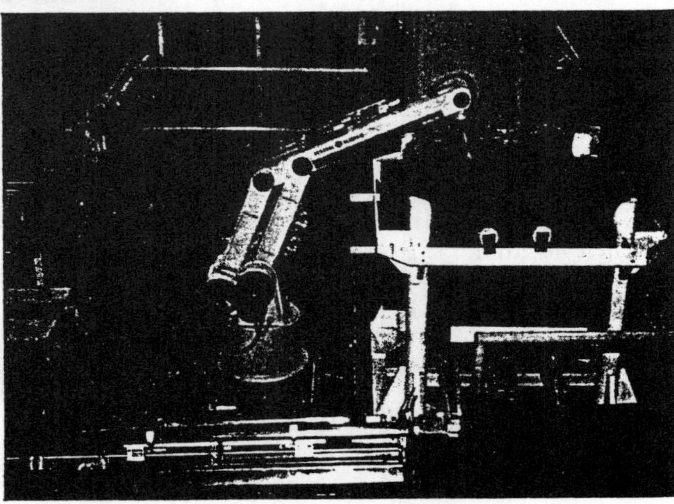

Robot work cell, at the GE Major Appliance Business Group in Louisville, Ky., was modeled with simulation software. Simulation allows the planner to check work cell design elements, such as interferences in the motion of the robot or end effector with other work cell components, before the hardware is installed.

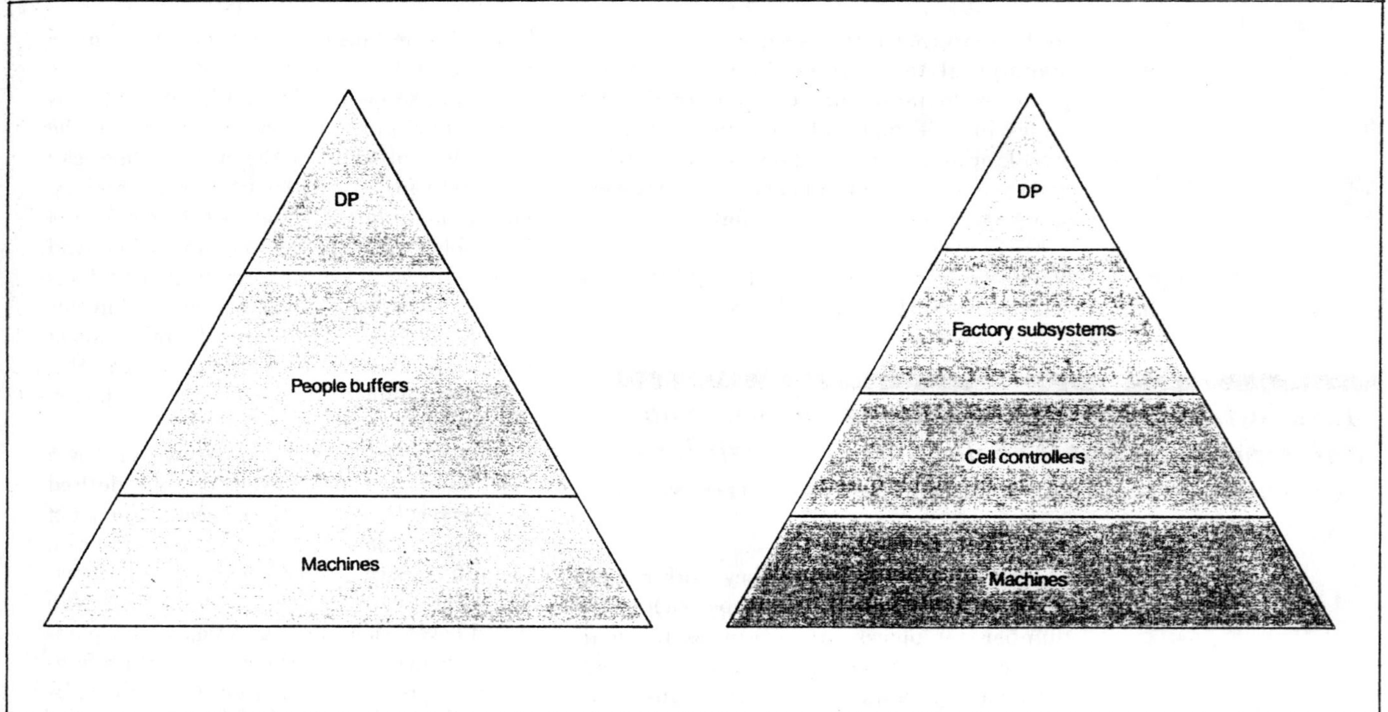

Today's factory consists of data processing capabilities at the top and machines—which incorporate more intelligence with the release of each generation of equipment—at the bottom. Between these two extremes are people—the buffers that have to perform the interfacing between the computers and the machines.

The need is for the intelligent machines to be more tightly linked with the computer capabilities. And for better feedback mechanisms between the top of the factory and the bottom to eliminate the people buffers. These feedback mechanisms can be provided by factory subsystems and cell controllers.

common batches. Modeling and simulation techniques can be used to reduce the design cycle and closely link the designer with the manufacturing engineer. On-line sensors can ensure a consistent level of quality throughout the product line, and common data base access reduces cycle time and impact by allowing simultaneous development of product and process information.

Networking is also an element of strategic planning. Planning and implementing a pilot engineering and manufacturing network based on coaxial cable technology will open new opportunities for moving from manual, paper-based systems to paperless electronic systems. And most networks available today provide adequate interfacing capabilities for the multivendor computer environment.

Many companies get their start in automation with a communications strategy built around an electronic data base designed to eliminate the people buffers. This is not the kind of data base most CEOs and data processing managers think about, however, which is little more than a list of part numbers. Rather, it is one that describes the geometry and the dimensions of the part. Once you have this, you have the ability to take the same information and automatically generate the computer-aided process plans, robotics programming, NC programming, tool and mold design, and other information needed for manufacturing the part.

A paperless factory, where information flows throughout the various operations automatically, eliminates not only the people buffers, but provides other, perhaps unexpected, benefits. For example, a moderate-size company typically accumulates 100 million sheets of paper a year, containing information that could be held in the computer. That much paper fills 4,200 file cabinets that occupy nearly an acre of floor space.

> ***"A paperless factory is a well-thought-out architecture for automation. It is what the leading companies today are working toward."***

A paperless factory is a well thought out architecture for automation. It is what the leading companies today are working toward. And it is the direction **you** want to head in.

Regardless of your entry point into factory automation, the most important recommendation still remains **start now.** The technologies and systems must be appropriate to the business strategy, but the technology is not a problem. Managing the application of the technology is the real issue. An FMS here, or a voice recognition system there, may be tempting. But they won't help you in the Red Queen's Race unless they fit into, and are basic elements of, the larger context of a total business system. And establishing the big picture takes top-down management commitment and involvement.

Financial justification. While everyone seems to agree that factory automation is necessary, the inevitable question is always, "Can I afford it?"

> ***"American businessmen have been trained by their financial brethren to look on investing in new equipment as a short-range tactical decision with limited objectives and financial implications."***

That, however, is the wrong question. The right question is, "Can I afford **not** to invest in factory automation?"

Unfortunately, American businessmen have been trained by their financial brethren to look on investing in new equipment as a short-range tactical decision with limited objectives and financial implications. An outstanding investment, based on traditional criteria, is one with an ROI of less than three years and preferably less than two years. But payback and ROI can be calculated in many different ways, and generally are, the method chosen depending on whether the person doing the calculation is in favor of or against the investment. For example, using only one of a number of commonly accepted financial procedures, you can calculate payback on a new machine of from 5 to 30 years; depending on the technique and method used, you can show an ROI as low as 3% or as high as 30%.

Rethinking the company's values in the business, and how they apply to investments in factory automation, is a necessary part of a corporate commitment to competitiveness. Management thinking needs to be switched from capacity-driven to productivity-driven. America has too much capacity, but a shortage of competitive capacity. We need to look at more than just a fast ROI because, in reality, the decision to automate is a strategic one—not a tactical one. It affects the long-range conduct of the entire business.

Right now, most businesses bend to overstate the justification when dealing with hard, quantifiable numbers, and to underestimate the value of soft attributes such as the value of having a more flexible shop, and the potential for markedly increasing sales because of better product quality and the ability to get to market quickly.

How, for example, do you quantify and amortize automation investments over products not even invented yet but that can be accommodated on your new flexible manufacturing system with only minor changes in the computer software? Most appropriation requests, following today's traditional procedures, dwell on direct labor costs and arbitrary fixed overhead rates. They also tend to overlook, or treat superficially, such areas as utility costs, inventory costs, tool and shop supervision, and labor variances.

Direct labor today accounts for only 10% of manufacturing costs. The big hits in automation come not from replacing a man with a machine, but in controlling materials, inventory and overhead costs, and—through better quality—reducing scrap and rework as well as warranty and liability costs. What we are doing with our inflexible and unrealistic accounting procedures is creating unnecessarily difficult hurdle rates for every major investment. With that kind of thinking—that kind of atmosphere—you are not only going to miss good opportunities for implementing factory automation systems, but you may also discourage people from even making suggestions about manufacturing automation.

Employee communication. One of the major barriers to successful automation is fear of change, and with it, possible elimination of jobs. There is only one way to deal with this fear, and that is by facing it head-on. From the moment your commitment to factory automation is made, you must communicate honestly and constantly with everyone involved—both sal-

Creating harmony between workforce and computer

Integrated manufacturing cannot succeed without a harmonious relationship between human workers and the computer systems. These six steps will go a long way toward assuring their peaceful and productive coexistence.

■ Define the problem. The higher the managerial level at which the problem is identified and defined, the greater the chance of success in gaining support from top management. However, the closer the origin of the problem to the end user, the greater the chance for successful implementation.

■ Organize the environment. New systems that create an environment in which employees monitor their own achievements and goals succeed faster than those where the systems control the employees.

■ Form a team. Select a team of dedicated, enthusiastic advocates to fill the roles of catalyst, entrepreneur, sponsor, technical gatekeeper, integrator, and manager for the factory automation program.

■ Install the plan. Every successful plan has these common denominators: Implementation proceeds in a realistic fashion, building on previous successes and known strengths, and progresses at a rate that the organization can cope with.

■ Transfer ownership. Ownership is the feeling of control; it applies to everyone involved in the new technology. The key is two-way communication. You have to talk with the workforce, and then truly listen to and act on the feedback you get from the conversation.

■ Integrate with the vendors. Make sure that outside vendors understand your problems and the need to tailor equipment to your specifications. Include vendors in the planning phase so they, too, have a vested interest in the achieving your goals through sale of their products.

Flexible machining system at GE's Erie, Pa. locomotive plant features nine CNC machine tools and a fixture setup station located on both sides of a 212-ft long chain-driven transporter. The automated transporter is controlled by a programmable controller interfaced with an executive computer, and delivers traction motor frames weighing up to 2,500 lb to the 21 load/unload stations in the system.

aried and hourly employees—telling them about what is going to happen and why it needs to happen. This sounds simple, but unless the communication program is carried out professionaly, with full backing of management— it can seriously damage prospects for success.

Once you have made everyone aware of why the factory automation program is needed, how it will increase output in terms of quality and quantity, and how it will make the business more competitive, you need to work on acceptance. Convince your staff, your workforce, and your community that it is necessary and that it is really going to happen, and show them the benefits.

Encourage the participation of your knowledgeable factory floor employees. Get them involved. Solicit their advice about equipment. And send them out to examine the competitive brands of equipment they will ultimately be working with.

Finally, work on developing ownership. Ownership is not simply another overworked cliche—it is a make or break factor in factory automation. In unionized plants with a cadre of professionals—the people who pride themselves on keeping the factory running in spite of its aging, out-of-date machinery—you are going to have a difficult time of automating unless you bring the workforce into the decision-making process. You must help them see these new systems as tools vital to their own interests.

Manufacturing management today needs to think strategically about how it will be able to compete five and tens years in the future. It needs to establish a long-term corporate architecture that will still be in place—and be a powerful driving force for improving the factory—even when the people who established it are no longer with the company. Top management also needs to understand that factory automation requires a substantial fixed investment which demands a different approach to financial justification, and that the major benefits to be derived from automation are not in direct labor savings.

Successful factory automation starts with the CEO. It is a business strategy that has to be effectively communicated through the organization to all levels, and employee ownership is a must.

Factory automation is nothing short of a commitment to stay competitive—to stay in business.

"Encourage the participation of your knowledgeable factory floor employees."

"Managers need not – and should not – abandon the effort to justify computer-integrated manufacturing on financial grounds. Instead, they need ways to apply the DCF approach more appropriately."

Must CIM be justified by faith alone?

Robert S. Kaplan

When the Yamazaki Machinery Company in Japan installed an $18 million flexible manufacturing system, the results were truly startling: a reduction in machines from 68 to 18, in employees from 215 to 12, in the floor space needed for production from 103,000 square feet to 30,000, and in average processing time from 35 days to 1.5.[1] After two years, however, total savings came to only $6.9 million, $3.9 million of which had flowed from a one-time cut in inventory. Even if the system continued to produce annual labor savings of $1.5 million for 20 years, the project's return would be less than 10% per year. Since many U.S. companies use hurdle rates of 15% or higher and payback periods of five years or less, they would find it hard to justify this investment in new technology – despite its enormous savings in number of employees, floor space, inventory, and throughput times.

The apparent inability of traditional modes of financial analysis like discounted cash flow to justify investments in computer-integrated manufacturing (CIM) has led a growing number of managers and observers to propose abandoning such criteria for CIM-related investments. "Let's be more practical," runs one such opinion. "DCF is not the only gospel. Many managers have become too absorbed with DCF to the extent that practical strategic directional considerations have been overlooked."[2]

Faced with outdated and inappropriate procedures of investment analysis, all that responsible executives can do is cast them aside in a bold leap of strategic faith. "Beyond all else," they have come to believe, "capital investment represents an act of faith, a belief that the future will be as promising as the present, together with a commitment to making the future happen."[3]

But must there be a fundamental conflict between the financial and the strategic justifications for CIM? It is unlikely that the theory of discounting future cash flow is either faulty or unimportant: receiving $1 in the future is worth less than receiving $1 today. If a company, even for good strategic reasons, consistently invests in projects whose financial returns are below its cost of capital, it will be on the road to insolvency. Whatever the special values of CIM technology, they cannot reverse the logic of the time value of money.

Surely, therefore, the trouble must not lie in some unbreachable gulf between the logic of DCF and the nature of CIM but in the poor application of DCF to these investment proposals. Managers need not – and should not – abandon the effort to justify CIM on financial grounds. Instead, they need ways to apply the DCF approach more appropriately and to be more sensitive to the realities and special attributes of CIM.

Technical issues

The DCF approach most often goes wrong when companies set arbitrarily high hurdle rates for evaluating new investment projects. Perhaps they believe that high-return projects can be created by setting high rates rather than by making innovations in product and process technology or by cleverly building and exploiting a competitive advantage in the marketplace. In fact, the discounting function serves only to make cash flows received in the future equivalent to

Mr. Kaplan is Arthur Lowes Dickinson Professor of Accounting at the Harvard Business School and a professor of industrial administration at Carnegie-Mellon University, where for six years he was dean of the business school. His first article for HBR, "Yesterday's Accounting Undermines Production" (July-August 1984), was a McKinsey Award winner.

cash flows received now. For this narrow purpose—the only purpose, really, of discounting future cash flows—companies should use a discount rate based on the project's opportunity cost of capital (that is, the return available in the capital markets for investments of the same risk).

It may surprise managers to know that their real cost of capital can be in the neighborhood of 8%. (See Part I of the *Appendix* at the end of the article.) Double-digit hurdle rates that, in part, reflect assumptions of much higher capital costs are considerably wide of the mark. Their discouraging effect on CIM-type investments is not only unfortunate but also unfounded.

Companies also commonly underinvest in CIM and other new process technologies because they fail to evaluate properly all the relevant alternatives. Most of the capital expenditure requests I have seen measure new investments against a status quo alternative of making no new investments—an alternative that usually assumes a continuation of current market share, selling price, and costs. Experience shows, however, that the status quo rarely lasts. Business as usual does not continue undisturbed.

In fact, the correct alternative to new CIM investment should assume a situation of declining cash flows, market share, and profit margins. Once a valuable new process technology becomes available, even if one company decides not to invest in it, the likelihood is that some of its competitors will. As Henry Ford claimed, "If you need a new machine and don't buy it, you pay for it without getting it."[4] (For a more realistic approach to the evaluation of alternatives, see Part II of the *Appendix* at the end of the article.)

A related problem with current practice is its bias toward incremental rather than revolutionary projects. In many companies, the capital approval process specifies different levels of authorization depending on the size of the request. Small investments (under $100,000, say) may need only the approval of the plant manager; expenditures in excess of several million dollars may require the board of directors' approval. This apparently sensible procedure, however, creates an incentive for managers to propose small projects that fall just below the cut-off point where higher level approval would be needed. Over time, a host of little investments, each of which delivers savings in labor, material, or overhead cost, can add up to a less-than-optimal pattern of material flow and to obsolete process technology. (Part III of the *Appendix* shows the consequences of this incremental bias in more detail.)

"I still think 'Buyout' is not a proper name for a dog."

Introducing CIM process technology is not, of course, without its costs. Out-of-pocket equipment expense is only the beginning. Less obvious are the associated software costs that are necessary for CIM equipment to operate effectively. Managers should not be misled by the expensing of these costs for tax and financial reporting purposes into thinking them operating expenses rather than investments. For internal management purposes, software development is as much a part of the investment in CIM equipment as the physical hardware itself. Indeed, in some installations, the programming, debugging, and prototype development may cost more than the hardware.

There are still other initial costs: site preparation, conveyors, transfer devices, feeders, parts orientation, and spare parts for the CIM equipment. Operating and maintenance personnel must be retrained and new operating procedures developed. Like software development, these tax-deductible training and education costs are part of the investment in CIM, not an expense of the periods in which they happen to be incurred.

Further, as some current research has shown, noteworthy declines in productivity often accompany the introduction of new process technology.[5] These productivity declines can last up to a year, even longer when a radical new technology like CIM is installed. Apparently, the new equipment introduces severe and unanticipated process disruptions, which lead to equipment breakdowns that are higher than expected; to operating, repair, and maintenance problems; to scheduling and coordination difficulties; to revised materials standards; and to old-fashioned confusion on the factory floor.

We do not yet know how much of the disruption is caused by inadequate planning. After investing considerable effort and anguish in the equipment acquisition decision, some companies no doubt revert to business as usual while waiting for the new equipment to arrive.

Whatever the cause, the productivity decline is particularly ill timed since it occurs just when a company is likely to conduct a postaudit on whether it is realizing the anticipated savings from the new equipment. Far from achieving anticipated savings, the postaudit will undoubtedly reveal lower output and higher costs than predicted.

Tangible benefits

The usual difficulties in carrying out DCF analysis – choosing an appropriate discount rate and evaluating correctly all relevant investment alternatives – apply with special force to the consideration of investments in CIM process technology. The greater flexibility of CIM technology, which allows it to be used for successive generations of products, gives it a longer useful life than traditional process investments. Because its benefits are likely to persist longer, overestimating the relevant discount rate will penalize CIM investments disproportionately more than shorter lived investments. The compounding effect of excessively high annual interest rates causes future cash flows to be discounted much too severely. Further, if executives arbitrarily specify short payback periods for new investments, the effect will be to curtail more CIM investments than traditional bottleneck-relief projects.

But beyond a longer useful life, CIM technology provides many additional benefits – better quality, greater flexibility, reduced inventory and floor space, lower throughput times, experience with new technology – that a typical capital justification process does not quantify. Financial analyses that focus too narrowly on easily quantified savings in labor, materials, or energy will miss important benefits from CIM technology.

Inventory savings

Some of these omissions can be easily remedied. The process flexibility, more orderly product flow, higher quality, and better scheduling that are typical of properly used CIM equipment will drastically cut both work-in-process (WIP) and finished goods inventory levels. This reduction in average inventory levels represents a large cash inflow at the time the new process equipment becomes operational. This, of course, is a cash savings that DCF analysis can easily capture.

Consider a product line for which the anticipated monthly cost of sales is $500,000. Using existing equipment and technology, the producing division carries about three months of sales in inventory. After investing in flexible automation, the division heads find that reduced waste, scrap, and rework, greater predictability, and faster throughput permit a two-thirds reduction in average inventory levels. (This is not an unrealistic assumption: Murata Machinery Ltd. has reported that its FMS installation permitted a two-thirds reduction in workers, a 450% increase in output, and a 75% cut in inventory levels.[6])

Pruning inventory from three months to one month of sales produces a cash inflow of $1 million in the first year the system becomes operational. If sales increase 10% per year, the company will enjoy increased cash flows from the inventory reductions in all future years too – that is, if the cost of sales rises to $550,000 in the next year, a two-month reduction

Example of an FMS justification analysis

With the following analysis, one U.S. manufacturer of air-handling equipment justified its investment in an FMS installation for producing a key component:

1
Internal manufacture of the component is essential for the division's long-term strategy to maintain its capability to design and manufacture a proprietary product.

2
The component has been manufactured on mostly conventional equipment – some numerically controlled – with an average age of 23 years. To manufacture a product in conformance with current quality specifications, the company must replace this equipment with new conventional equipment or advanced technology.

3
The alternatives are:
Conventional or numerically controlled stand-alone.
Transfer line.
Machining cells.
FMS.

4
FMS compares with conventional technology as Table A shows.

5
Intangible benefits include virtually unlimited flexibility for FMS to modify mix of component models to the exact requirements of the assembly department.

6
The financial analysis for a project life of ten years compares the FMS with conventional technology (static sales assumptions, constant, or base-year, dollars) as Table B shows.

7
With dynamic sales assumptions showing expected increases in production volume, the annual operating savings will double in future years and the financial yield (still using constant, base-year, dollars) will increase to more than 17% per year.

On the basis of this analysis and recognizing the value of the intangible item (5), which had not been incorporated formally, the company selected the FMS option.

Table A

	Conventional equipment	FMS
Utilization	30 %-40 %	80 %-90 %
Number of employees needed (including indirect workers, such as those who do materials handling, inspection, and rework)*	52	14
Reduced scrap and rework	–	$ 60,000 annually
Inventory	$ 2,000,000	$ 1,100,000†
Incremental investment	–	$ 9,200,000

*Each employee costs $36,000 a year in wages and fringe benefits.
†Inventory reductions because of shorter lead times and flexibility.

Table B

Year	Investment	Operating savings	Tax savings ITC and ACRS depreciation	After-tax cash flow 50 %
0	$ 9,200	$ 900‡	$ 920	$ −7,380
1		1,428§	1,311	1,370¶
2		1,428	1,923	1,675
3		1,428	1,835	1,632
4		1,428	1,835	1,632
5		1,428	1,835	1,632
6		1,428		714
7		1,428		714
8		1,428		714
9		1,428		714
10		1,428		714

After-tax yield: 11.1 %.
Payback period: during year 5.

‡$ 900 = Inventory reduction at start of project.

§$ 1,428 = 38 fewer employees at $36,000/year + $60,000 scrap and rework savings.

¶$ 1,370 = (1,428) (1 − 0.50) + (1,311) (0.50).

in inventory saves an additional $100,000 that year, $110,000 the year after, and $121,000 the year after that.

Less floor space

CIM also cuts floor-space requirements. It takes fewer computer-controlled machines to do the same job as a larger number of conventional machines. Also, the factory floor will no longer be used to store inventory. Recall the example of the Japanese plant that installed a flexible manufacturing system and reduced space requirements from 103,000 to 30,000 square feet. These space savings are real, but conventional financial accounting systems do not measure their value well–especially if the building is almost fully depreciated or was purchased years before when price levels were lower. Do not, therefore, look to financial accounting systems for a good estimate of the cost or value of space. Instead, compute the estimate in terms of the opportunity cost of new space: either its square-foot rental value or the annualized cost of new construction.

Many companies that have installed CIM technology have discovered a new factory inside their old one. This new "factory within a factory" occupies the space where excessive WIP inventory and infrequently used special-purpose machines used to sit. Eliminating WIP inventory and rationalizing machine layout can easily lead to savings of more than 50% in floor space. In practice, these savings have enabled some companies to curtail plant and office expansion programs and, on occasion, to fold the operations of a second factory (which could then be sold off at current market prices) into the reorganized original factory.

Higher quality

Greatly improved quality, defined here as conformance to specifications, is a third tangible benefit from investment in CIM technology. Automated process equipment leads directly to more uniform production and, frequently, to an order-of-magnitude decline in defects. These benefits are easy to quantify and should be part of any cash flow analysis. Some managers have seen five- to tenfold reductions in waste, scrap, and rework when they replaced manual operations with automated equipment.

Further, as production uniformity increases, fewer inspection stations and fewer inspectors are required. If automatic gauging is included in the CIM installation, virtually all manual inspection of parts can be eliminated. Also, with 100% continuous automated inspection, out-of-tolerance parts are detected immediately. With manual systems, the entire lot of parts to be produced before a problem is detected would need to be reworked or scrapped.

These capabilities lead, in turn, to significant reductions in warranty expense. When General Electric automated its dishwasher operation, for example, its service call rate fell 50%. Designing manufacturability into products, making the production process more reliable and uniform, and improving automated inspection can all contribute to major cash flow savings. Although it may be hard to estimate these savings out to four or five significant digits, it would be grossly wrong to assume that the benefits are zero. We must overcome the preference of accountants for precision over accuracy, which causes them to ignore benefits they cannot quantify beyond one or two digits of accuracy.

We can estimate still other tangible benefits from CIM. John Shewchuk of General Electric claims that accounts receivable can be reduced by eliminating the incidence of customers who defer payment until quality problems are resolved.[7] Consider too that because improved materials flow can reduce the need for forklift trucks and operators, factories will enjoy a large cash flow saving from not having to acquire, maintain, repair, and operate so many trucks. All these calculations belong in a company's capital justification process.

Intangible benefits

Other benefits of CIM include increased flexibility, faster response to market shifts, and greatly reduced throughput and lead times. These benefits are as important as those just discussed but much harder to quantify. We may not be sure how many zeros should be in our benefits estimate (are they to be measured in thousands or millions of dollars?) much less which digit should be first. The difficulty arises in large part because these benefits represent revenue enhancements rather than cost savings. It is fairly easy to get a ballpark estimate for percentage reductions in costs already being incurred. It is much harder to quantify the magnitude of revenue enhancement expected from features that are not already in place.

Greater flexibility

The flexibility that CIM technology offers takes several forms. The benefits of economies of scope – that is, the potential for low-cost production of high-variety, low-volume goods – are just beginning to flow from FMS environments as early adopters of the technology start to service after-market sales for discontinued models on the same equipment used to produce current high-volume models. We are also beginning to see some customized production on the same lines used for standard products.

Beyond these economy-of-scope applications, CIM's reprogramming capabilities make it possible for machines to serve as backups for each other. Even if a machine is dedicated to a narrow product line, it can still replace lost production during a second or a third shift when a similar piece of equipment, producing quite a different product, breaks down.

Further, by easily accommodating engineering change orders and product redesigns, CIM technology allows for product changes over time. And, if the mix of products demanded by the market changes, a CIM-based process can respond with no increase in costs. The body shop of one automobile assembly plant, for example, quickly adjusted its flexible, programmed spot-welding robots to a shift in consumer preference from the two-door to the four-door version of a certain car model. Had the line been equipped with nonprogrammable welding equipment, the adjustment would have been far more costly.

CIM's flexibility also gives it usefulness beyond the life cycle of the product for which it was purchased. True, in the short run, CIM may perform the same functions as less expensive, inflexible equipment. Many benefits of its flexibility will show up only over time. Therefore, it is difficult to estimate how much this flexibility will be worth. Nonetheless, as we shall see, even an order-of-magnitude estimate may be sufficient.

Shorter throughput & lead time

Another seemingly intangible benefit of CIM is the great reductions it makes possible in throughput and lead time. At the Yamazaki factory described at the beginning of this article, average processing time per work piece fell from 35 to 1.5 days. Other installations, including Yamazaki's Mazak plant in Florence, Kentucky, have reported similar savings, ranging from a low of 50% reduction in processing time to a maximum of nearly 95%. To be sure, some of the benefits from greatly reduced throughput times have already been incorporated in our estimate of savings from inventory reductions. But there is also a notable marketing advantage in being able to meet customer demands with shorter lead times and to respond quickly to changes in market demand.

Author's note: Especially helpful comments on the preliminary draft were made by Robin Cooper and Robert Hayes (Harvard Business School), Alan Kantrow (*Harvard Business Review*), George Kuper (Manufacturing Studies Board), and Scott Richard and Jeff Williams (Carnegie-Mellon).

Increased learning

Some investments in new process technology have important learning characteristics. Thus, even if calculations of the net present value of their cash flows turn up negative, the investments can still be quite valuable by permitting managers to gain experience with the technology, test the market for new products, and keep a close watch on major process advances.

These learning effects have characteristics similar to buying options in financial markets. Buying options may not at first seem like a favorable investment, but quite small initial outlays may yield huge benefits down the line. Similarly, were a company to invest in a risky CIM-related project, it could reap big gains should the technology provide unexpected competitive advantages in the future. Moreover, given the rapid pace of technological change and the advantages of being an early market participant, companies that defer process investments until the new technology is well established will find themselves far behind the market leaders. In this context, the decision to defer investment is often a decision not to be a principal player in the next round of product or process innovation.

The companies that in the mid-1970s invested in automatic and electronically controlled machine tools were well positioned to exploit the microprocessor-based revolution in capabilities – much higher performance at much lower cost – that hit during the early 1980s. Because operators, maintenance personnel, and process engineers were already comfortable with electronic technology, it was relatively simple to retrofit existing machines with powerful microelectronics. Companies that had earlier deferred investment in electronically controlled machine tools fell behind: they had acquired no option on these new process technologies.

The bottom line

Although intangible benefits may be difficult to quantify, there is no reason to value them at zero in a capital expenditure analysis. Zero is, after all, no less arbitrary than any other number. Conservative accountants who assign zero values to many intangible benefits prefer being precisely wrong to being vaguely right. Managers need not follow their example.

One way to combine difficult-to-measure benefits with those more easily quantified is, first, to estimate the annual cash flows about which there is the greatest confidence: the cost of the new process equipment and the benefits expected from labor, inventory, floor space, and cost-of-quality savings. If at this point a discounted cash flow analysis – done with a sensible discount rate and a consideration of all relevant alternatives – shows a CIM investment to have a positive net present value, well and good. Even without accounting for the value of intangible benefits, the analysis will have gotten the project over its financial hurdle. If the DCF is negative, however, then it becomes necessary to estimate how much the annual cash flows must increase before the investment does have a positive net present value.

To see how one manufacturer justified its investment in FMS, turn to the insert entitled "Example of an FMS Justification Analysis."

Suppose, for example, that an extra $100,000 per year over the life of the investment is sufficient to give the project the desired return. Then management can decide whether it expects heightened flexibility, reduced throughput and lead times, and faster market response to be worth at least $100,000 per year. Should the company be willing to pay $100,000 annually to enjoy these benefits? If so, it can accept the project with confidence. If, however, the additional cash flows needed to justify the investment turn out to be quite large – say $3 million per year – and management decides the intangible benefits of CIM are not worth that sum, then it is perfectly sensible to turn the investment down.

Rather than attempt to put a dollar tag on benefits that by their nature are difficult to quantify, managers should reverse the process and estimate first how large these benefits must be in order to justify the proposed investment. Senior executives can be expected to judge that improved flexibility, rapid customer service, market adaptability, and options on new process technology may be worth $300,000 to $500,000 per year but not, say, $1 million. This may not be exact mathematics, but it does help put a meaningful price on CIM's intangible benefits.

As manufacturers make critical decisions about whether to acquire CIM equipment, they must avoid claims that such investments have to be made on faith alone because financial analysis is too limiting. Successful process investments must yield returns in excess of the cost of capital invested. That is only common sense. Thus the challenge for managers is to improve their ability to estimate the costs and benefits of CIM, not to take the easy way out and discard the necessary discipline of financial analysis.

References

1 This example has appeared in several articles on strategic justification for flexible automation projects. Clifford Young of Arthur D. Little has traced the example to *American Market/Metalworking News*, October 26, 1981. Other examples of the labor, machinery, and throughput savings from flexible manufacturing system installations are presented in Anderson Ashburn and Joseph Jablonowski, "Japan's Builders Embrace FMS," *American Machinist*, February 1985, p. 83.

2 John P. Van Blois, "Economic Models: The Future of Robotic Justification," Thirteenth ISIR/Robots 7 Conference, April 17-21, 1983 (available from Society of Manufacturing Engineers, Dearborn, Michigan).

3 Robert H. Hayes and David A. Garvin, "Managing As If Tomorrow Mattered," HBR May-June 1982, p. 70.

4 Quoted in John Shewchuk, "Justifying Flexible Automation," *American Machinist*, October 1984, p. 93.

5 See Robert H. Hayes and Kim B. Clark, "Exploring the Sources of Productivity Differences at the Factory Level," in *The Uneasy Alliance: Managing the Productivity-Technology Dilemma*, ed. Kim B. Clark, Robert H. Hayes, and Christopher Lorenz (Boston: Harvard Business School Press, 1985), and Bruce Chew, "Productivity and Change: Understanding Productivity at the Factory Level," Harvard Business School Working Paper (1985).

6 "Japan's Builders Embrace FMS," *American Machinist*, February 1985, p. 83.

7 John Shewchuk, "Justifying Flexible Automation."

Appendix

Getting the numbers right

Part I
The cost of capital

A company always has the option of repurchasing its common shares or retiring its debt. Therefore, managers can estimate the cost of capital for a project by taking a weighted average of the current cost of equity and debt at the mix of capital financing typical in the industry. Extensive studies of the returns to investors in equity and fixed-income markets during the past 60 years show that from 1926 to 1984 the average total return (dividends plus price appreciation) from holding a diversified portfolio of common stocks was 11.7% per year. This return already includes the effects of rising price levels. Removing the effects of inflation puts the real (after-inflation) return from investments in common stocks at about 8.5% per year (see *Table A*).*

These historical estimates of 8.5% real (or about 12% nominal) are, however, overestimates of the total cost of capital. From 1926 to 1984, fixed-income securities averaged nominal before-tax returns of less than 5% per year. Taking out inflation reduces the real return (or cost) of high-grade corporate debt securities to about 1.5% per year. Even with recent increases in the real interest rate, a mixture of debt and equity financing produces a total real cost of capital of less than 8%.

Many corporate executives will, no doubt, be highly skeptical that their real cost of capital could be 8% or less. Their disbelief probably comes from making one of two conceptual errors, perhaps both. First, executives often attempt to estimate their current cost of capital by looking at their accounting return on investment – that is, the net income divided by the net invested capital – of their divisions or corporations. For many companies this figure can be in the 15% to 25% range.

There are several reasons, however, why an accounting ROI is a poor estimate of a company's real cost of capital. The accounting ROI figure is distorted by financial accounting conventions such as depreciation method and a variety of capitalization and expense decisions. The ROI figure is also distorted by management's failure to adjust both the net income and the invested capital figures for the effects of inflation, an omission that biases the accounting ROI well above the company's actual real return on investment.

The second conceptual error that makes an 8% real cost of capital sound too low is implicitly to compare it with today's market interest rates and returns on common stocks. These rates incorporate expectations of current and future inflation, but the 8.5% historical return on common stocks and the less than 2% return on fixed-income securities are *real* returns, after the effects of inflation have been netted out.

Now it is possible, of course, to do a DCF analysis by using nominal market returns as a way of estimating a company's cost of capital. In fact, this may even be desirable when you are doing an after-tax cash flow analysis since one of the important cash flows being discounted is the nominal tax depreciation shield from new investments. I have, however, seen many a company go seriously wrong by using a nominal discount rate (say in excess of 15%) while it was assuming level cash flows over the life of their investments.

Consider, for example, the data in *Table B*, which is excerpted from an actual capital authorization request. Notice that all the cash flows during the ten years of the project's expected life are expressed in 1977 dollars, even though the company used a 20% discount rate on the cash flows of the several investment alternatives. This assumption of a 20% cost of capital most likely arose from a prior assumption of a real cost of capital of about 10% and an expected inflation rate of 10% per year. But if it believed that inflation would average 10% annually over the life of the project, the company should also have raised the assumed selling price and the unit costs of labor, material, and overhead by their expected price increases over the life of the project.

It is inconsistent to assume a high rate of inflation for the interest rate used in a DCF calculation but a zero rate of price change when you are estimating future net cash flows from an investment. Naturally, this inconsistency – using double-digit discount rates but level cash flows – biases the analysis toward the rejection of new investments, especially those yielding benefits five to ten years into the future. Compounding excessively high interest rates will place a low value on cash flows in these later years: a 20% interest rate, for example, discounts $1.00 to $.40 in five years and to $.16 in ten years. If companies use discount rates derived from current market rates of return, then they must also estimate rates of price and cost changes for all future cash flows.

Table A **Annual return series** 1926-1984

Mean annual returns

Series	**1926-1984**	**1950-1984**	**1975-1984**
Common stocks	11.7%	12.8%	14.7%
Long-term corporate bonds	4.7	4.5	8.4
U.S. Treasury bills	3.4	5.1	9.0
Inflation (CPI)	3.2	4.4	7.4

Real annual returns net of inflation

Series	**1926-1984**	**1950-1984**	**1975-1984**
Common stocks	8.5%	8.4%	7.3%
Long-term corporate bonds	1.5	0.1	1.0
U.S. Treasury bills	0.2	0.6	1.6

Part II
Measuring alternatives

Look again at the capital authorization request in *Table B.* The cash flows from alternative 1 assume a constant level of sales during the next ten years; the cash flows from alternative 5 show a somewhat higher level of sales based on a small increase in market share. The difference in sales revenue as currently projected, however, is not all that great. Only if managers anticipate a steady decrease in market share and sales revenue for alternative 1, a decrease occasioned by domestic or international competitors adopting the new production technology, would alternative 5 show a major improvement over the status quo.

Obviously, not all investments in new process technology are investments that should be made. Even if competitors adopt new technology and profits erode over time, a company may still find that the benefits from investing would not compensate for its costs. But either way, the company should rest its decision on a correct reading of what is likely to happen to cash flows when it rejects a new technology investment.

Table B **Example of a capital authorization request***

Alternative 1 **Rebuild present machines**

Year	**1977**	**1978**	**1979**	**1980**	**1981**	**...**	**1986**
Sales	$6,404	$6,404	$6,404	$6,404	$6,404	...	$6,404
Cost of sales:							
Labor	168	168	168	168	168	...	168
Material	312	312	312	312	312	...	312
Overhead	1,557	1,557	1,557	1,557	1,557	...	1,557

Alternative 5 **Purchase all new machines**

Year	**1977**	**1978**	**1979**	**1980**	**1981**	**...**	**1986**
Sales	$6,404	$6,724	$7,060	$7,413	$7,784	...	$7,784
Cost of sales:							
Labor	167	154	148	152	152	...	152
Material	312	328	344	361	380	...	380
Overhead	1,557	1,440	1,390	1,423	1,423	...	1,423

*Adapted from Robert S. Kaplan and Glen Bingham, *Wilmington Tap and Die,* Case 185-124 (Boston: Harvard Business School, 1985).

Part III
Piecemeal investment

Each year, a company or a division may undertake a series of small improvements in its production process – to alleviate bottlenecks, to add capacity where needed, or to introduce islands of automation based on immediate and easily quantified labor savings. Each of these projects, taken by itself, may have a positive net present value. By investing on a piecemeal basis, however, the company or division will never get the full benefit of completely redesigning and rebuilding its plant. Yet the pressures to go forward on a piecemeal basis are nearly irresistible. At any point in time, there are many annual, incremental projects scattered about from which the investment has yet to be recovered. Thus, were management to scrap the plant, its past incremental investments would be shown to be incorrect.

One alternative to this piecemeal approach is to forecast the remaining technological life of the plant and then to enforce a policy of accepting no process improvements that will not be repaid within this period. Managers can treat the money that otherwise would have been invested as if it accrued interest at the company's cost of capital. At the end of the specified period, they could abandon the old facility and build a new one with the latest relevant technology.

Although none of the usual incremental process investments may have been incorrect, the collection of incremental decisions could have a lower net present value than the alternative of deferring most investment during a terminal period, earning interest on the unexpended funds, and then replacing the plant. Again, the failure to evaluate such global investment is not a limitation of DCF analysis. It is a failure of not applying DCF analysis to all the feasible alternatives to annual, incremental investment proposals.

*Roger G. Ibbotson and Rex A. Sinquefield, *Stocks, Bonds, Bills and Inflation: The Past and the Future* (Charlottesville, Va.: Financial Analysts Research Foundation, 1982). The author has updated this study for returns earned during 1982-1984.

This estimate should be adjusted up or down, depending on whether the project's risk is above or below the risk of the average project in the market. A detailed discussion of appropriate risk adjustments is beyond the scope of this article. Good treatments can be found in David W. Mullins, Jr., "Does the Capital Asset Pricing Model Work?" HBR January-February 1982, p. 105, and in chap. 7-9 in Richard Brealey and Stewart Myers, *Principles of Corporate Finance,* 2d ed. (New York: McGraw-Hill, 1984).

Reprinted from ***Proceedings of the Third International Conference on FMS****, 1984.*

CONDITIONS UNDER WHICH FLEXIBLE MANUFACTURING IS FINANCIALLY VIABLE

P.L. Primrose
and
R. Leonard
University of Manchester Institute of Science and Technology, UK

SUMMARY

In order to overcome the inherent difficulties of evaluating whether a specific F.M.S. is financially viable, two computer programs have been developed at UMIST, and extensively used both in industry and to re-evaluate the published work of previous authors. The first program relates to single machine installations, such as Flexible Manufacturing Modules (F.M.M.), whilst the second, more complex program, embraces the criteria for viability of a full F.M.S. After explaining why the two situations have to be dealt with in markedly different ways, the results from the programs are analyzed over a range of situations and the conditions under which both F.M.M. and F.M.S. give an adequate D.C.F. return are stated. The paper then goes on to demonstrate how the use of a comprehensive range of computer programs, using D.C.F. principles and taking full account of all the parameters involved in a specific manufacturing situation, results in the conclusion that Flexible Manufacturing is applicable for a much wider range of applications than traditional methods of evaluation have suggested in the past.

NOMENCLATURE

C.N.C.	Computer Numerical Control
D.C.F.	Discount Cash Flow
F.M.C.	Flexible Manufacturing Cell
F.M.M.	Flexible Manufacturing Module
F.M.S.	Flexible Manufacturing System
I.R.R.	Internal Rate of Return
M.R.P.	Materials Requirement Planning
N.P.V.	Net Present Value
W.I.P.	Work in Progress

INTRODUCTION

A major research project is being conducted at UMIST regarding the financial evaluation of advanced manufacturing technology such as C.N.C. and F.M.S. The main objectives are:

1. To provide a methodology for companies who wish to evaluate potential applications and identify areas that are likely to give the best return on investment.

2. To use financial viability as a basis for identifying the most suitable applications of C.N.C. and F.M.S. This will enable technical development to be concentrated in the most important areas.

3. To develop methodologies to enable Engineers to conduct investment appraisals such that the results are acceptable to Accountants.

To achieve these objectives, 2 computer programs have been developed, the first is for single machine applications, whilst the second relates to major projects such as F.M.S. Initially the paper explains why the two situations have to be dealt with in markedly different ways, it then goes on to describe why, if condition 3 above is to be achieved, non-discounting techniques of investment appraisal, such as Payback and Accounting Rate of Return, must be ignored and in their place, I.R.R. and N.P.V. should be used.

DEFINITIONS

Many different concepts exist with regard to what constitutes a F.M.S.; for this paper, the definitions used are:

Flexible Manufacturing Module (F.M.M.):-	1 machine
Flexible Manufacturing Cell (F.M.C.):-	2 or 3 machines, linked by a management control computer
Flexible Manufacturing System (F.M.S.):-	3 or more machines, with a control computer and interlinked by a suitable material transport system.

METHODOLOGY

Two computer programs have been used in a variety of applications, both within engineering companies and for research. Although the programs have a common objective, namely the financial evaluation of capital investments, they are markedly dissimilar in philosophy and content. For example, an implicit assumption exists in the single machine program that apart from any factors directly related to the machine being evaluated, the operation of the company will remain unaffected by the purchase. With F.M.S., however, potentially the entire operation of the company can be altered by the introduction of such a system. In fact the complexity of an F.M.S. evaluation is reflected in the number of factors which had to be included in the program. Specifically, while the single machine situation could be conducted by evaluating 31 factors (1), 85 specific factors required consideration for F.M.S.

Because of the 'company wide' implications of introducing F.M.S., many authors have suggested that the benefits are considerable but intangible, with the word 'intangible' being regarded as 'synonymous' with 'unquantifiable'. Nevertheless, much effort has been devoted to identifying these potential areas of 'intangible' savings, with Ingersoll Engineers (2) writing a book containing a list of 18 advantages which may be achieved from the introduction of F.M.S. Because the inclusion of the 'intang-ibles' significantly influences the viability of a proposed F.M.S., it was necessary to develop techniques to enable all the potential benefits of such a system to be both quantified and included in an evaluation. Unfortunately the complexity of the models prevent them being rigorously described in this paper.

Concurrent with the technical considerations of evaluating F.M.S., a major study had to be conducted into the accounting implications of such systems. This revealed that in addition to the normal objection that accountants make regarding non-discounting techniques, such as Payback, the evaluation clearly demonstrated that the Payback concept was completely impractical for F.M.S. Figure 1 shows the way that cash flows are considered to change in conventional appraisals, with a fixed point in time being defined as the date when the investment actually takes place. Figure 2 however shows the way that Revenue cash-flows will actually occur with F.M.S. It is thus self-evident that capital expenditure, which is not included in the figures, is much more complex with F.M.S. because of the time scale of the project. Thus, because of the large number of factors requiring consideration, and the complexity of the accountancy procedures used, traditional evaluation procedures, using discount tables, are completely impractical. This is especially true for a company wishing to evaluate a range of possible F.M.S. applications and design configurations. Fortunately, the use of a comprehensive computer program enabled these problems to be overcome. Similarly, the use of simulation techniques, together with the results of the detailed design work associated with F.M.S. applications, provides data for the financial appraisal which is on a higher level of accuracy than normally exists for most major capital projects.

AREAS OF SAVINGS

Dempsey (3) suggests that the real success of F.M.S. exists in its way of thinking and application, not in its technology, and that many of the savings claimed for F.M.S. can be achieved without actually installing a F.M.S. system. It has also been suggested (4) that the biggest change a company can make is the move from manual machines to N.C., with the subsequent evolution to F.M.S. producing only a minor effect. Similarly, it is stated that although F.M.S. will reduce lead times, the company will have already achieved much of this reduction with the installation of C.N.C. If the previous opinions are valid, it follows that when considering the financial advantages of either F.M.M. or F.M.S., the nature of the savings must be identified and an investigation made regarding whether those savings could be achieved without the need for the major capital outlay associated with F.M.S. Fortunately, there is a clear division between the categories of savings achievable with each type, or category, of manufacturing process evolution, i.e., N.C., C.N.C., F.M.M. and F.M.S.

It is necessary at the outset of a financial evaluation to identify, in detail, any savings that may ensue and then to ascertain whether these savings could equally be achieved without the need for capital expenditure. Specifically, are the savings the result of a real change in technology or is the investment being used as an excuse to eliminate bad working practices. It is important to realize that this analysis is both different from, and additional to, that described in (1) where it is being decided if the potential saving that is being claimed will actually occur in practice. For example, there is a dichotomy in the claims regarding the savings attributable to C.N.C., with some authors suggesting that batch sizes will inevitably reduce, thus saving W.I.P., while other writers advocate an increase in batch sizes, thereby improving the relative utilization of the machine. Therefore, having first identified a specific saving that can be achieved, the question must be posed:- "Would this saving take place anyway if working practices were rearranged, rather than by making the investment?" Often the answer to this question is not a categorical 'Yes' or 'No' but a belated awareness that without the stimulus of the major investment forcing the company to change its practices, there may be insufficient motivation for the improvements in procedures to occur.

Dempsey (3) is correct with regard to the way that he dismisses some types of savings with F.M.S. on the grounds that they can be attained without investment, however, the fact remains that for many companies these savings will NOT be achieved unless a F.M.S. is installed. This statement is verified by the fact that the savings have not already been made, nor do specific plans exist to achieve the improvements in the foreseeable future. A similar situation occurred in companies who were installing Group Technology. Specifically, where major savings were made these did not result from the introduction of G.T. but rather from the concentration of management effort which brought about the necessary changes in operating procedures. Similarly with the introduction of M.R.P. systems, many of the benefits of M.R.P. could be achieved with efficient manual systems, however the strict disciplines imposed by the computer operating the M.R.P. system ensures that the required improvements in procedures are adhered to.

CATEGORIES OF SAVINGS

Companies can evolve through 3 distinct stages in the pursuit of F.M.S., namely C.N.C., F.M.M. and F.M.S., with each stage being characterized by differing categories of savings, namely:-

1. C.N.C. - Machine efficiency
2. F.M.M. - Machine utilization
3. F.M.S. - Production system changes

Flexible Manufacturing Modules (F.M.M.) are normally based on C.N.C. machines, therefore to justify the additional capital cost of F.M.M. over C.N.C., they must generate sound savings from the flexibility which results from the F.M.M. operation being independent of its operator. Similarly, because a F.M.S. can be regarded as a series of F.M.M.'s linked together, the additional capital cost of the F.M.S. must be equated with any additional savings which result from the modules being joined instead of operating in 'stand alone mode. This approach is consistent with other forms of investment appraisal, where two alternatives are mutually compared, even if one alternative is to do nothing.

The use of the F.M.S. program, acting upon actual company data, demonstrated that the complexity of the evaluation rendered it difficult to decide which cash flows should be included in the overall analysis. Thus it was only by systematically proceeding through the stages of the evaluation (1 to 4), that the alternative patterns of capital expenditure, together with their resulting incremental cash flows, could be correctly identified. It is thus important, when evaluating a F.M.S., to progress through the various levels of investment decisions in a logical sequence, namely:-

1. Does the company intend to retain a manufacturing facility?
2. Could the existing machines be replaced by C.N.C.?
3. Would it be better to invest in F.M.M. rather than C.N.C.?
4. Are the overall returns and benefits to the company improved by investing in F.M.S. rather than F.M.M.?

Assuming that the answer to question 1 is YES, then by progressing analytically through stages 2 and 3, before evaluating a full F.M.S., those savings which can be achieved by C.N.C. and F.M.M. will be identified and quantified, thus preventing them being used erroneously to strengthen stage 4 unfairly. It has been shown (5) that under certain circumstances F.M.M. may be viable yet C.N.C. could not generate an adequate return on investment. Thus a negative answer to either stages 2 or 3 should not preclude stage 4 being evaluated as the nature and extent of the savings made by each level of increasing technology are different.

MACHINE EFFICIENCY

The machining technology used in F.M.S. is normally comparable with that employed in C.N.C., thus a manufacturer (6) may use the same basic flexibility required with F.M.S. For this situation, any savings which might result from improved speeds/feeds/depth of cut, machine (e.g. a machining centre) could equally be obtained by introducing C.N.C. on its own. Likewise, because the technology of C.N.C. is well documented and understood, the savings in machining efficiency can be accurately estimated.

MACHINE UTILIZATION

C.N.C. machines are similar to manual machines in that their operation is directly related to the attendance of an operator. Thus, when the operator leaves the machine to go to the stores, or to book on/off etc., traditionally, the machine will be switched off. Hence an activity diagram for the C.N.C. machine is highly interrelated with that of its operator. The problems of the machine being operator dependent are increased when one man is running 2 machines because of the difficulties associated with one machine being stopped while the other is being set. Thus a major advantage of F.M.M. is that it offers the potential to sever the link between the 2 activity diagrams, with a measure of flexibility being the extent of which this separation occurs.

One of the problems relating to achieving the full potential of F.M.M. and F.M.S. concerns set-up. If a F.M.M. requires a conventional set-up at the end of a batch of components, the operator must spend time changing the machine or tooling before commencing manufacture of the new parts. For this situation, the activity diagrams are no longer independent, consequently, utilization will be reduced significantly. It will be shown that with F.M.S., savings are highly dependent on reducing the size of batches, potentially down to 'one-off'. This approach to economically manufacturing small batches cannot be achieved if the system has to remain idle while an operator changes set-up. Fortunately F.M.M., based on machining centres, encompasses much of the technology required to eliminate operator dependent set-up. For this situation, machines can, in theory at least, economically machine a series of one-offs. Naturally, for single machines, there will be limitations to the variety of components which can be produced due to the size of tool magazines and the range of different tools.required, this being an area where a full F.M.S. can increase the variety of components produced. F.M.M., based on turning machines, do not possess the same level of technology for eliminating Set-Up. Although Nordstrom (7) describes the adaptation of C.N.C. lathes into F.M.M., the problems of changing chuck jaws are not discussed. Thus the impression is given that considerable development is needed if lathes are to achieve the same capability as machining centers to make component Set-Up independent of the operator. F.M.S. which are designed for rotational, rather than prismatic, parts comprise different types of machines in addition to lathes, e.g. gear cutting and grinding. Full flexibility and maximum utilization will only be achieved when all the machines in the system achieve operator independence.

OPERATING SYSTEM CHANGES

Although the introduction of individual F.M.M. will achieve improvements in machine efficiency and utilization, and some systems may even be deployed in multiple machine/operator configurations, such modules will still tend to perform as conventional machines when viewed from the standpoint of the flow of work through the factory. It is only when the models are integrated within an efficient management control system, together with automated work/tool movements, that significant improvements can be made in terms of much reduced lead times and stock levels. Franks and Scholefield (8) point out the "any reduction in the uncertainty of lead times, enables lower stocks to be maintained". One objective of F.M.S. is to reduce lead time uncertainty to the minimum possible, with a measure of the success of a system being just how close to zero uncertainty is in pushed. When developing the computer program for F.M.S. evaluation, it was necessary to quantify all potential benefits. In doing this, it was found that most of the benefits which led to reduced stock levels and increased sales were directly attributable to a major shortening of product lead time which, in turn, derived from the reduction in uncertainty.

A major requirement for reducing uncertainty is the ability to schedule work through a system with the maximum of flexibility. To achieve this, it is necessary to eliminate any 'operator dependent' Set-Ups, so that work can be scheduled without reference to an 'operator interface'. Potentially, this enables components to be sequenced on a 'One-Off' basis. Basically, Flexible Manufacturing Systems can either be categorized as a 'Complementary type' (Figure 3) or as an 'Interchangeable type' (Figure 4). For the Interchangeable type, where several similar machines exist, the problem of scheduling is considerably reduced, in addition, the utilization of the system is not adversely affected by minor changes in product mix. With 'Complementary systems', however, conflicting objectives exist, namely to balance the advantages of being able to schedule components with significantly differing work content, against the need to achieve a high utilization of the expensive machines in the system. Only by being able to rigorously evaluate the financial implications of these conflicting objectives can the correct balance be achieved between flexibility and utilization.

Dempsey (3) suggested that, "the fault with current F.M.S's is that they do not cover a wide enough area of manufacture". This statement implies that to obtain the full benefits of F.M.S., it is necessary to embrace the whole manufacturing process. However, taking due account of the current levels of technology and costs, the financial returns from F.M.S. would progressively worsen as the scale of application was widened. This primarily occurs because most factory-wide applications would inevitably be of the Complementary type, generating major utilization problems. F.M.S. exhibits many of the characteristics of Group Technology, it should therefore be noted that when the first G.T. groups were selected in a factory, these tended to be successful, however, as attempts were made to extend the families, less viable 'cells' were created and the overall economies of the system deteriorated.

In addition to reducing component lead times, F.M.S. also has the potential to reduce product lead times. It is thus important at the system design and evaluation stage to differentiate between these two types of 'lead times'. In general, the overall lead times of a company's products are determined by a relatively small number of key components. If these components are produced on the F.M.S., with significant reductions in lead times, then the time needed for product delivery will be correspondingly shortened. In addition, when considering the magnitude of savings which might ensue from the introduction of a F.M.S., those resulting from reduced product delivery times are potentially far greater than just the algebraic savings from individual components.

POTENTIAL APPLICATIONS FOR F.M.S.

When considering the financial viability of C.N.C., the concepts involved are relatively simple, thus the cost of C.N.C. for a typical application can be contrasted with a standard machine and general conclusions derived. For example, it was possible to show (9) that by using correct evaluation methodologies, C.N.C. machines are viable in a much wider range of applications than previous authors had suggested. With F.M.M., although the technology involved is still at an early stage of development, it can be regarded conceptually as being similar to C.N.C., with additional savings being generated by improved efficiency and utilization. Thus a standard F.M.M. can be postulated, working in a typical application, and general conclusions may therefore be derived (5). However, for full F.M.S., no such a thing as a standard system or typical application currently exists. In addition, the financial program referred to in this paper has yet to be used in a sufficient number of companies to enable the conditions under which F.M.S. is viable to be clearly defined. However, sufficient experience does exist to give companies a set of general guidelines to assist them in deciding if potential applications for F.M.S. exist which will ultimately prove viable.

Although, at present, no such installation exists that can be classified as a standard F.M.S., an approximate guide to relative cost, defined as the price per machine center spindle, can be stated as:-

C.N.C	150,000 pounds
F.M.C.	250,000 pounds
F.M.S.	500,000 pounds

Thus, the cost of a full F.M.S. is approximately double that of a similar number of unconnected Modules, however, as development work continues on F.M.S., this difference will reduce considerably. In addition, at present within the U.K., Department of Industry grants are available for F.M.S but not for F.M.M., thereby significantly reducing the cost difference. The availability of grants, however, is unlikely to be a permanent feature of F.M.S. installations and therefore it is important to consider the intrinsic viability of such systems in the absence of Government inducements.

It has been shown (5) that the improved utilization of F.M.M. enables a Module to produce more work than a C.N.C. machine. However, when progressing from F.M.M. to F.M.S., it may occur that a 4 machine F.M.S. is LESS productive than 4 single Modules, this being due to the loss of utilization which, as suggested earlier, can be a feature of a Complementary type system. In order to minimize the loss of utilization, a balanced work load is essential for a viable F.M.S., not only in terms of work volume but also a work feature distribution which enables all the machines in the system to attain a high level of utilization, even if the component mix changes. In order to achieve the required product mix balance, it may be necessary to select certain components whose role is to act as a buffer, with the 'buffer' components being equally capable of production on either the F.M.S. or by conventional machining.

The viability of any manufacturing system is significantly influenced by the Taxation policies of the country concerned. For example, because of the nature of the U.K. tax system, any reductions in Work in Progress and Finished Stock will have a considerable effect on the tax liabilities of the cash flows included in the evaluation. The F.M.S. application should enable lead times to be reduced, thus allowing the company to operate at lower stock levels, not only with respect to component lead times and stocks, but also product lead times. Although the detailed calculations are complex, savings ensue from Materials, Variable overheads and Fixed overheads.

The ability to significantly reduce lead times can however result in a conflict between design and manufacture, components with long lead times are normally those associated with multiple operations on a variety of machines. If such components are to be produced on a F.M.S., a Complementary type system will be required; however this will exhibit the inherent utilization problems already discussed. Conversely, components which can be produced on an Interchangeable type F.M.S. system are those which exhibit an observable degree of simplicity in terms of machining operation variety; in consequence, such components offer less advantage for lead time reductions. The conclusion must thus be drawn that a viable F.M.S. application must have the potential to achieve significant reductions in lead times without the need for an unnecessary proliferation of machine types within the system.

In the same way that a standard F.M.S. does not, as yet, exist, equally there is no such part as an 'average component'; however, for purposes of discussion, a typical component can be assumed whose cost comprises:-

Material	40%
Labour + Variable Overhead	20%
Fixed Overhead	40%

The introduction of C.N.C. or F.M.M. have the capability to reduce the 20% Labour and Variable Overhead element contained within the total cost of the component, however, even a 50% improvement in this category would only induce a 10% fall in overall costs. Therefore it is by progressing from F.M.M. to F.M.S. that reductions in the other elements of component costs become possible. For example, a reduction in lead time, with its consequent fall in stock levels, allows a saving to be made in, all 3 areas of cost. Unfortunately, the saving is of a 'one-off basis within the D.C.F. calculations and therefore does not generate an annual return. Within the British tax system, the stock reductions are taken as occurring at the start of the project, therefore their D.C.F. significance can be considerable in the evaluation. For example, in one evaluation carried out in a company, the inclusion of stock reduction raised the I.R.R., after tax, from 6.1% to 15.6%. This calculation took no account of the D.O.I. grant and was for a project whose potential investment exceeded 2M pounds.

The reduction in lead times, and thus product delivery times, achieved with F.M.S., can result in increased company competitiveness and hence sales. Because this benefit recurs annually, thus reducing the 40% Fixed Overhead element of the component cost, the financial advantages from increased sales can exceed the savings made by the other factors in the analysis. An F.M.S. application which either results in increased sales or, equally importantly, prevents sales declining, can give a contribution to overhead recovery which significantly improves the overall return on investment of the system, offsetting any loss of utilization which results from providing the flexibility needed to reduce delivery time.

DIRECT LABOUR SAVINGS

A comparison of F.M.S. with existing, low level technology, based on manually operated machines, may initially indicate that F.M.S. can be justified. However, a subsequent evaluation of C.N.C. where the relationship with manual machines is assumed as a 2.5:1, will reveal that most of the direct labour reductions with F.M.S. can equally be achieved by the much lower investment needed for C.N.C. This example highlights the difficulty of attempting to justify F.M.S. on traditional cost reduction criteria, such as Direct Labour savings.

EXAMPLE APPRAISAL

To ensure that the potential benefits of F.M.S. are not overstated within the following analysis, very conservative assumptions have been made. For example, the F.M.S. will operate 24 hours a day, 5 days a week with weekends being used for maintenance etc. Equally conservative estimates are made for F.M.M. against C.N.C. For example, when considering the % of total time that is spent machining, it has been shown (5) that the utilization of C.N.C. is likely to be 58% compared with 81% for F.M.M. Similarly, the available hours per week of a module can correspondingly be increased by running during break periods, for the 78 available for C.N.C., to 85 for the F.M.M. It is also assumed that the F.M.M. are worked on the basis of 1 man/1 machine on a conventional 2 shift system. For each situation, the cost of an operator is taken as 10,000 pounds, which includes Basic Pay, Shift Premium, National Insurance and Welfare Costs. Within the above framework, a comparison of the 4 potential approaches to manufacture results in the cost relationship shown in Table 1. (Amounts marked with * are given in pounds.)

	Manual	C.N.C.	F.M.M.	F.M.S.
Cutting hours (1 F.M.M.)	4,860 x 2.5	4,860	4,860	4,860
Cutting hours (4 M/c F.M.S.)	48,600	9,440	19,440	14,440
Efficiency	58%	58%	81%	81%
Machine Hours	83,788	33,516	24,000	24,000
Number of Machines	23	9	6	4
Number of Operators	45	18	12	4
Operator Cost	450,000*	180,000*	120,000*	40,000*

	Manual	C.N.C.	F.M.M.	F.M.S.
Capital cost per M/c	Existing	150,000*	250,000*	500,000*
Total Capital Cost	Existing	350,000*	1,500,000*	2,000,000*
Incremental Savings	-	270,000*	60,000*	80,000*
Incremental Cost	-	350,000*	150,000*	500,000*

Table 1. Cost and Saving Comparison

Within the above table, any loss of utilization of the F.M.S. beyond that stated, or the F.M.M. being operated in a more efficient manner for longer hours, will correspondingly reduce the calculated advantage of F.M.S. Conversely, as the Direct labour reduces from 45 to 4 for the F.M.S., there will also be a marked reduction of Indirect Labour, such as Inspection and Supervision. However, because the size of the reduction will relate to a particular company and type of product, it has been excluded from the analysis. Although much of the saving in Indirect Labour might possibly be achieved by introducing C.N.C. or F.M.M., these lesser changes would be made on a piecemeal basis over a number of years, thus positive changes in the operation of the machine shop are unlikely to be as far reaching as when F.M.S. is introduced. Specifically, with F.M.S., the whole manufacturing organization can be considered from first principles, allowing the case for a radical restructuring of the factory to be evaluated.

CONCLUSIONS

By using a methodology based on D.C.F. principles, it has been shown (9) that the advantages of deploying advanced technology, such as C.M.C., are much greater than previous authors have indicated. Similarly, when considering the purchase of F.M.S., the marked benefits resulting from reduced lead times and improved sales, reveals these systems to be financially effective when deployed in the correct circumstances. However, firms which do not carry out a detailed financial appraisal of a potential F.M.S. application, run the following risks:-

1. They might invest several million pounds in a project which is incapable of generating an adequate return on capital.

2. They might invest in a project which does NOT represent the best potential application of F.M.S. within their company, with the project offering the greatest potential remaining unidentified.

3. They may refrain from investing in F.M.S. even though the correct application would have provided a much greater return than simply taking the easy option of replacing single machine tools.

REFERENCES

1. Primrose, P.L. and Leonard, R. "The development and application of a computer-based appraisal technique for the structural evaluation of machine tool purchases". Proceedings of I.Mech.E., Issue 3, 1984 Part B.

2. Ingersoll Engineers. "The F.M.S. Report." I.F.S. (Publications) Limited, 1982.

3. Dempsey, P.A. "New Corporate Perspectives in F.M.S." Proceedings of 2nd International Conference on F.M.S. 1983, pp.3-17.

4. "How to Justify Installing F.M.S." The Production Engineer April, 1982, pp. 30-32.

5. Primrose, P.L. and Leonard, R. "The Financial Evaluation of Flexible Manufacturing Modules (F.M.M.)" Proceedings of the International Machine Tool Conference, Birmingham, June, 1984.

6. "KTM Flexi-Matic FM 100". Sales literature from Kearney & Trecker Marwin Ltd.

7. Nordstrom, C. "The Complete Turning Cell for F.M.S.: Proceedings of 2nd International Conference on F.M.S. 1983, pp.169-182.

8. Franks, J.R. and Scholefield, H.H. "Corporate Financial Management", Gower Press, 2nd Edition 1977.

9. Primrose, P.L. and Leonard, R. "Optimising the financial advantage of using C.N.C. machine tools by use of an integrated suite of programs". Proceedings of I.Mech.E. Issue 3, 1984, Part B.

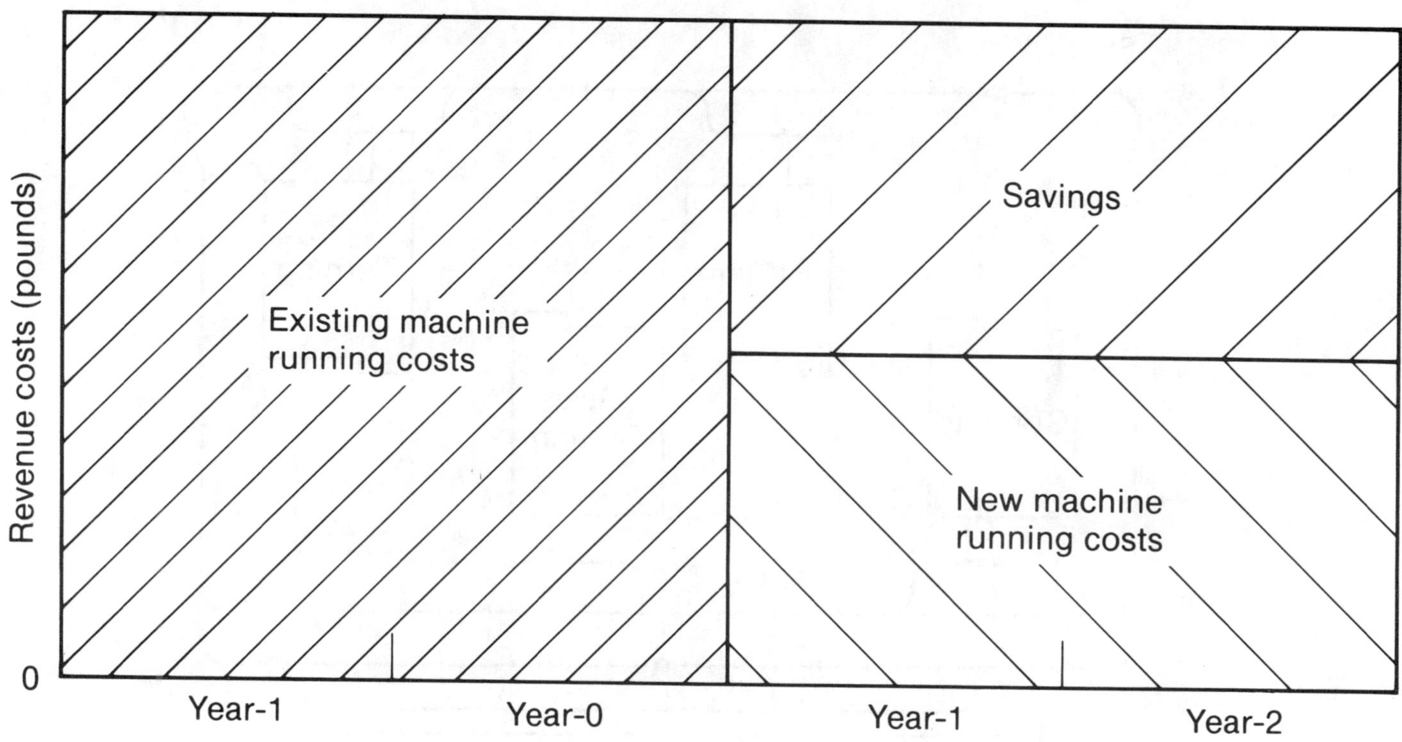

Figure 1. Conventional Appraisal

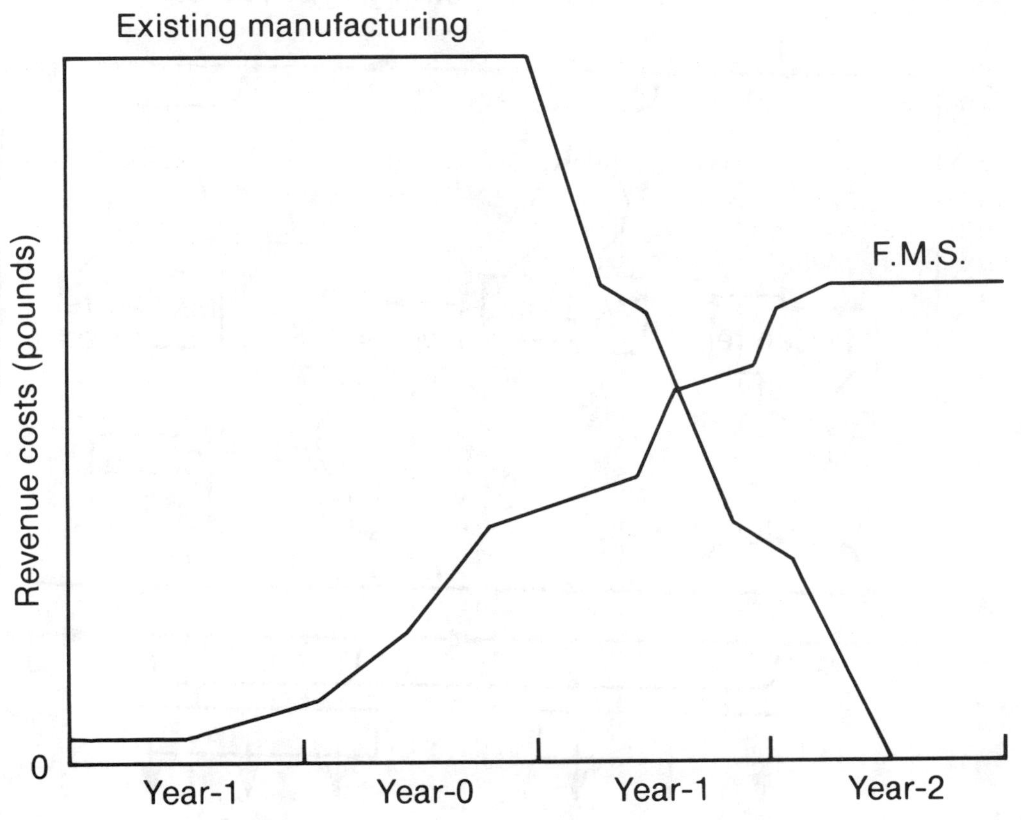

Figure 2. F. M.S. Revenue Cash Flows

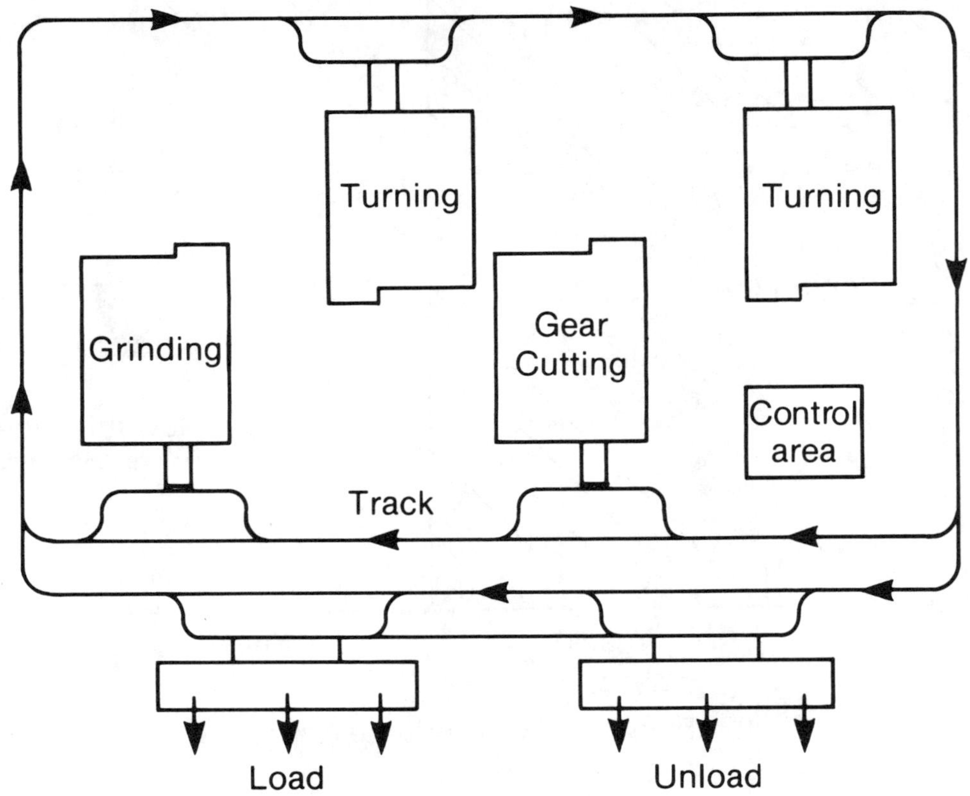

Figure 3. Complementary Type F.M.S.

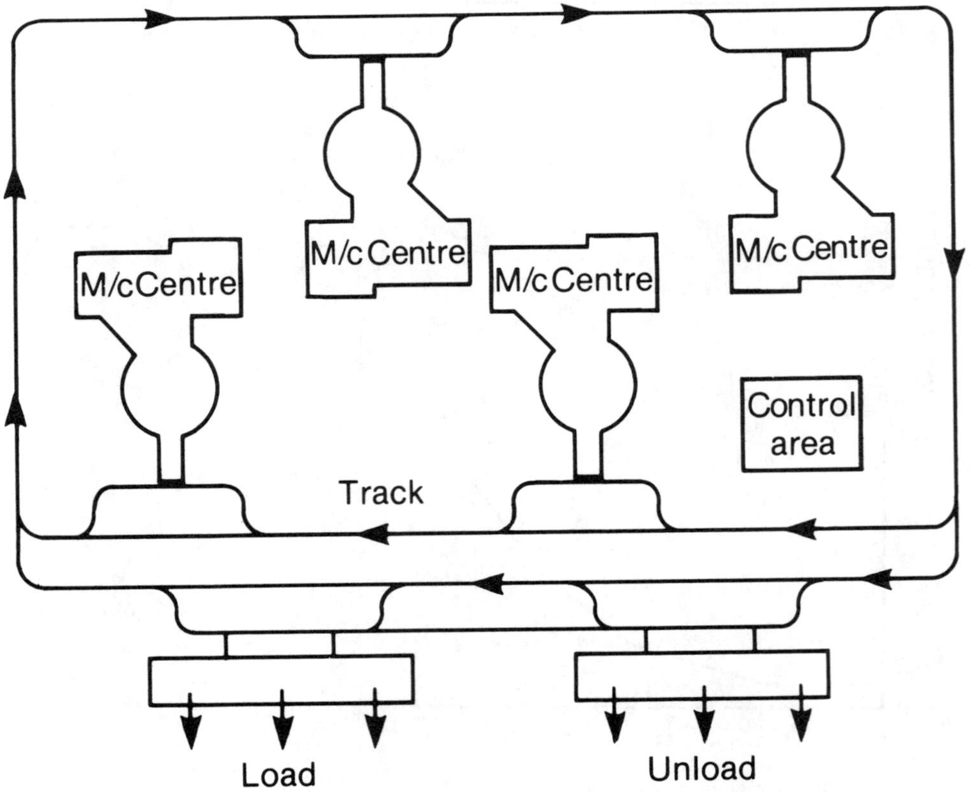

Figure 4. Interchangeable Type F.M.S.

Do's and don'ts of computerized manufacturing

Careful planning is a must to make new advances in process technology fulfill their promise to batch manufacturers

Donald Gerwin

A decade ago it was common for American managers, flush from their successes in the market, to think they had licked the problem of production for good. Today the sterling performance of global competitors has with a vengeance burst that bubble of complacency. In fact, to hold their own against world-class competition, American manufacturers have had to relearn how to compete on the basis of excellence in production. When they turned for help to automated production machinery, more often than not they found that lower unit costs and higher quality could be achieved only by sacrificing flexibility. For companies in flow- or mass-production industries, this was not a crippling problem; for batch manufacturers, however, it was a nightmare. They could get their costs in line only by giving up the ability to shift machinery quickly from one task to another–the very essence of their production system.

No wonder, then, that batch manufacturers look on recent developments in flexible computerized production technology as if they were pieces of the true cross. But in their haste and excitement to bring this equipment on-line, they are apt to overlook the many questions it raises for the rest of a production system. It is precisely to these questions that the author addresses himself.

Mr. Gerwin is professor of business administration at the University of Wisconsin-Milwaukee. His research interests are in the management of technology, especially the problems of adopting and implementing new manufacturing equipment. Mr. Gerwin is also an associate editor of Management Science *and is on the editorial board of* Human Systems Management.

Illustrations by Karen Watson.

The lagging growth in productivity of American industry has at last captured public attention. Many observers attribute the problem, at least in part, to management's reluctance to invest in the capital equipment necessary to automate production systems at fully competitive levels. Not all such investment, however, carries equivalent benefits. Of the three primary methods of manufacture – flow production of liquids and gases, mass production of discrete parts, and batch production of discrete parts – industry probably stands to gain the most from stimulating automation in the last category: batch manufacturing operations.

In both flow and mass production, which are appropriate to large-volume single products or a few standardized products, operations are continuous, follow a cost-efficient predetermined sequence, require specialized equipment, and are already heavily automated. By contrast, batch production, which applies to the manufacture of several different products each with relatively low volume and low standardization, is intermittent, follows no invariable sequence, requires general-purpose equipment, forces work to remain in process for considerably longer intervals, has higher unit costs – and is much less automated.

Improvement here would translate quickly into productivity gains, for batch manufacturing represents more than 35% of the U.S. manufacturing base and constitutes 36% of manufacturing's share of GNP. Moreover, that improvement is almost at hand. Recent developments in the technology of computer-aided manufacturing (CAM) may well provide batch manufacturers with the efficiencies long enjoyed by flow- and mass-production systems, although many companies in non-defense-related industries still do not know enough about CAM technology – its potential or its limitations – to justify the investment it requires.[1] Thus, for managers concerned with the productivity of their batch manufacturing operations, a number of critical issues require attention. What exactly is CAM? For whom is it appropriate? By what criteria is it best to evaluate a proposal to adopt CAM technology? What problems are likely to arise after installation? Do effective strategies exist with which to solve these problems? From intensive interviews in companies both here and abroad, I have begun to obtain answers to these and similar questions. Some of the answers are quite surprising.

CAM–what is it?

Normally, batch producers had two kinds of equipment from which to choose. The first, dedicated machinery such as transfer lines, is best suited for mass production of a single part at an annual volume of 20,000 units or more. This process specialization permits low unit costs, but it also inhibits flexibility. The second–unautomated general-purpose machine tools such as conventional lathes, milling machines, and drill presses–is best suited for one-of-a-kind or very small batch production of many different parts at an annual volume of, say, 200 units or less. Costs per unit tend to be high, but the flexibility of the process can accommodate engineering changes, fluctuations in demand, and shifts in product mix.

For those batch producers making several parts at annual volumes between 200 and 20,000 units each, neither alternative is quite right: general-purpose equipment is too costly, and dedicated machinery lacks flexibility. Today, however, computer-aided manufacturing offers batch producers a third choice–one with more flexibility than transfer lines and lower unit costs than general-purpose machine tools.

Forerunners

To understand the potential of CAM technology, it is important to view this new equipment as but the latest step in the evolution of numerical control (NC) machine tools. Developed after World War II, numerical control was first applied to milling and other metal-cutting operations–drilling, boring, turning, grinding, and sawing–in the production of jet aircraft. More recently, its range of application has grown to include tube bending, shearing, torch cutting, and fabric-cloth cutting.

The significance of an NC system is that it runs according to a detailed set of coded instructions on punched paper tape. Compared with conventional equipment, NC machines offer increased accuracy, flexibility, and uniformity even with highly complex parts. Design changes and special adjustments require only a change of instructions, nothing more. One helicopter manufacturer, for example, used NC to machine a transmission housing whose geometry was such that tolerances could not be met with conventional equipment. Another company using NC made in one year more than a hundred alterations in the design of a new product. These experiences are not unusual.

Problems do exist, however. NC equipment usually costs more than conventional machinery that performs the same function. It also affects the control of factory operations. Because NC systems employ parts programmers to translate engineering drawings into punched-tape code, machine operators no longer actually control machine motion, and thus less-skilled workers can do the job. Foremen, as always, are caught in the middle. They lose some control over the production process to staff and service personnel at the same time that the cost of NC machines puts them under increased pressure to boost manufacturing performance.

Machining centers, which first appeared in 1959, combine very efficiently NC operations that previously occupied separate machines; in fact, such centers can reduce setups and floor space by as much as two-thirds, direct labor by half, and in-process inventories by nine-tenths. With tool changing automatically controlled by instructions on the tape, as many as 90 tools can be changed in a few seconds.

In 1969 computer numerical control (CNC) replaced the hard-wired control unit of the NC system with a stored-program minicomputer. By 1974 less expensive microcomputers were also available to act as control units. With its programs stored in the computer's memory and not on fragile paper tape, a CNC machine or machining center is more reliable than NC equipment. It is also more flexible. Once a program is in memory, editing and revision are simple tasks, as is the addition of new systems options. The minicomputer can also be used, of course, for machine monitoring, scheduling, and the reporting of performance data.

1 Comptroller General of the United States, *Report to the Congress: Manufacturing Technology– A Changing Challenge to Improved Productivity* (Washington, D.C.: U.S. General Accounting Office, 1976).

Exhibit **Layout of a flexible manufacturing system**

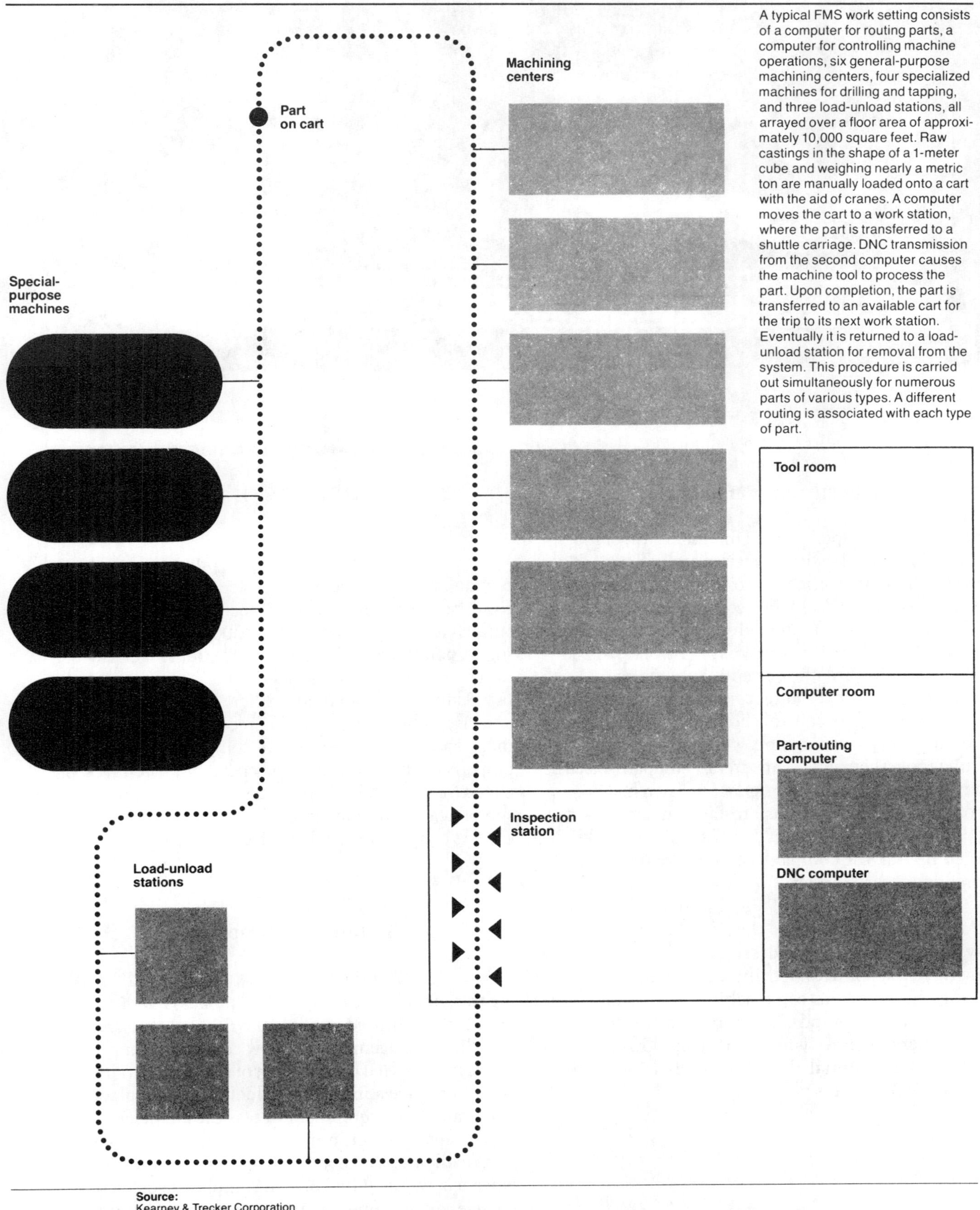

A typical FMS work setting consists of a computer for routing parts, a computer for controlling machine operations, six general-purpose machining centers, four specialized machines for drilling and tapping, and three load-unload stations, all arrayed over a floor area of approximately 10,000 square feet. Raw castings in the shape of a 1-meter cube and weighing nearly a metric ton are manually loaded onto a cart with the aid of cranes. A computer moves the cart to a work station, where the part is transferred to a shuttle carriage. DNC transmission from the second computer causes the machine tool to process the part. Upon completion, the part is transferred to an available cart for the trip to its next work station. Eventually it is returned to a load-unload station for removal from the system. This procedure is carried out simultaneously for numerous parts of various types. A different routing is associated with each type of part.

Source:
Kearney & Trecker Corporation

The current version

Computer-aided manufacturing proper, the CAM system, also first appeared around 1969. Known as direct numerical control (DNC), it consists of a battery of NC and/or CNC equipment that is connected to a central computer, which usually controls from 5 to 20 machine tools but may control up to 250.

By providing a centralized source of information and by extending computational ability, DNC helps managers control shop operations. When, for example, its parts programs were stored at the machine tools themselves, an aircraft company faced with frequent changes in component design had no way of knowing whether the latest, updated program for a part was in use. When it switched to a central DNC computer, the company could make sure that all the programs were correct.

In the early 1970s, a second generation of CAM system technology – the flexible manufacturing systems (FMS) – appeared (see the *Exhibit*). Combining DNC capabilities with automated materials handling, these systems can machine parts in any sequence at any time and can automatically reroute parts to other machines when one breaks down. This further reduces materials-handling time – in one company, from 160 minutes per part to 20.

Problems of adoption

Given the immense promise of this technology, why have American companies been slow to adopt it? One major reason, according to the Comptroller General's survey of some 200 U.S. metalworking corporations, is that one in five simply lack the necessary understanding of advanced process technology. Sadly, American industry here faces a chicken-and-egg problem. If, as another recent study suggests, many small and medium-sized companies cannot properly evaluate whether they need NC, their lack of expertise will likely keep them from buying the sophisticated equipment on which their technical experts could acquire the needed experience.[2]

Finding a champion

At the companies I interviewed, breaking this vicious circle required a "process champion" – a committed individual willing to take chances in selling management on a new concept.[3] When slumping sales led one division of an American manufacturer to invest in a new product line and replace antiquated shop equipment, it was the engineering head's tenacious commitment to purchasing an FMS that finally overcame internal resistance. Similarly, when a newly acquired British company decided to replace aging equipment and increase capacity, it was the previous head of manufacturing development (who

had just retired) whose plan for modernization sold the DNC concept to management.

But even when a process champion wins his campaign, the task force commonly appointed to investigate designs and vendors is often led astray by a company's bureaucratic structure and power centers. Usually composed of manufacturing engineers, the task force will likely treat CAM adoption as just another capital budgeting decision, which it is not. CAM's inevitable effects on accounting procedures, production scheduling, quality control, maintenance, foundry and assembly operations, plant management, and job structure make it important that workers and functional experts participate in the task force or at least consult with it.

The task force is also likely to discover that financial tools like discounted cash flow analysis do not help much. Either not enough information is available to support estimates of future net returns or the problem of how to quantify the benefits of flexibility – especially where new products are involved – makes very difficult a comparison between, say, a transfer line that machines specified parts at low cost and a CAM system that machines the same parts at higher cost but that can also produce other unspecified parts in the future.

2 George P. Putnam, "Why More NC Isn't Being Used," *Machine and Tool Blue Book*, September 1978, p. 98.

3 Alok K. Chakrobarti, "The Role of Champion in Product Innovation," *California Management Review*, Winter 1974, p. 58.

Difficult, too, is quantifying the benefits of centralizing operating information in a computer as opposed to collecting it at various points in the shop. If the task force is under the gun to demonstrate short-term returns, these advantages will not receive a proper weighting. One sales manager for a leading American vendor wondered aloud to me whether he should de-emphasize flexibility in his selling efforts because so many prospective customers were concentrating only on short-run returns.

When faced with pressures for concrete, short-term results, a project champion on a task force should:

1 **Reduce the visibility** of CAM proposals by including them in capital improvement programs involving scores of machines.

2 **Project a confident image.** Unswerving commitment can influence managers who lack information on costs and benefits.

3 **Maintain credibility.** Once managers lose faith in the recommendations of their technical experts, they will review all details intensively and more than likely wind up rejecting proposals.

4 **Appear to be rational.** Preparing rudimentary financial analyses of CAM and a written evaluation of its merits – however inexact such statements must be – helps satisfy corporate guidelines for rationality in capital budgeting.

Avoiding the numbers trap

As Robert H. Hayes and William J. Abernathy, two authorities on process technology, have warned, an undue emphasis on analytic techniques biases decisions against the purchase of advanced manufacturing equipment that promotes long-term technical superiority.[4] Among the companies I've studied, those that have adopted CAM technology did so on the basis of experience rather than estimated profits. At least three of them did not even conduct sophisticated financial analyses.

One made its selection on the basis of the technology's flexibility, reasoning that a new product line was bound to involve considerable design change and that an FMS could accommodate these changes more readily than could transfer equipment. Another was influenced by its favorable experience with stand-alone NC equipment, by the proven superiority of the NC concept in the aerospace industry, and by a work-piece analysis that showed that an integrated system would reduce waiting times.

Implementing CAM technology

The unfamiliar complexity of an integrated CAM system is virtually guaranteed to produce novel problems when it comes to implementation. My interviews clearly show that bringing a CAM system on-line has major effects on the blue-collar workers who operate it, the staff who control it, and the managers responsible for strategy development.

Worker attitudes

My colleague Mel Blumberg and I found that most blue-collar participants in the American company's FMS – especially those who were loaders and operators – felt that their jobs were stressful and offered little motivation. In particular, they regretted the lack in their work of both autonomy (the opportunity to exercise discretion) and task identity (the chance to complete an identifiable piece of work). At the same time, they had a great need for personal growth and development that was not met at work. Not surprisingly, then, most workers (with the exception of foremen and repairmen) were dissatisfied with several important aspects of their jobs: comfort (as well as challenge and security), co-worker relations, adequacy of resources, clarity of expectations about performance, promotion opportunities, and financial rewards.

If these findings are typical, they indicate a pressing need to redesign the jobs in CAM systems. One possibility is a joint work group that does away with separate jobs for operators and loaders. Given the self-contained nature of tasks in an integrated system, the group can as a whole be responsible for loading, monitoring, unloading, routine repairs, and tool setting. When each member can participate in several tasks and in decisions on such matters as job

rotation, motivation is apt to increase and dissatisfaction to decline.

Management & control

Another common effect of CAM system technology is to call into question the validity of the traditional standards for monitoring quality and financial performance. One quality control unit, for example, wanted quickly to identify the source of defects in order to limit damage to the smallest number of expensive parts. But because quality checks could be made only at the end of the machining process or between machining sequences, two hours might elapse before the unit caught a defective part. Moreover, because defects could arise from problems in the machine tools, the computers, the loading and materials-handling systems, or the parts themselves, the unit had such a hard time knowing where to look that it finally lowered the original quality standards.

CAM system implementation can also play havoc with a factory's accounting standards. Since direct labor hours do not vary with the cost of the part being processed, cost standards must be restated in terms of machining hours while the rest of the shop maintains direct labor hours. Manufacturing managers cannot, therefore, rely on their informal procedures (based on experience with direct labor hours) to control the operation of the new system.

Further, little of the usual data for calculating standard costs is available. For so new a process, no factory in the country can yet provide all the needed historical information. Consequently, standards have to be set largely by intuitive estimate, which experience has shown to be an unreliable benchmark for such major cost components as rework and maintenance. By the same token, revised accounting procedures and radically new process capabilities can make a comparison between data on pre- and post-adoption costs virtually useless.

Living with flexibility

Those batch manufacturers who switch from transfer lines to CAM systems invert the normal developmental course of productive units (as sketched by Abernathy and others) from a fluid to a rigid state – that is, from customized products made on general-purpose equipment to standardized products made on specialized equipment.[5] As a result, the primary strategic significance of CAM lies in its potential for reversing the trend toward more cost-efficient but inflexible productive units. This potential stems from its ability to loosen the ever-tighter integration of product with process.

Unfortunately, manufacturers have to date only a limited understanding of the problems

4 Robert H. Hayes and William J. Abernathy, "Managing Our Way to Economic Decline," HBR July-August 1980, p. 67.

5 William J. Abernathy, *The Productivity Dilemma* (Baltimore: Johns Hopkins University Press, 1978); Robert H. Hayes and Steven C. Wheelwright, "Link Manufacturing Process and Product Life Cycles," HBR January-February 1979, p. 133; and Robert H. Hayes and Steven C. Wheelwright, "The Dynamics of Process-Product Life Cycles," HBR March-April 1979, p. 127.

Suggested reading

Advisory Committee on Industrial Innovation, Final Report (Washington, D.C.: Department of Commerce, 1979).

A.J. Burge and R.E. Goforth, *Survey of Numerical Control Equipment Application in Texas Manufacturing Plants* (College Station, Tex.: Texas Engineering Experiment Station, Texas A&M University, 1976).

Nathan H. Cook, "Computer-Managed Parts Manufacture," *Scientific American*, vol. 232, 1975, p. 22.

John J. Hughes, George K. Hutchinson, and Kenneth E. Gross, "Flexible Manufacturing Systems for Improved Mid-Volume Productivity," in *Understanding Manufacturing Systems* (Milwaukee, Wisc.: Kearney & Trecker Corporation, 1976).

George K. Hutchinson, "Advanced Batch Machining Systems" (Working paper, School of Business Administration, University of Wisconsin at Milwaukee, 1979).

Alan M. Kantrow, "The Strategy-Technology Connection," HBR July-August 1980, p. 6.

involved in moving backward to more fluid – and thus more uncertain – conditions.[6] With transfer lines, for example, calculating production time per unit is a fairly easy matter; with a CAM system, production depends on a host of interrelated factors, the effects of which are hard to know in advance.

Moreover, managers are not even agreed among themselves about what flexibility means. My interviews show they use the term in at least five different senses:

1 **Mix flexibility.** The processing at any one time of a mix of different parts loosely related to each other.

2 **Parts flexibility.** The addition of parts to the mix and removal of parts from the mix over time.

3 **Routing flexibility.** The dynamic assignment of parts to machines – that is, the rerouting of a given part if a machine used in its manufacture is incapacitated.

4 **Design-change flexibility.** The fast implementation of engineering design changes for a particular part.

5 **Volume flexibility.** The accommodation of shifts in volume for a given part.

The potential for conflict among the different sources of flexibility makes it imperative for a company to decide early on which kinds of flexibility it values most. Setting these priorities accurately, however, requires a prior strategic consensus formulated by top management. Whether, for example, a company needs parts flexibility will depend to some extent on its plans to engage in new-product development. Without a firm decision on a new-product orientation, top management simply cannot give its manufacturing operations a well-defined mission.

One manufacturer in my sample made a strategic decision to invest in a new product line for which it needed new process equipment. Believing that engineering changes and perhaps even abandonment of the line might be necessary, it put a high priority on both design-change flexibility and parts flexibility.

A second company, planning for a gradual buildup in production, stressed volume flexibility. A third, which intended to produce a large number of products at different volume levels, sought to minimize the costs of frequent setups by focusing on mix flexibility. Each purchased a CAM system well suited to its needs.

Consider, by contrast, an American company that adopted CAM technology to manufacture a new part without understanding the equipment's potential for machining several parts. When demand for the new part fell off, management was unprepared and had to rush to come up with new tooling, fixtures, and parts programs while idle time mounted. Had the company's strategy identified mix flexibility as an important ingredient in its business environment, the problem would have been far less serious.

Curiously enough, the very flexibility of CAM may introduce some undesirable rigidities to decision making. Its ability to machine a variety of parts often makes it indispensable to a company's manufacturing process. Once CAM becomes indispensable, however, it tends to displace all other production equipment and to make subcontracting difficult. These developments, in turn, render any disruption in its normal functioning intolerable.

Building an infrastructure

All these considerations underscore the importance of the fit between CAM and the manufacturing system it serves. A good fit is the result of a company's having developed not only a coherent strategy but also a human and technical infrastructure to support its manufacturing equipment. This infrastructure includes:

Skills. Quality control personnel must have experience in diagnosing the defects that occur in CAM machinery.

Attitudes. Skilled maintenance people must be willing to work second and third shifts.

Systems and procedures. Accounting systems must adapt to the peculiarities of CAM machinery.

The state of a company's support system is closely related to its stage of manufacturing development. The more experience a company has with NC – converting its cost standards from direct labor to machine hours, for example – the more its support activities will be able to cope with additional advances.[7]

The companies I studied varied markedly in their exposure to NC and therefore in their degree of infrastructure development. This had a profound impact on the nature of their implementation plans. One British company, with almost no NC experience except with a balky NC lathe installed in one of its shops, had to forge from scratch a comprehensive human-development and technical-development plan to accompany installation of CAM equipment. A German manufacturer, by contrast, was able to install its CAM system immediately in its main machine shop because the shop already had experience with NC. Quite different was the second British company in my sample, which planned to establish its CAM system in an autonomous unit staffed with new personnel in a new location. With virtually no NC and no infrastructure, the company thought it necessary to insulate the system from the attitudes of existing foremen and workers, who were dedicated to hand skills.

Developing an infrastructure is an easier task when management installs CAM equipment in stages. This approach keeps problems to a manageable size and allows the company time to gain experience in solving problems. In fact, all the companies that I studied have installed or are installing both machinery and software in stages. As a rule, companies starting from scratch may want to prove out one machine at a time, while companies with NC experience may prefer to install an integrated system in modules.

No experience with stand-alone equipment, however, can fully prepare a manufacturer to cope with the complexities of a CAM system. Indeed, my interviews suggest that there is a qualitative leap in complexity between stand-alone machines and integrated systems. Novel problems in control and strategy are compounded by the lack of opportunity to prove out a large integrated system before installation. Later, when manufacturing people press to use the equipment, support personnel must learn about the new system while it is in daily use.

This leap in complexity also prompts some companies to rely too heavily on vendors to solve problems, even though vendors are not completely knowledgeable about what new technology can and cannot do. Turnkey projects, in which vendors assume all responsibility for making the system work, are particularly dangerous. Building up in-house experience is critical. How to avoid the pitfalls of dependence? Begin to worry the very moment your manufacturing people tell you with reference to some problem or other, "Don't worry, that's the vendor's responsibility."

Research methodology

My research included detailed interviews with 35 managers – in the United States, Great Britain, and West Germany. Meanwhile, my colleague Jean-Claude Tarondeau obtained written responses to the same interview questions from a French company. Also, a questionnaire tapped workers' reactions in one American company. H. Thomas Klahorst of Kearney & Trecker provided helpful advice throughout the project.
Among the companies studied were:

1
A diversified American manufacturer (1980 sales, $2 billion) that has a division making a product line for which the major housings are machined on a flexible manufacturing system.

2
A British producer of medium-sized and large electrical motors and generators (1980 sales, $86 million) that was in the process of installing a direct numerical control system for the machining of prime components.

3
A British manufacturer of plain bearings for industrial engines that has a rudimentary direct numerical control system for machining thin wall bearings.

4
A German aircraft manufacturer (1980 sales, $420 million) that was installing a flexible manufacturing system designed to machine over 200 different parts.

5
A German producer of transmission systems for industrial and agricultural vehicles (1979 sales, $750 million) that is developing its own flexible manufacturing system to machine gears and other rotary parts.

6
A French manufacturer of industrial vehicles that began installing a flexible manufacturing system for gear boxes in 1981.

6 William J. Abernathy, Kim B. Clark, and Alan M. Kantrow, "The New Industrial Competition," HBR September-October 1981, p. 68.

7 John Ettlie, "Technology Transfer – From Innovators to Users," *Industrial Engineering*, June 1973, p. 16.

Do's & don'ts

From my interviews I have compiled a summary list of "do's" and "don'ts" for management to consider when adopting and implementing CAM technology:

Do

Identify knowledgeable technical people and provide them with the time and resources to evaluate alternatives.

Bring all potentially affected parties–including staff, operating managers, and workers–into the adoption decision.

Develop an infrastructure that will be ready to support a CAM system as it is installed.

Install in stages.

Consider organizing a CAM system work force on a group basis, with job switching allowed and encouraged.

Don't

Bank on analytic financial techniques to provide the last word on whether to adopt.

Expect that the standards used to control CAM operations will be as trustworthy as those for conventional equipment, at least for the first few years.

Consider CAM just another collection of equipment with no implications for strategy.

Allow CAM to become indispensable in the production process.

Expect vendors to be completely knowledgeable about operating problems in a CAM system.

This checklist cannot, of course, provide all the answers, but it can give you a sensible starting point for dealing effectively with advanced manufacturing technology. The rest is up to you.

Strategic Planning for Factory Automation by the Championing Process

JACK R. MEREDITH

Abstract - The implementation of "factory of the future" manufacturing technologies has tested to the limit our ability to implement strategic change--and largely failed. The traditional concept of strategic planning conducted through an executive analysis and decision process and passed down the managerial hierarchy is clearly inadequate for these advanced, computerized technologies. A different, although not new, strategic planning process is required for the complexity of implementing changes on the scale of factory automation. Borrowing from the concept of the "product champion," a process championing procedure for strategic change is postulated. This approach is then described and examples are given to illustrate the procedure. Last, some implementation guidelines are given to facilitate the use of this process for achieving strategic level change.

I. FACTORY AUTOMATION'S IMPLEMENTATION PROBLEMS

The difficulties and complexity of implementing the much-heralded "factory of the future" are so well publicized that the major problems now have their own terminologies: "islands of automation," "short-term management attitudes," "automating your mistakes," "equipment incompatibility," conducting the "as is" study, "automate, emigrate, or evaporate," "justification dilemmas," and so on. As one example, the failure rate for implementing manufacturing resource planning systems is commonly quoted [1] as 80 percent.

It is clear that achieving a strategic goal of implementing a computer-integrated manufacturing (CIM) facility will be accomplished neither quickly nor soon. But what is it about CIM that makes factory automation so much more difficult than other, seemingly similar manufacturing projects such as quality circles, statistical process control, or even implementing an MRP (material requirement planning) program?

A recent study [2] indicates that the broad extent of change required for CIM goes far beyond that required for previous manufacturing projects, such as installing a robot or NC machine. The coordination needed, not just for the manufacturing functions, such as purchasing quality control, and scheduling, but also all the other company functions--engineering, finance, marketing, accounting, human resources--is at least an order of magnitude greater than ever needed before, particularly for manufacturing projects. As stated recently [3]:

> "Tomorrow's manufacturing managers will need the broader skills necessary to design and operate manufacturing systems, rather than manufacturing plants. ... Managers also are going to need the skills to work with computer-based information systems that integrate the manufacturing information systems with those of other functional areas, such as marketing, finance, and strategic planning."

Clearly, achieving computer-integrated manufacturing must be viewed as a strategic task, with companywide requirements and ramifications. But strategic level planning, whether for manufacturing automation or some other strategic goal, has been under attack recently for continually being a planning process but never attaining the implementation stage. Firms are experimenting with a number of solutions: changing the strategic planning staff, placing strategic planning in the hands of those who must implement the plans, dropping the strategic planning process, and so on.

Based on the experience of firms that have successfully implemented factory automation such as General Electric, Ingersoll Milling Machine, and John Deere and Co., these are probably the wrong solutions for achieving CIM and, quite likely, for achieving any other type of strategic change. It appears that any strategic level change as extensive and complex as CIM requires the spearheading of a project "champion" or else it will fail.

II. THE CHAMPIONING PROCESS

The concept of a "champion" is largely attributed to Schon who states [4] that this champion

> "identifies with a new development (whether or not he made it), using all the weapons at his command, against the funded resistance of the organization. He functions as an entrepreneur within the organization, and since he does not have official authority to take unnecessary risks ... he puts his job in the organization (and often his standing ...) on the line. ...He (has) great energy and capacity to invite and withstand disapproval."

Schol is speaking here of _product_ champions, but we believe that his description fits that of a _process_ champion, or any other kind of project champion, as well. As with the product champion, the process champion perceives the potential future in a new process and adopts the successful implementation of the process as a personal crusade, in spite of the possible risk, not only to the project but also to his own career. The process champion typically has a "vision" of the firm's future were it to adopt the process innovation and strives to make this vision a reality.

It is difficult to capture the fervor that champions have for their idea. Tom Peters (of _In Search of Excellence_) describes them as "fanatics," and notes that Peter Drucker calls them "monomaniacs," referring to their single-minded drive for project success. Peters further notes [5] that "The people who are tenacious, committed champions are often royal pains in the neck. ... They must be fostered and nurtured--even when it hurts."

A. Factory Automation Champions

In our studies of successful automation projects, we have been intrigued by the consistent presence of a "champion" for the project, typically at the vice-presidential or similar level. It became clear that it was this person who had the "vision" of where the firm should be headed and took the responsibility to see that this vision materialized. The task included considerable salesmanship, of both superiors and subordinates, and significant risk since the extent of change was so extreme, particularly compared to previous manufacturing projects.

Others have also noticed this phenomenon; some examples [5]-[9] are quoted in Table 1. In our experience, and based on the above, it would seem appropriate to conclude that for strategic level manufacturing systems change (that is, large, extensive, complex change), a "process champion" is a virtual necessity.

B. Championing Roles

But are champions always the same people, or is it possible, as one article [5] implies, just a "role" that someone happens to be filling? In fact, successful champions do not seem to have usually been a champion before. The champion's belief is more of an all-consuming passion than a bandwagon interest. Thus, the champion's position is indeed akin to a "role," but it is a deadly serious one.

For that matter, research [9]-[11] indicates that there may be a number of roles that are necessary for successful implementation of strategic level changes in organizations. Some of these are commonly known by the following terms:

-- Creative originator - an engineer, scientist, or similar such person who is the source of the idea.

-- Entrepreneur - the person who adopts the idea and "sells" the project throughout the organization, eventually pushing it to success.

-- Sponsor - a senior level manager who obtains the resources, coaches the team when problems arise, and protects the project when necessary.

-- Project Manager - the administrator of the project who handles the operational planning and day-to-day details.

However, the most common realization of the championing effect is seen in small- to medium-sized firms where the champion fills all three of the bottom roles listed above (the role of "creative originator" is frequently external to the firm). Examples of this situation abound where "true believers" have championed the concept of JIT (just-in-time production), quality circles, CAD/CAM (computer-aided design and manufacturing), or some other such systems created or discovered elsewhere.

What is important to note here is that when all four of the roles are diffuse, as commonly occurs in large firms because people from different functional areas fill them, the championing effect does not seem to materialize to a significant extent. In these situations, the "entrepreneur" role--which is the role usually identified with the champion--does not seem to have the force or "drive" that is associated with championing. But when the entrepreneur, sponsor, and project manager roles get combined in the same person, a new chemistry seems to occur. It is this phenomenon that is being described here and the one that has significant implications for an alternative approach to strategic planning.

A note of caution is appropriate, however. The role of the idea generator (creative originator) has been much overemphasized, to the detriment of successful strategic level change in most organizations. As noted [9],[12] in the literature:

> "Our culture seems to be fixated beyond reason with the 'great idea' syndrome. We search for and reward heavily the person who devised the great idea, but ignore those who struggle to implement it. ...As long as the idea is seen as the focus of the process, rather than implementation, we will be plagued with 'implementation failures'."

"The role of creative scientist seems to be overemphasized; organizations tend to assume that having creative people on the payroll guarantees effective development of new products, new processes, and product improvements."

C. Championing in the Dynamics of the Firm

One may well question why strategic change is so difficult. But a more appropriate question here is: Why does championing succeed in strategic change when other methods fail? Schon [4] indicates, first, that people need and try to maintain stability but strategic level changes upset that stability. Thus, to maintain equilibrium, people tend to resist change or, failing that, take alternative actions that inhibit the effectiveness of the change. Even without the countervailing efforts of people, operating systems tend to maintain their own stability and changes in one part of the system automatically bring about "coping" changes in other parts of the system. Were this not the case, every change in the environment could bring the many systems by which we run our society, and our lives, to an immediate halt.

Second, strategic level change is apparently more difficult these days than previously, such as in the 1950's when markets and processes were more stable. This is particularly the case for large firms [5]: "The product champion is a creation of the age of large organizations and of the special problems of innovation which have become so noticeable in the last decade or two." More explicitly: "No ordinary involvement with a new idea provides the energy required to cope with the indifference and resistance that major technical change provokes." Thus, growth in firms automatically tends to create bureaucracy and resistance to change, thereby requiring a special force, in the form of the process champion, to push the change through.

Schon [4] also reveals why strategic level changes tend to be resisted in bureaucracies: "The formal chain of command ... is wedded to the major committments of the firm and treats radical innovation as a threat." Hence, the procedures put in place to originally encourage innovation in the firm unwittingly grown into the official obstacles to prevent innovation. The champion is thus forced to use the "informal" system in the organization, bootlegging funds and promoting the project through his or her personal contacts.

Of course, the champion may be wrong and the price of failure in such a case, more than likely, is professional suicide. It is not unknown for such champions to eventually even become martyrs to their idea [4]. The point is, the championing process does not guarantee successful projects; it only helps assure that change has a fair chance of success in the firm. Let us now look closer at how the championing approach can be incorporated into the strategic planning process.

III. STRATEGIC PLANNING THROUGH CHAMPIONING

The essence of the championing approach to strategic planning is that high-level managers knowledgeable about specific functions of the business bring their functional projects to the executive management of the firm for consideration. Thus, a Vice President of Technology, or the Manufacturing VP, may propose an automation project, or a Marketing VP a sales project. Innovation thus begins, as shown in Fig. 1, with a highly placed knowledgeable manager, rather than either a low-level expert or a top-level generalist. The

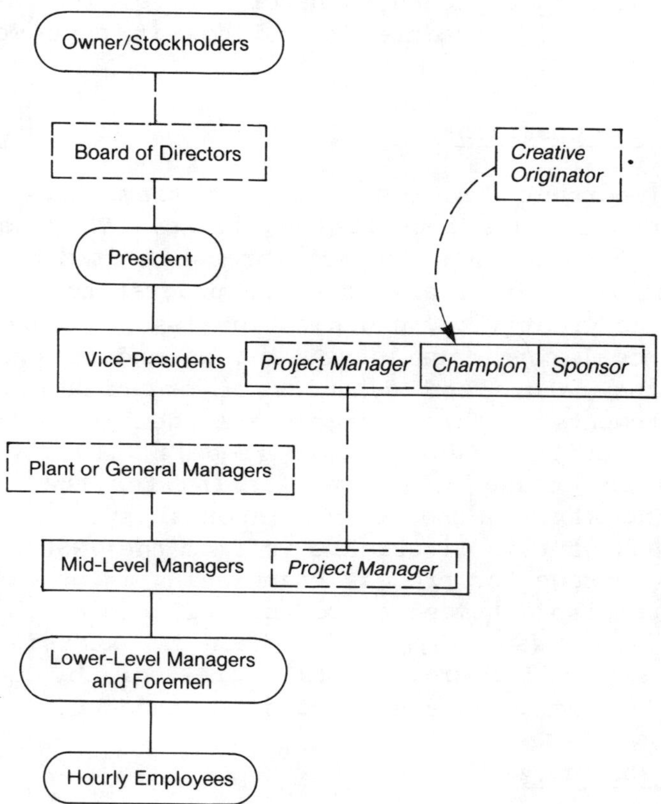

Fig. 1. The appropriate level for championing roles. Notes: (1) in smaller firms there may not be some of these levels, particularly the heavy dashed boxes; 2) the creative originator is often external to the firm, as indicated; and 3) the champion, sponsor, and project manager may or may not be the same person, depending on the firm's size.

point is, only this level of the firm, typically vice-presidential, has both the technical knowledge and the administrative knowledge to know what will be a success and what will not. Following top management approval, this person then becomes the champion of the idea and "sells" it throughout the rest of the company.

This was the approach taken at Peerless Saw Co. [13], a very small firm that conceived and implemented a computer-driven laser cutting system to drive the front end of their production process. In this case, all four roles (the "creative originator" included) were played by the champion, who happened to be the president of the firm. This was also the basic style adopted by Ingersoll of their "factory of the future" which won the Society of Manufacturing Engineers' prestigious LEAD award for automation.

But for such an approach to be effective, all parties in the process must understand what the procedure is and how it best works. This is described below.

A. Training the Champion

It is important for the champions themselves to understand the championing process and its benefits and risks. They need to be educated in how to successfully champion a project through a firm and what their roles should be. The champion should also clearly understand the "soft" aspects of championing such as creativity, organization issues, managerial control, and other such nontechnical considerations. And an understanding of both human nature as well as the nature and dynamics of organizations is needed. The major issues and elements in this category are identified in Table II.

There are quite a few articles (see [4], [9], [14]) on the product championing idea and these give some feeling for the tasks facing the champion such as making maximum use of the informal system and the value of personal contacts. Much of this literature is from the 1960's and need not be repeated here but the message for process championing is no different today.

There are also a number of books available on the project management process itself (e.g., see [15]). And there is extensive literature on implementation and project failure. A full bibliography is given in [16] that can direct the champion to issues of specific concern.

B. Training Executive Management

Top management also must understand the championing process in order to use it effectively for strategic change in their organizations. They must understand too the role they have to play and, particularly in large organizations, the importance of a "sponsor" at their level, as described earlier. To use a championing process for strategic change means foregoing the notion that the top managers will be able to create a strategic plan by themselves and then pass it down the hierarchy for implementation. Strategic change simply does not seem to occur that way.

Top management also must realize that grass roots innovation and change will not occur through the systems that have become part of the operating arms of the organization. For example, standard justification procedures consistently fail to pass the concept of computer-integrated manufacturing. Again, strategic change does not occur that way either.

And last, strategic change does not seem to come about through the efforts of strategic planning groups.

Top management should also be aware that with this type of approach to strategic change, it is often wiser to back the person, the champion, rather than the particular project [9]. That is, project costs and benefits, particularly with the widespread change they entail, are often virtually impossible to forecast accurately. This is especially true of factory automation. Thus the project go-ahead decision often rests on the reputation and vision of the champion. Approving a plan based on the reputation of its champion is not a dereliction of duty but rather a reiteration of the age-old wisdom that a firm's most valuable assets are its people.

C. Training the Subordinates

A major portion of the champion's tasks is "selling" the project to his or her subordinates, and their colleagues throughout the firm. When it comes to major change in an organization, nothing can be mandated--the people must be treated like volunteer workers and "catch" the project fever from the champion. Normally, the champion's enthusiasm, the challenge of an important task, and the chance for fulfillment can be sufficient inducements.

The champion's task will again be easier if their subordinates and the hourly and clerical employees also understand the championing process and what is expected of them. Workshops, development seminars, and training can be advantageous here.

They should also understand the pros and cons of the process. Strategic level change is extensive and impacts employees throughout the firm. This naturally engenders resistance and fear. Layoff fears must be put to rest immediately, and job changes must be developed in a positive, supportive fashion. The project must be seen as an opportunity, not a threat; otherwise it will never be a success. With more knowledge of the championing process and the typical changes that occur when this process is successful, the chances of project success are that much greater.

IV. CONCLUSION

For change as comprehensive and complex as factory automation, it has been found that a championing procedure is mandatory. This procedure, used for process changes as well as more traditional product changes, has been described in the previous pages. But possibly of more significance than just factory automation, it may well be that no strategic level change can occur in an organization without following a championing process. If this is true, it carries major implications for how we run and modify our organizations to keep them competitive and contemporary.

REFERENCES

[1] G. Hoyt, "Successes and failures in MRP user involvement," in Proc. 20th APICS Ann. Conf. (Cleveland, OH), Nov. 2-4, 1977.

[2] J. R. Meredith, "Results of the manufacturing management council study on justification procedures," Soc. Manuf. Eng., June 1985.

[3] J. Lynch and D. Orne, "The next elite: Manufacturing supermanagers," Manag. Rev., Apr. 1985.

[4] D. A. Schon, Technology and Change, Delacorte Press, 1967.

[5] "A passion for excellence," Fortune, May 13, 1985.

[6] A. Ashburn, "Flexibility, patents, and champions," editorial for Special Report: Reexamining flexible manufacturing systems, Amer. Machinist, Mar. 1985.

[7] L. Everitt, Jr., "Profile," Operations Manag. Rev., Summer 1985.

[8] S. T. Parkinsons and G. V. Avlonitis, "Management attitudes to flexible manufacturing systems," in Proc. 1st Int. Conf. FMS, 1982.

[9] E. B. Roberts, "Generating effective corporate innovation," Tech. Rev., Oct.-Nov. 1977.

[10] M. B. W. Graham and D. Leonard-Barton, "Technology transfer within the firm: A framework for R&D implementation," Synergy Conf. Proc., Soc. of Manufacturing Engineers, 1984.

[11] M. A. Maidique, "Entrepreneurs, champions, and technological innovation," Sloan Manag. Rev., Winter 1980.

[12] J. R. Meredith, "Belling the cat: An implementation failure," Operations Manag. Rev., Spring 1984.

[13] J. R. Meredith, "Peerless laser processors," case study developed under Cleveland Foundation grant Management Issues in High Technology Manufacturing Industries, Univ. of Cincinnati, Cincinnati, OH, Dec. 1984.

[14] R. M. Kanter, The Change Masters: Innovations for Productivity in the American Corporation. New York: Simon and Schuster, 1983.

[15] J. R. Meredith and S. J. Mantel, Jr., Project Management: A Managerial Approach. New York: Wiley, 1985.

[16] J. R. Meredith, "Implementation of computer-based systems," J. Oper. Manag., Oct. 1981.

Jack Meredith received the Bachelor degrees in math and mechanical engineering from Oregon State University, and received an MBA and Ph.D. degrees in business administration from the University of California, Berkeley.

He has been employed at TRW Corp., Douglas Aircraft Co., Hewlett-Packard, and Ampex Corp. Presently, he is an Associate Professor of Production/ Operations Management Programs at the University of Cincinnati, He is the former Editor of Operations Management Review, Associate Editor of the Journal of Operations Management, and is also a Series Editor in Production/ Operations Management for John Wiley & Sons as well as being an Advisory Editor. He is the author or coauthor of a number of books in the project/operations management field.

Dr. Meredith is the chair of the Publications Committee of the Manufacturing Management Council of Society of Manufacturing Engineers and a Director of the Operations Management Association.

TABLE 1
COMMENTS RECOGNIZING THE NECESSITY OF A CHAMPION

* "There was a champion behind every FMS in operation today."
* "There must be someone at the 'top' willing to try these (automation technologies), a champion for manufacturing.
* ". . . many projects would not have been successful were it not for the subtle and often unrecognized assistance of such senior people acting in the role of sponsors."
* "Formal in-house studies of research project successes at IBM always unearth the presence of a champion."
* ". . . the adoption of (the FMS depended on) the existence of a key figure in the adoption process, that of the 'process champion' . . . For instance, we found cases where the adoption process ceased. . . because the project champion had left the organization."
* "Where radical innovation is concerned, the emergence of a champion is required. . . the new idea either finds a champion or dies."
* "It is, we believe, the 'process champion' factor, or lack of it, which may explain why numerous large organizations. . . have not moved even beyond the interest stage of the adoption process of such a system."
* "(For) the truly innovative enterprise, people called champions are a necessity."

TABLE II
TYPICAL ISSUES A CHAMPION NEEDS TO UNDERSTAND

Technical	**Organizational**	**Risks**
Project elements	The championing process	Project risk
Existing process	Championing roles	Political roadblocks
Economic impacts	Human behavior	Career risk
Financial analysis	Project organizational impacts	Potential pitfalls
Nature of innovation	Capital authorization process	Implementaiton difficulties
Market environment	Group behavior	
Budget process	The informal system	

Manufacturing today needs to be viewed holistically...and that includes justifying automation

Winning Your Case for Automation

BY JAMES A. BAKER

One of the characters in the James Michener book *Space* says, "Scientists are men who dream about doing things. Engineers do them." From an author known for clarity, drama, and incredibly thorough research, this is a thought well worth holding.

No disrespect intended to scientists; they play an essential role in developing the theory and the tools. But, it's engineers who apply the theory and the tools in a hands-on way.

This applies very directly to the subject of automation, because for all that's been said about it and written about it, precious little has been done about it. The theory and the tools exist; now they need to be applied.

It's not enough for a business to decide it needs to automate. It's like losing weight. *Deciding* to lose weight doesn't do any good. You have to *do* something about it. And you can't just "lose a little around the middle"...you can't send your stomach out for a jog around the block while the rest of you sits in the office and works. You must take a holistic approach to weight loss.

By the same token, automation needs a holistic approach. New technologies must fit together with new management disciplines to make the total payback greater than the sum of the parts. Manufacturing problems must be looked at as integrated problems, not isolated problems that go away with a little technology.

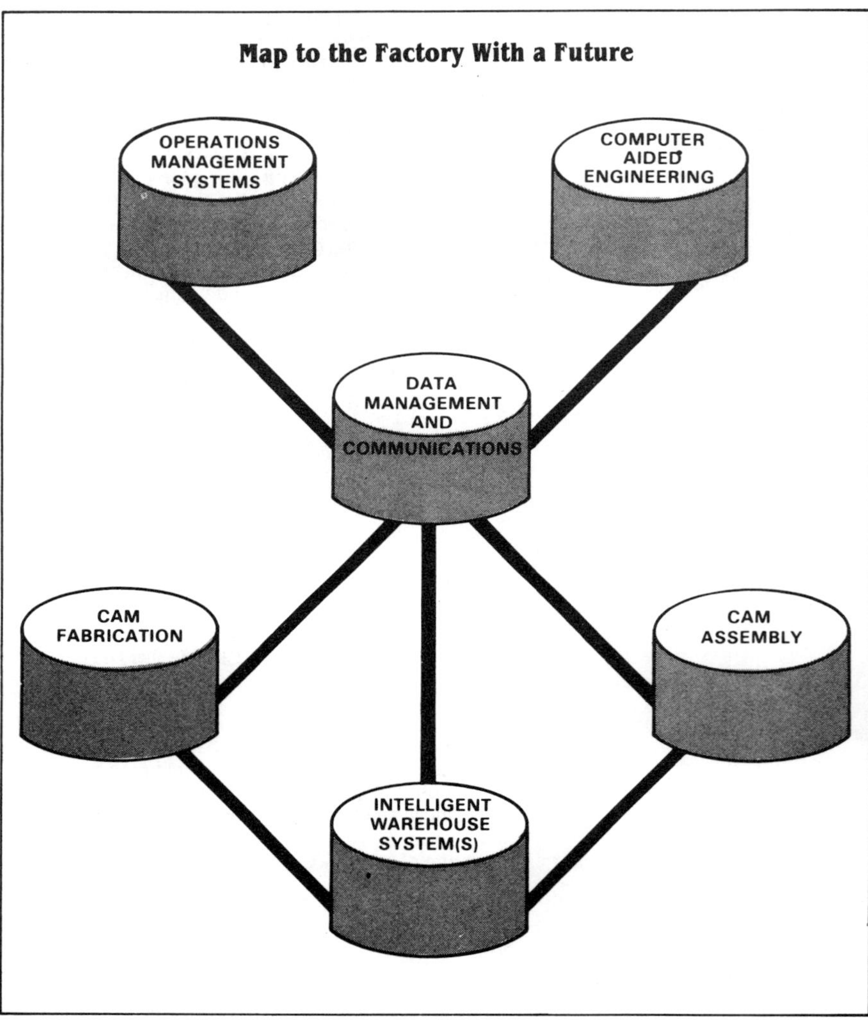

The GE approach to the factory with a future: Integrated problems getting integrated solutions.

Demand for integrated solutions

What is needed are cost/benefit formulas and measurements that go beyond the usual ROI evaluations and take into account the total impact of automation. Manufacturing engineers hold the key to making this happen.

Engineers, being technical creatures by nature, tend to think and talk in very technical ways. To win a case for automation with management, engineers need to show how the technologies that will help in manufacturing or financial areas will also help in other areas, areas such as inventory control, materials handling, market response, and so on.

A recent A. D. Little "Impact" report cites a major Japanese manufacturer who installed an FMS. The financial return over the first two years was only $6.9 million on an investment of $18 million. On the basis of conventional accounting principles, this kind of ROI could be difficult to justify.

However, by looking beyond the short-range financial implications, it makes sense. For one thing, the system reduced space requirements from 103,000 to 30,000 ft^2 (9569 to 2790 m^2). Seventy-three thousand square feet (6789 m^2) of manufacturing floor space is worth a lot and ought to be factored into the equation. Moreover, the new system reduced processing time from 35 days to 1½ days. How much is this kind of response to market demand worth? That depends on the application, but it's more than just a technical or ROI consideration.

At General Electric, it's been discovered that a carefully thought-out, planned systems approach is the only way to get at the interlinking issues involved in automation. But that is not an easy matter. It's difficult to do the planning and to factor in all the considerations for an automated system if that task has never been done before.

GE has been fortunate. It has 377 plants, and some pretty extensive automation projects have been completed in some of them. This automation systems experience is so valuable, in fact, that a large part of it has been pooled into an operating component—the Industrial Automation Systems Dept. (Charlottesville, VA).

And, what's been learned? First, there are no simple solutions; no easy answers. A well-thought-out manufacturing strategy must be planned. Second, any one part of a factory cannot be looked at in isolation. Third, organizations are *not* linear and rigid in nature, rather they are fluid and flexible. At GE the view of manufacturing is holistic, taking in the manufacturing operation in its entirety.

GE's James Baker says it's not enough for America to just be a contributor to the rest of the world's manufacturing success.

Justifying investment in automation

Here are some specific recommendations about justifying investment in flexible automation based on GE's experience.

First, don't try to justify automation through direct labor take-out. In the past, direct labor accounted for as much as 50% of the total product cost, but no longer. Today, on the average, direct labor accounts for about 12% of the cost of a product. Parts and material costs are about 50%, indirect labor 26%, and overhead costs about 12%.

Opportunities exist to cut inventory costs through techniques such as just-in-time systems and improved material handling methods. GE manufacturing engineers have reduced inventory needs and materials usage by millions of dollars.

For example, when the just-in-time inventory concept was piloted on a shaft line at one of GE's motor plants in Hendersonville, TN, the concept eliminated 60% of in-process inventories and 35% of the material handling. In addition, the manufacturing cycle was shortened by 60%.

Second, don't allow individual pieces of automation equipment to be considered in isolation from the rest of the process. Tactically, it may be impossible to justify the cost of a robot, but strategically, it may make good economic sense. If a robot can increase the percentage of time the inventory is being worked on, then a tremendous potential to reduce material and overhead costs exists.

Third, factor in the positive effects that improved product quality has on the productivity of the business. For example, since automating its dishwasher plant in Louisville, GE has experienced a one-third improvement in product quality as measured by service calls, and scrap and rework costs have dropped 40%. Also, a higher quality product substantially helps the overall marketing effort.

Fourth, don't just design an automation process to make a product; design the product so that it can be efficiently made with automation techniques. CAD/CAM is a must to integrate design and manufacturing. It not only ferrets out production problems before they occur, it also allows for evaluation of alternative materials quickly and efficiently through CAE.

Fifth, if flexibility is built-in, then it will be used. GE's flexible machining center in Erie, PA, was originally designed to make six different motor frames. Today, less than two years later, it's been expanded to handle 10 different motor frames.

Finally, know that success in this new field of automation depends heavily on a close alliance between those who know the business to be automated and those who know the business of automation. Automation is a long-term undertaking that requires long-term relationships between functions that previously had only a vague awareness of each other. Manufacturing, design, finance, order entry, sales, shipping and receiving, and the personnel department...all will be affected by automation, and all can make a contribution in making it work. A question manufacturing industries need to ask is: Do we need separate design engineering and manufacturing engineering functions, or could they be combined?

If engineers are to take technology from the laboratories to the factory floors, they cannot just sell the sizzle of bits and bytes and bauds. If the case for automation is to be won, management will have to be sold on the solid benefits that automation can bring to the business: benefits like reduced manufacturing costs, higher margins, better market response, better quality assurance, improved product design, reduced work-in-process inventory costs, and better product mix. The bottom line is not reduced labor or manufacturing costs, it's better market share.

And, this needs to be done quickly. If America does not move soon with major automation investments in its bread-basket industries, all the Silicon Valley technology in the world will only make us a contributor to the rest of the world's manufacturing success. **ME**

James A. Baker is executive vice president and sector executive of the Technical Systems Sector, General Electric Co. (Fairfield, CT).

Reprinted from ***Industrial Management****, September-October 1984.*

For Major Investment Decisions, How Much Does the Board Need to Know?

James B. Weaver

The first article of this series outlined the role of the Board of Directors in setting strategies. The second dealt with the type of information typically provided to a board in connection with a major capital expenditure. The final article of this series will propose the ways in which internal management and the Board of Directors could work together to obtain a sound, independent review of major capital expenditures. There will be two scenarios. Board A will represent the current operating posture of a typical board. Board B will be an idealized board working for the stockholders to guide management toward sound long-range decisions.

Board A

Board A likes to work by full consensus. It is provided a well-ordered presentation of material and usually has access to presentation details well in advance of each meeting. An overall strategic plan is presented once a year, covering all the company's businesses, and each business is also reviewed annually in more detail as a feature of one board meeting. Board review is required for capital investments of over $5 million. Approval is based on a brief, fairly standard presentation. The format is a shorter version of the document used for review and approval inside the company.

Board members are assumed to be familiar with the business area for which the project is proposed, as well as the current company assumptions about the future national economic environment, so the presentation deals only with the proposed investment. Price and volume forecasts are provided by a single set of best-estimates on all key forecasts for the "proposed case." Profitability is calculated as a discounted cash flow rate of return. Sensitivity is shown, graphically, to 10% and 20% individual variation in each of several key variables.

CURRENT BOARD PRESENTATIONS ARE USUALLY TALKING TO, NOT DISCUSSION WITH.

The major intent is to present a persuasive, synopsized version of expected benefits from the proposed project. At the board meeting, 90% of the time on the proposal is spent on presentation. Board members may make comments from their special backgrounds or experience, but generally there is little discussion and they are successfully persuaded to approve the projects. Everyone is pleased at the obvious breadth of consensus.

Board B

The Chief Executive Officer at Company B is more anxious to have his Board of Directors review with him the major aspects of such important projects. He feels their backgrounds can add to the resources inside the company, and feels the board is ideally situated to participate in such critical decisions. However, to achieve the benefits of an independent review, the board must be given a different sort of presentation.

The advance mailing must first provide a discussion of the *problem* which the recommended project is expected to solve. This discussion would cover the expected general economic climate and then the business environment for the specific markets involved. It should also detail the company's competitive position in the business, both commercially and technically, and provide a brief review of the firm's desired strategy in this market. To show the need for action, the company's situation would be forecast assuming no major investments were to be made. Profitability of the proposal would be calculated on a "with" or "without" basis, not just "before" and "after" the proposed investment comes on stream. The proposed investment would then be described, emphasizing the impact on the corporation within the project's economic life. The expenditure's commercial and technical advantages and disadvantages would be discussed. The discussion should show how the project fits into the company's strategy.

The next section of information provided to the board would review *alternative* ways of solving the same problem which were considered and rejected, with reasons for the recommended choice. This discussion of other solutions helps let board members know enough about the critical variables – the "gut issues" – to contribute to the decision.

To this point the presentation would be almost all text. The final section would contain the profitability calculations. The key forecasts would be discussed, and the profitability of a best-estimate case would be provided, along with the cash flows supporting it. Sensitivities would be discussed on variables both internal and external to the project, such as inflation and GNP growth rate. If significant variables usually change together, such as price and volume, the *combined* impact would be assessed. After reading this section, the board should know which variables are felt by operating management *most likely* to change, and what the impact of that change might be.

The board meeting would review the ground covered in the advance mailing, by means of an oral summary. The CEO would try to balance simplicity with frankness, while still making a firm recommendation. The points brought up orally, to *elicit* discussion and input from board members, would not be solely to persuade them to the prior recommendation. The board's total discussion time on critical projects would often be longer than the original presentation. Length of discussion would depend on board input, internal management's confidence, and the degree of unanimity expressed early. Some projects are more debatable than others.

Commentary

Both manners of presentation are completely honest – extremes are described to emphasize the differences. The sins are of omission, more than commission, and in different treatment of the relation between management and the board. The image of full agreement is much higher in Board A. Discussion in Board B may leave some members uncertain of the wisdom of the chosen outcome, but Board B probably ends up with a broader understanding of the project and its real prospects for success. Board A may lack full understanding of the risks in the investment – but they do go home relishing the feeling of agreement.

Neither board is likely to hear about such projects again – *except* in the cases of booming successes or dismal failures. When failures occur, Board A can only resort to personal recriminations with internal management. Board B should be able to hear the causes, relate them to the discussion at the proposal meeting, and perhaps gain feedback to improve subsequent analyses and presentations.

I'd like to suggest a difference in purposes of the alternative approaches, which may not be apparent to internal management or the Board of Directors. Consider the following:

PURPOSE IN COMPANY A: Prepare the most persuasive document to gain consensus approval from a relatively isolated Board of Directors.

- Present enough about a situation to convince Board of Directors that funds should be granted the project
- Make a case that will assure priority for this project over others from other parts of the company.
- Emphasize a "best-estimate" case, based on the best judgment of the CEO, without making it clear that the one set of cash flows may have a low probability of being achieved.

MORE PRODUCTIVE BOARD PRESENTATIONS SEEK TO REVIEW, NOT JUST PERSUADE.

- Calculate various sensitivities to change, but don't discuss probabilities of deviation from best-estimate.

PURPOSE IN COMPANY B: Seek out the best decision for the future of the company, considering various possible competing definitions of "best."

- Achieve the highest long-term profitability for the stockholders, considering new project together with current assets and businesses.
- Allocate available resources among current and prospective businesses.
- Protect the company's current proprietary advantages, of any sort for future benefit.
- Broaden proprietary skills, as appropriate to developing trends.
- Elicit board "advice and consent" to current management on major current issues and decisions, so long as current management warrants such support.
- Maintain useful working relationships among the board members, as well as between the board and management.
- Document the proposed decision for future reference within the company, for outside legal use if necessary, and for possible improvement of later decisions.

Both lists suggest the purposes of the presentation to the board, not the purposes of the project being proposed. The latter set could be taken as an ideal listing of purpose for *any* board presentation, while the former indicates that *internal management* in Company A feels it has *already* reached the best decision.

Board B really reaches a decision, although they usually support the recommendation from within the company. In Company A, the real decision point on an investment is very fuzzy; veto power exists at any level, but is seldom exercised. Most projects accepted for presentation are eventually implemented. If delays occur, new ventures and speculative acquisitions may disappear from the lists; but projects in current businesses are seldom rejected.

Will the Real Board of Directors Please Rise?

Most companies operate their boards more like Board A than Board B. Board B is strictly a figment of my imagination – no company I know has a board sharing decisions to that extent. Furthermore, internal management doesn't want a change. Major capital investment approval procedures in all these companies are more like the Company A extreme, and no change is felt desirable by internal management. I feel procedures should be modified more toward the Board B type. While my industry survey found managers who said they'd welcome improved procedures for reaching investment decisions, they also said they wouldn't share these complexities with the board. I think this is wrong. Avoiding help from the board is a mistake. The reputation of American business management might be higher if they had used the skills available in the Board of Directors to help their investment decisions.

FMS - CONVINCING THE BOARD

G.R. STAPLES, URWICK, ORR & PARTNERS (UK) LTD., UK

ABSTRACT

Every board of directors is a unique combination of personalities, interests, disciplines and experience and this fact must be recognized when trying to convince it that an investment in FMS would be a good investment. A FMS proposal which advances only the manufacturing benefits is likely to get short shrift and so it should. If there ever was a case for showing how manufacturing aims can be supportive of, integrated with, those of the business as a whole, then FMS is that case. Therein lies the key to convincing the board about the merits of FMS.

INITIATING THE FMS IDEA IN A COMPANY

Typically, the people in a company who have heard of FMS ahead of most others are Engineers--Production or Mechanical, Electrical or Electronic. Most often, those Engineers are in the manufacturing or design function. Any progressive and inquisitive manufacturing engineer should by now know in some detail what a FMS is and will almost certainly have considered, if only casually, whether the products his company make might lend themselves to FMS treatment.

If the practice (or even theory) of Group Technology means anything to him, he will be able to picture whether sufficient similar batch-made components are produced to justify a number of linked processes where each would be reasonably loaded. There are, of course, many other criteria which might prompt the initial thought that FMS is worth examining (high set-up/run relationships, scrap etc.) but that which gives rise to the thought is less important than the need for top management to encourage it. This is the first duty of the Board of a company with manufacturing involvement, mainly to encourage the expression and examination of ideas from manufacturing people. Probably the biggest difference between Japanese and British manufacturing companies is in the status and hence influence accorded to manufacturing engineers. The sooner engineers are given more scope and encouragement to express themselves, the better.

However, for his part, the engineer must not believe that all he needs is a good technological case. He needs important allies and these need to be respected and lobbied well ahead of finalizing and advancing a FMS justification. These allies are considered in turn.

THE MARKETING AND SELLING DIRECTOR

This person usually has problems. He will say they are problems not of his own making and will list them:

- product costs too high
- delivery lead time too long and unreliable
- quality variable
- too long getting new products tested/launched.

He needs to be shown that a FMS can reduce these problems significantly and he should be persuaded to do his own assessments to show how the company's competitive position would be altered if the potential improvements available from FMS were realized.

In our most recent FMS project, for instance:

- product cost will fall by 16%
- delivery lead time will be two weeks instead of six weeks
- products, as delivered, will be consistent and almost certainly faultless in quality terms
- new designs will be processed along with existing designs and take no longer to produce once designs are put into "machine understandable" language.

The prospect of these levels of gains will almost certainly harness the support of the Marketing and Selling person--at least sufficiently for him to approve the idea of a detailed Feasibility Study.

THE FINANCIAL DIRECTOR

The next ally needed is the Financial Director. Most Financial Directors are accustomed to giving support to capital investment but only if profitability projections show that cash will be available for investment and then, only to the level needed to replace aging facilities.

A FMS investment might need more than that; in a recent case of ours, double the previous year's average investment was needed for the succeeding three years. Few Financial Directors would be pleased with that prospect but when other likely implications are advanced, a different attitude can emerge. For instance, major reductions in working capital become possible by lower stock levels of raw materials, work-in-progress and finished goods, and in a recent case, the following were achieved:

- 30% reduction in Raw Materials stores

- 93% reduction in Work-In-Progress

- 65% reduction in Finished Goods.

In this same case these reductions funded 40% of the total capital needed and, after deducting the normal replacement capital requirements and the Department of Trade and Industry FMS grant, a 2½ year payback period was achievable. Whatever Financial Directors' general attitudes to spending might be, most would prefer investment in hardware rather than in stocks.

Figures such as these cannot be obtained reliably until a fully detailed feasibility study is completed. Even so, an experienced and fertile mind can develop "ball-park" figures sufficiently impressive to gain support for the view that an independent, soundly-based study should be undertaken.

In depression-ridden UK manufacturing companies, many have been unable to maintain on their payroll the number and quality of qualified staff to handle a major project like a FMS Feasibility Study within an acceptable timescale. This fact is recognized by the Department of Trade & Industry as a consequence of which, a 50% grant is available against external consultant fee costs, when a feasibility study is entrusted to them.

Reaching the Feasibility Study stage represents the first major hurdle that the initiator of the FMS idea has to surmount. These studies may appear costly but when compared with the annual savings that a FMS could generate, their costs are puny--less than 1% is not uncommon.

The prospect subsequently of up to a one-third grant to cover development and capital costs can cause the average Financial Director to modify his traditional conservatism.

THE MANAGING DIRECTOR

The Managing Director is frequently the most important person to be persuaded that a feasibility study is desirable and, while he might be impressed by the Marketing and Financial arguments, there are other, less tangible but nonetheless real, arguments he should ponder:

- A Managing Director ought to be concerned to push or otherwise take his company forward to ensure that he doesn't suddenly find that the competition has overtaken him. By adopting a soundly-based FMS he would know he is at the edge of technological development and is therefore unlikely to be overtaken for a year or two at least.

- By examining the FMS possibilities in detail, he is showing shareholders and employees alike that his is a progressive business and worthy of their continued support.

- The successful and speedy adoption of modern technology is vital to the survival of manufacturing in this country and FMS is a significant form of this with potentially wide application. The sooner staff and work people adjust themselves to accepting and exploiting this technology, the sooner the sleepy, comatose reputation much of British industry has will disappear. The top men in every company must accept their share of the reputation this country has, and likewise, they should take responsibility for putting it right.

- Inflexibility in all its forms delays progress. Inflexibility is born of the problems that change would cause. FMS coupled with CADCAM can handle change as well as it can handle continuity and treat both imposers just the same. Showing that change is manageable will help in the essential re-education of the people in his company.

- The difference between forecast sales and actual sales accounts for more adverse variances in manufacturing than any other single cause, and the executive time taken to explain and control these variances is usually enormous and disproportionate. With FMS, manufacturing lead times can be so short that they become acceptable as sales lead-times, often eliminating the "make for stock" ideas and enabling, for the first time in many companies, the possibility of introducing "make to order" systems. Moreover, with FMS, it is possible to vary volume within very wide limits without the penalty of excess labour costs which, with FMS, are usually very low.

- Although FMS can displace some so-called skilled jobs, its main effect is to replace the menial tasks. FMS operating and maintenance jobs become more meaningful, flexible and responsible, thus giving greater job satisfaction and benefiting the whole employee relations climate in the business.

The strength of intangible points such as these will vary with each organization. Nevertheless, they should be thought out and presented if the Managing Director's support is to be secured.

THE MANUFACTURING DIRECTOR

We should be prepared for the possibility that it might not be the manufacturing people who initiated discussions about a FMS. It could have come from a dozen other sources, in which case an equally difficult job of selling the idea to the manufacturing people could present itself. Indeed, they could be the most reluctant people of all if they have, until now, been the custodians of the technological state of the art in the company.

The problems could be even more difficult if the manufacturing people have not already adopted NC/CNC and other computer aids in their company. In this case, there is a job, not merely of convincing them of the manufacturing benefits, there is also fear to be overcome--fear that they will be unable to understand, let alone manage, a development as sophisticated as FMS.

To deal first with the fear problem. There is only one way to overcome this and it isn't exhortation, misinformation or threats. It is in fact training--a combination of a little formal classroom work, a visit or two, but principally it is involvement right from the outset, well before a decision to proceed is reached.

When Urwick, Orr carries out a preliminary survey, and follows this by the detailed Feasibility Study, they go to great lengths to involve various members of the production staff in different parts of the study. They hold regular progress review meetings, not least to ensure that everyone sees their part of the Project in context. Slowly, but surely, they come to recognize that they will be capable of handling the development if in due course it is authorized.

Naturally enough, their involvement in the study is also invaluable for reasons of both speed and credibility. Although Urwick, Orr would expect to assign consultants reasonably knowledgeable of the industry and its processes and products, there is no substitute for the intimate knowledge possessed by the people who have been in the business for years.

Turning now to the tangible benefits which could accrue to the manufacturing people. There can be no doubt that production managers are plagued by problems such as:

- shortage of materials
- short runs
- excessive set-ups and other indirect activities
- poor workmanship, high scrap and re-work
- loss/shortage of cutting tools, jigs etc.

- progress chasing of batches between processes

- regular criticism of their high WIP and poor delivery performance.

A FMS can ease most of these problems and relieve the production director and his staff of much of the frustration and dissatisfaction they experience at work. They will have to learn new ideas and unlearn some of the old ones which earned them the jobs they currently hold, but so long as patience, participation and thoroughness characterize their development, they will respond to and enjoy the new challenge that FMS will provide.

THE PERSONNEL DIRECTOR

The Personnel Director shares with the Production Director some of the heaviest burdens. When all opportunities for job enrichment, job enlargement, enhancement of pay, and status for FMS operatives, etc., have been fully exploited, FMS means fewer employees for a given volume of work.

Maybe, as in many companies, joint agreements on the introduction of new technology exist, but not all have been put to the test and fewer have anticipated the large proportion of job losses that a FMS might require. Nor can we assume that the present mood of acquiescence and compliance by workforces will continue indefinitely.

Some crumbs of comfort can apply however. First, from the time expenditure approval is given to the FMS being up and running can take up to, and sometimes longer than, two years. Conscientious manpower planning, retraining and redeployment efforts over such a period can produce some unexpectedly successful results and these, coupled with enhanced severance terms for those who just cannot be retained, can go a long way to solving the residual problem.

The next possibility, which should not only be recognized but positively sought, is an expansion of business. After all, an ambitious FMS can and should transform the company's competitive position. It most certainly should improve its lead time and delivery performance; enable much quicker response to design change demands; it could if necessary, enable a wider product range to be produced without increasing inventory or the likelihood of stock redundancy losses, and finally, it could herald the end of the "Friday Car" quality image because FMSs don't get tired towards the end of the week. All of these influences should help to create circumstances in which there is scope for expansion with accompanying new job opportunities.

THE TECHNICAL DIRECTOR

For those companies which design the products they manufacture , a FMS offers both a challenge and an opportunity. The challenge lies in the need to undertake the long overdue review of design, with a new factor in mind which might not have figured when the product was designed originally. That additional factor is ease of manufacture. Designers in companies which have previously adopted stand-alone CNC equipment will already have experienced the obvious need to re-examine, with their production engineering colleagues, component design for purposes of fixture location and registration. Extension into FMS adds further challenge to that form of teamwork.

The opportunity lies not so much in FMS itself but in the stimulus it provides to take a fresh look at design for other reasons--for instance, to rationalize materials, shapes and sizes in order to reduce component varieties, stock levels and weights. Moreover, the Technical Director can begin tackling these design reviews using other Department of Trade & Industry support schemes such as the recently introduced CADCAM scheme and the Design Advisory Service.

If FMS can help inspire work along these lines, it will be performing another enormously important development in British manufacturing industry. It will also demonstrate, in no uncertain terms, the real interdependence of marketing, design and production and emphasize, again, the contention that FMS is a concern of the whole business, not just a production concern.

CONCLUSION

Experience to date shows that it is not easy to persuade a board to authorize a Feasibility Study which could cost between 5,000 pounds and 25,000 pounds, but paradoxically they find it comparatively easy to agree to spending a massive sum on the resulting FMS. This paradox is not too difficult to understand. At the time the FMS idea is new to them, its possibilities are difficult to grasp, its potential costs appear enormous and, unless great care is taken, it can be written off as just another consultancy gimmick like Quality Circles or Zero-based budgeting.

To authorize a Feasibility Study therefore becomes an act of faith rather than a firm conviction about the soundness of the step.

By the time the Feasibility Study is completed, all concerned are very much more aware of what a FMS is and can do. After that, persuasion becomes unnecessary because the potential effects on costs, delivery, quality and flexibility simply speak for themselves.

CASES IN JUSTIFICATION

This final part is devoted to case studies. It is one thing to read about the theory of an approach and how it should be applied but it is quite another to attempt to apply it, as many of these case studies show.

Cochran describes an extensive study to justify the cost of a computer-aided process planning system for Garrett Turbine Engine Company. He uses no sophisticated models, only cash flows, but the required analysis is complex nonetheless and the solution requires some special analytical tools.

In the second case, Sullivan uses a weighted scoring model to consider both non-monetary factors as well as the direct and indirect costs and benefits of replacing a conventional manual system with a robot. This case originated with the Air Force ICAM project; Sullivan extended the analysis to include additional factors.

The case study by Barr reports on the Upjohn Company's use of risk analysis. The task was to evaluate the economics of a new production process renovation. Barr describes the major parameters in the study and their effect on the results. He also discusses some possible extensions of the approach.

Rolston reports on the simulation of an FMS to determine its equipment requirements and utilization characteristics. The problem situation is first described, and then the resulting throughput and utilization graphs are given for various configurations.

The final case, by Gerwin, describes the activities of a large manufacturer considering the acquisitions of an FMS. The case describes the process the FMS project planning team went through in trying to justify the equipment, the extensive analyses they conducted, and the eventual emergency of a project champion. The case is especially candid concerning the human and organizational issues involved in this firm's automation planning, justification, and implementation.

An Approach to CAPP Cost Justification

abstract

There are many different methods to analyze and justify the cost and benefits of a proposed system. This paper will discuss one approach used by the Garrett Turbine Engine Company to justify the cost of a computer-aided process planning (CAPP) system. Among the tools used for analysis were group technology and a form of production flow analysis. This method of cost justification can be adapted by other companies in a variety of proposals.

author

WILLIAM D. COCHRAN
Production Systems Engineer
Garrett Turbine Engine Company
Phoenix, Arizona

conference

AUTOFACT 6 Conference
October 1-4, 1984
Anaheim, California

index terms

CAM
Process Planning
Costs
Group Technology

Introduction

An important part of any manufacturing enterprise is the making of its process plan. The process plan dictates to the shop floor such vital information as: step by step procedures to be used in making a part, machine tools used, perishable tooling requirements, jigs, fixtures, cut parameters, inspections, special handling call-outs, tolerance stacking, dimensions, etc. The quality of these plans directly affect the company's profits. Poor planning produces scrap and rework as well as wasted time and labor. For this reason most companies use their more experienced machinists or engineers to produce these plans.

The clerical nature of this manual method is very time consuming and prone to human error and individual subjectively of the planner. In addition, experienced planners are becoming scarce necessitating the use of younger less seasoned planners. For this reason there has been a growing trend over the past decade to switch to some type of a Computer-Aided Process Planning (CAPP) system. These systems are or should be characterized by their quick response in producing accurate plans that are both consistent and free of error. The term "should be" is used in that present CAPP systems are still some what dependent on art as well as science.

Though we still are not intelligent enough to capture the seemingly infinite number of variables that are associated with this key step in manufacturing, CAPP systems are maturing and becoming more of an economically useful tool rather than an expensive high tech curio. As these systems grow in power and sophistication they are becoming recognized as a "necessity" for those companies who wish to stay competitive in the future. Many large companies have delved into CAPP and found it to be economical in spite of its current limitations and cost. As an increasing number of larger companies use CAPP systems from a growing number of vendors, a trend takes place that drives the cost and availability of such systems down within the grasp of medium and small size companies.

Regardless of size, there comes a time when a company in the market for a CAPP system needs to justify the cost of such a system. Before management commits significant dollars or resources in obtaining and implementing a CAPP system they want to know if the benefits will outweigh costs. For this reason two questions need to be answered. How much is it going to cost and what savings can be expected?

Questions pertaining to acquisition costs of hardware, software, maintenance, training, etc., are straight forward. These can be readily obtained from a vendor. If the CAPP system is to be home grown then software development cost, dedicated human resources, implementation lead time, etc. are less concrete. Capturing cost of an in-house developed system is a bit more elusive. Some form of a project management system is usually employed. The scope of work is drawn up and the resources needed to support the work are estimated. From these estimated resources a ball park figure can be arrived at for the total system cost.

However, the question of savings becomes even more elusive than defining the cost of an in-house developed system. Some fortunate companies have a cost tracking system that is well developed and supported. This makes the job of estimating and tracking savings much easier. A good cost tracking system is used as a solid standard on which savings can be based and projected. A number of companies however, are not as advanced in their accounting methods and do not separate out as many cost categories. Thus, they struggle to find an exact cost in their existing system.

A real problem with projecting savings is to determine how much work a piece of software as complex as a CAPP system will save in an area as all encompassing as process planning. Will there be a 10%, 20%, 30% overall reduction? After installation, system performance can be measured and compared with historical data if that data exists. In this case foresight is needed first to justify the system. Many approaches have been taken to answer this question. The following is just one of those approaches taken.

Background

The Garrett Turbine Engine Company (GTEC) of Phoenix, Arizona is a manufacturer of gas turbine engines. It produces a wide variety of parts most generally characterized by complex shapes and tight tolerances. An example of this is a commonly used blue print geo-tolerance call-out of run-out within .0002 for gears. Due to the inherent quality and reliability requirements for aircraft propulsion engines and the complexity of design, process plans can be very lengthly, complex and meticulous. A typical process plan will range between 30 to 50 operations. Each operation will contain the following: machine tool used, perishable tooling, jig and fixture tool numbers, operation description, machining dimensions and tolerances, caution notes, specifications in a condensed form, speeds, feeds, depth of cut, and graphics.

Another characteristic of GTEC is its use of group technology. All parts are produced in one of four shops, namely, wheels, gears, cases, and sheet metal. Though this is not a complete breakdown of parts into "families of parts" as pertaining to group technology it does significantly reduce the magnitude of searching needed to develop the complete process plan as well as benefit the shop floor in facilitating production.

GTEC has for sometime been examining the detailed requirements for a CAPP system and defining the modules needed to satisfy its design. Because of the interest aroused by this study, it became necessary to provide to management as soon as possible a complete project analysis with associated costs and benefits. The specific assignment derived from this was to:

> "project as accurately as possible, given a three week time period, the savings of an in-house designed CAPP system."

In addition, this was to be a one person job. The only other human resources dedicated were a few hours each week of an experienced process planner.

This in-house CAPP development at GTEC is known as CAPE, or Computer Aided Production Engineering. Parts of the system are purchased from outside vendors, but for the most part it is an in-house design. CAPE is an enhanced variant process planning system which will interface with the existing company data base and be integrated with all other departments and functions concerned with process planning.

Approach

The strategy used for cost analysis was simple: select a representative family of parts and develop a standard routing for that family. Next, compare and contrast the standard routing with the existing routings and carefully note all differences. This would be followed by an analysis of those differences to see what, if any, the cost impact would be. Lastly, the findings of the analysis would be applied to actual 1983 cost data for a projected cost savings.

Due to the time constraints only one family was chosen for analysis. The family chosen was that of spur gears with a trepan configuration. This choice was made for two reasons. First, there was an existing classification and coding scheme in the gear shop with a significant number of the parts coded where as no such code existed in the other shops at that time. Second, it was felt that spur gears would be about as representative of a family as could be considering the wide variety of part types. Seven parts fell within this spur gear family and were used for the analysis.

After the process plans were gathered a production flow analysis was performed on the seven parts of this family. A matrix was constructed consisting of a vertical listing of the routing, or in other words, a list in descending order of each machine tool used in the routing of the part. The horizontal axis of this matrix consisted of the respective part numbers of the spur gear family, seven columns in all (see figure 1). This matrix had to be built one column at a time. Each operation for a part had to be as closely fitted and aligned to the previous part routing as possible. This was made fairly simple due to the great number of common operations that are a natural outcome of the group technology families. At the intersection of each part number and machine tool the operation number was placed. To avoid confusion it should be pointed out that GTEC does not assign a unique operation number to a specific type operation. Operation numbers are assigned on strictly a sequential basis regardless of machine type or process. The load center number which appears beside each operation takes the place of what many companies call an operation number.

Following the matrix construction another vertical column was added to the right. This column was a machine class or process type column. An example of this would be calling out an operation as using a medium horizontal N/C lathe as opposed to a specific company name and model type N/C lathe. This column was added for use in generating a graphics flow chart of the production flow on a higher level to avoid a chart with too much detail and clutter.

At this point the experienced process planner was called upon to go through each routing in detail to create a standard routing. This was a time consuming task taking one of the three weeks allotted to this project. It must be appreciated that a process planner makes hundreds of calculations, judgments, and decisions whether conscious or subconscious to create a standard routing. It was discovered that though the seven parts of the spur gear family were nearly identical in geometry and material their routings differed in some cases considerably. It was judged that the differences came from a number of reasons. Among these reasons were obsolete machinery, changes in methods, new equipment, different approaches and preferences of the planner who created the routing, changes in tooling, and changes in company policy. From this careful examination of each routing a standard process plan was created that would cover all seven spur gears.

This standard routing was then compared with the matrix created earlier. The matrix was adjusted to reflect the standard routing. Operations were deleted, combined, and in some cases added to this matrix. The net outcome of all these changes was a decrease in the number of operations. Before the analysis there were a total of 221 operations among the seven parts studied. After the standard routing was produced and compared there were 201 operations among the seven parts. This was a net decrease of 20 operations or 9.1% (see Table 1).

CHANGE IN NUMBER OF OPERATIONS

PART #	PRESENT # OF OPER	CHANGES IN # OF OPER	% CHANGE
364199-1	23	-2	-8,7%
364199-2	24	-2	-8.3%
896803-1	41	-3	-7.3%
3101535-1	32	-1	-3.1%
3501687-1	33	-3	-9.1%
3860012-1	37	-5	-13.5%
3860038-917	31	-4	-12.9%
TOTAL	221	-20	-9.1%

TABLE 1

Upon examining this number it was realized that this could be applied in a generic sense to set-ups, tool design, perishable tooling, fixtures and gages (see table 2). N/C in particular was reduced from 14 operations to 9, a 35.7% decrease in the number of N/C operations. Though many operations were deleted there were a number that were added. The number of operations added, 9, was subtracted from the number deleted, 18, for a net run-time reduction of 9 operations or 4.1%. It was felt that those operations that were moved or combined still required the same amount of run time and thus would have no effect on this 4.1%. Two categories remained to be examined: process planning production time and scrap. These figures could not be directly derived from this method. For this reason commonly cited figures in current literature were used as found by other companies who have installed CAPP systems. In the case of reduced process planning production time a frequently cited figure is 58%. In the area of scrap 6% reduction is found frequently.

For demonstration purposes and as a visual aid a flow chart was made of the spur gear family routing flow through the shop. This simple line and circle flow diagram was used to visualize the magnitude of changes that had taken place on the matrix (see figure 2). It is easier to verify a standard routing when all possible paths are laid out for easy visual access. In addition, this chart was also used in presenting the CAPP system to management. A network, if not too complex visually, is more easily understood in a meeting atmosphere than volumes of tabular data such as the matrix.

PERCENT REDUCTION OF OPERATIONS

9.1%	SET-UP
9.1%	TOOL DESIGN
9.1%	PERISHABLE TOOLING
9.1%	FIXTURES & GAGES
35.7%	N/C
4.1%	RUN-TIME
58.0%	PROCESS PLANNING
6.0%	SCRAP

TABLE 2

After the standard routing was created, the changes were transferred to the matrix. This resulted in moving, deleting and in some cases adding operation numbers at the intersections of the part number and the machine or operation description. When this process was finished all vacant lines were removed. With the machine group in the right column it was noted that a number of operations were performed in the same sequence from part to part with the only difference being the brand name of the machine it was routed to. For this reason the left hand column was ignored and the right hand column was used to further condense the matrix. An example of this would be four lines of medium horizontal N/C lathes with different manufacture names being combined into one line. To have every specific machine on the flow chart would add a degree of complexity that is unwanted clutter for a presentation. Suffice it to say that all the parts go to a medium horizontal N/C lathe.

After the estimated percentage savings were calculated, costs were gathered. Baseline cost data in this case was from the previous calendar year, 1983, for each of the respective areas shown in table 2. This was another time consuming task. Of the three weeks allocated one full week alone was dedicated to the task of cost gathering. No one group or individual had access in any detail of all the fields concerned. Cost data was gathered from a number of individuals and reports. In most cases the information was present and accurate. Only in a few instances was a "best guess" made by an experienced person who had a good handle on the situation of that department.

With the cost information collected it was decided to format the data on a personal computer using a spreadsheet software package. The spreadsheet lends itself nicely to this application (see figure 3). The format used for this spread sheet was to group each major section (eg. run-time, N/C, scrap). The first line of each group would contain the group name such as set-up or tool design, followed by the percent savings expected, followed by the total dollars spent in

1983 in that area. The following four lines were for each of the respective manufacturing shops, namely, gears, sheet metal, wheels, and cases, followed by their percent of impact in that area. The percent of impact was calculated by the amount that shop spent for that particular area compared with the total for all four shops. The remaining columns for these four lines contain a series of numbers representing thousands of dollars saved per quarter. These values were calculated by multiplying the dollar amount spent per quarter in that area for 1983 by the percent of impact for that shop. The results were multiplied by the percent savings expected for that area which then was multiplied by an implementation phasing factor.

This phasing factor took into consideration the time each shop would require to move from the manual system to the CAPP system up to the twentieth quarter. The phasing was linear and reached 97% at its highest point in anticipation that a full implementation would not be realized due to various unforeseen conditions and circumstances. Savings for each shop were totaled as well as saving by each column or quarter for the different areas. This pattern was repeated for each of the areas with a grand total of savings tallied at the bottom.

Following total savings were total costs. These were a combination of purchased software, hardware requirements, and personnel needed to design, implement, and maintain the total system over a period of five years. Wages and salaries were adjusted using a 6% per year inflation factor. Net savings per quarter were derived from the total savings less total costs. These net savings were then summed to a total net savings over five years.

Results

Substantial savings were projected from this study. A rate of return was calculated at greater than 50% with the pay back occuring in the eleventh quarter. With these figures it was deemed a worthwhile project to undertake. A presentation was made to management using these figures as well as the graphical flow chart of the spur gear routing as mentioned earlier. The outcome was a general approval of the CAPE system with further deliberation on the technical requirements of the software. Human resources for the remaining system design have been approved and are presently dedicated to the project.

Considerations

Although the general method is sound there are a few factors that should be considered before using this approach. Since time was a major limiting factor a number of estimates or assumptions were used. One of the most notable shortcomings of this study was that only one family of parts, spur gears, was used. This does not invalidate the data that was found, but there could be a greater degree of confidence in the results if several families could have been studied and combined. It would have been best to have one and possibly two families from each of the shops as a representative sample.

Another area of enhancement would be to build into the spreadsheet a discounted cash flow model. This or any other financial analysis tool that a company sees prudent could be used. Though this might not dramatically affect the numbers that are generated, it could change the magnitude of the return on investment.

Conclusion

The goals for cost analysis of CAPP at GTEC were met. A fairly accurate cost savings study was performed and accepted by management. For the time and resource constraints it was felt to be a viable method, though recognized as not being entirely complete. An inexpensive method was explored for cost/benefit analysis that can be used by any company with similar needs. It can be as in-depth as desired if time and resources are available or as simple as was this case without tying up a lot of money on labor.

OPERATION DESCRIPTION	LOAD CENTER	364199-1	364199-2	856803-1	3101535-1	3501687-1	3860012-1.	3860038-917	MACHINE GROUP
FAB CONT SER NO	260			015	015				MISC
AUTO DO ALL SAW	585			016			015	020	SAW
SHOP OVER LOAD	666	020	020		020	020			MISC
PRE INSP 1201	901	025	025		030	030			INSPECTION
#5 W&S TL	1LB			018					MED HZ TUR LATHE
J&L 312 CNC	1TD			020					MED HZ N/C LATHE
#3 W&S TL	3LC	030	030				020		MED HZ TUR LATHE
#3 W&S TL	3LC						030		MED HZ TUR LATHE
H-S NC CHUCKER	3TD			030		040			SML HZ NC LATHE
L&S NC BAR CHUCKER	2TC							030	MED HZ N/C LATHE
H-S NC CHUCKER	3TC							040	SML HZ NC LATHE
L&S NC BAR CHUCKER	2TD				040		040		MED HZ N/C LATHE
LAPMASTER	BL7					050		050	ROTARY LAP TABLE
HEALD ROTARY	IND					060	050	060	ROT SUR GRINDER
HEALD ROTARY	IND						060		ROT SUR GRINDER
SUNNEN HONE	2PC	040	040						HORZ HONE
KARSTENS 9X18	3PD				050				O.D. GRINDER
PFAUTER 10 HOB	6HD	050	050	040					SMALL GEAR HOB
LIEBHERR HOB	5HD						070	070	SMALL GEAR HOB
8C HOB 10-20	2HD				060				SMALL GEAR HOB
PFAUTER 16 HOB	1HE					070			MED GEAR HOBBER
OSBORN BRUSH	1YD	060	060	050	070	080	080	080	SPINDLE DEBURR
TUMBLE	023	070		060					TUMBLE DEBURR
VIBRATION SWECO	024		070			090	090	090	TUMBLE DEBURR
PRE INSP 1201	901			070	075		100	100	INSPECTION
H.T. DIM INSP	173					100	110	110	HEAT TREAT
COPPER PLATE	183	080	080	080	080	110	120	120	PLATING

(FIGURE 1) PART/OPERATION MATRIX

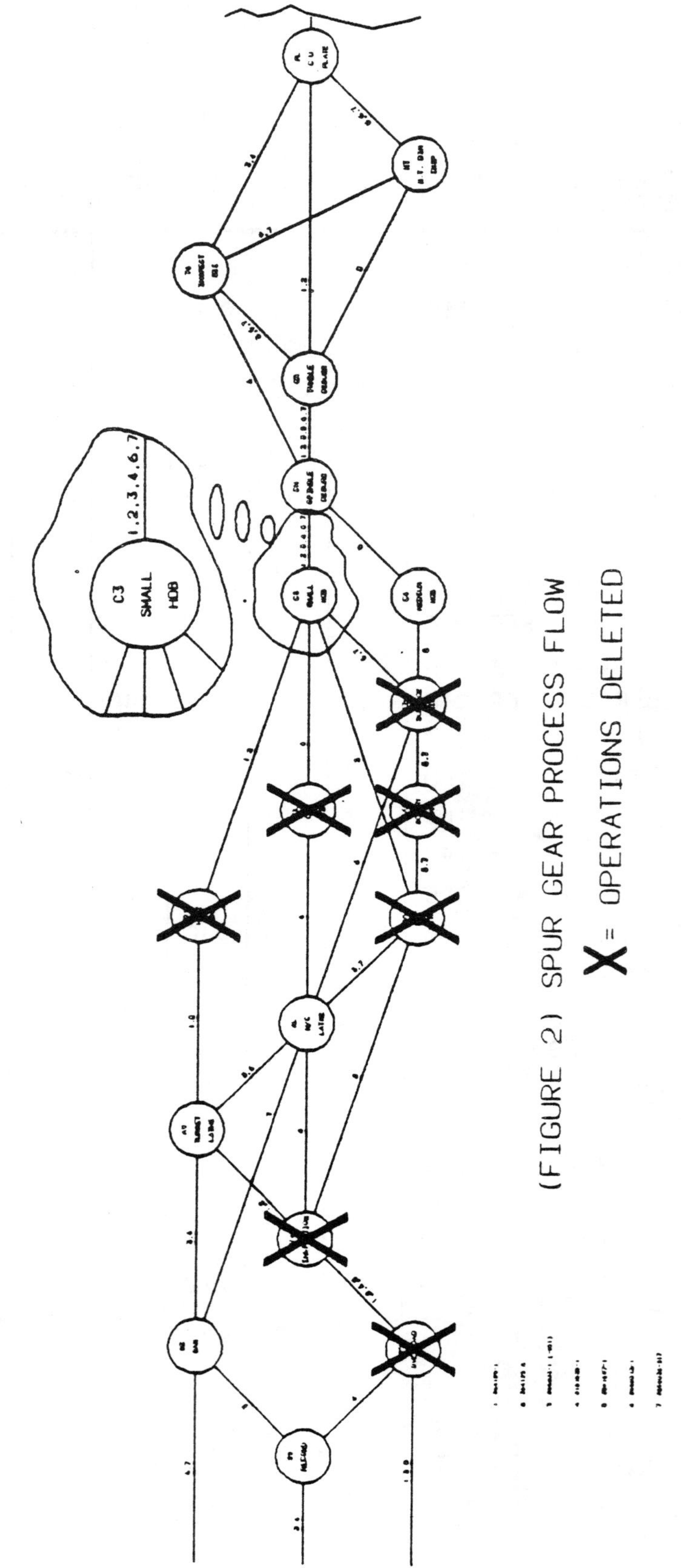

(FIGURE 2) SPUR GEAR PROCESS FLOW

X = OPERATIONS DELETED

COMPUTER AIDED PRODUCTION ENGINEERING
COST ANALYSIS

(ALL FIGURES IN 1000'S OF $'S)

AREA	SAVINGS PERCENT	1983 $'s	SHOP	IMPACT	ESTIMATED SAVINGS PER QUARTER								TOTAL
SET-UP	9.1%	346		QUARTER	1	2	3	4	5	6	7		
			GEAR	17%	0	1	2	4	9	15	29		130
			SHEET METAL	38%	0	0	0	0	1	3	5	7	270
			WHEELS	31%	0	0	1	2	3	4	5	6	245
			CASES	14%	0	0	0	1	2	3	4		112
			TOTAL		0	1	3	7	15	25	43		757
RUN-TIME	4.1%	2,500		QUARTER	1	2	3	4	5	6	7		
			GEAR	23%	0	2	4	7	13	21	32		320
			SHEET METAL	41%	0	0	0	0	1	3	7		602
			WHEELS	27%	0	0	2	3	6	8	13		346
			CASES	9%	0	0	0	1	2	4	8	14	105
			TOTAL		0	2	6	10	22	36	60	83	1,373

	QUARTER	1	2	3	4	5	6	7		TOTAL
GROSS SAVINGS		0	6	23	45	57	89	99	1	6,421
TOTAL COSTS		56	23	24	25	26	27	28	3	2,704
NET SAVINGS		-56	-17	-1	20	31	62	71		3,717

(FIGURE 3) SPREADSHEET EXAMPLE

Reprinted from ***1984 Annual International Industrial Engineering Conference Proceedings.***

REPLACEMENT DECISIONS IN HIGH TECHNOLOGY INDUSTRIES -- WHERE ARE THOSE MODELS WHEN YOU NEED THEM?

William G. Sullivan
Arizona State University

ABSTRACT

The aim of this paper is to examine the role that industrial engineers will play in replacement studies of existing capital assets with those incorporating improved technology. Most investment decisions that involve CAM/CAM and robotics applications are essentially replacement decisions, but for a variety of reasons conventional analytical models appear to be inadequate in effectively dealing with them. A broader decision analysis consisting of several levels of evaluation criteria is suggested in which strategic technological objectives of the firm are taken into account. To illustrate the more global perspective needed in such situations, an example involving an industrial robotics application is presented.

"The ability to prove convincingly a robot or automation system is a good investment, and prioritizing its capital needs within the constraints of a limited capital budget, seems to require the talents of a 'super engineer' [17]."

1. INTRODUCTION

It is common knowledge that productivity in the United States has not fared well since 1973. In fact output per man-hour from 1969 to 1973 increased at a compound rate of 2.9 percent, while these gains dwindled from 1973 to 1979 to an embarrassing 1.6 percent a year [14]. It is also widely acknowledged that reverses in this downward trend rely heavily on the continued development of advanced technology <u>and its successful application</u> [1].

Factory automation, which includes computer-aided design (CAD) and computer-aided manufacturing (CAM), will assume increased importance in productivity improvement programs in manufacturing industries. A subset of CAM is industrial robots, broadly defined as "microprocessor-controlled mechanical devices that perform a function or provide an intelligent interface between machines and processors" [14]. Industrial robots represent a substitution of capital for labor and are ideal for hostile environments and/or batch production shops where <u>flexibility</u> can be economically traded off for efficiency [1]. Thus, a basic goal in many manufacturing firms is to increase the capability of effectively processing small lot sizes by drastically reducing setup times with the help of industrial robots.

A subtle benefit of flexibility is that it encourages technological change and innovation. However, flexibility is initially destructive of capital - whether in the form of labor skills, technological processes, or capital equipment - and tends to create rapid obsolescence and uncertainty regarding existing investments and their known replacements. In the face of these adversities, it's no wonder that a "wait and see" attitude develops when large investments in fixed capital assets are involved!

At the moment, it is generally difficult to justify robotic applications on purely economic grounds [12,14,17]. "Why install a $100,000 to $150,000 robot to perform a $25,000-a-year job?" It is clear that concerns besides dollars and cents provide the driving force in today's implementation efforts. For robots to become practical during the next decade, their size, cost and mechanical complexity must be reduced. This will occur primarily through the expanded use of computer and control technology [14].

The aim of this paper is to explore the talents needed by Meyer's "super engineer" (quoted above) in justifying capital investments in high technology* manufacturing industries. A robotics application from the Air Force's ICAM project is used to illustrate the broader perspective required of industrial engineers in such analyses.

2. BACKGROUND

Let's consider the present situation. The United States is the leader in robotics technology, yet Japan is acknowledged as the world leader in robot applications [10]. Will U.S. manufacturing companies continue to be reluctant to replace obsolete capital assets when it is apparent that hundreds of thousands of manually operated machine tools, for example, should be replaced in the foreseeable future by computer-integrated systems?

A variety of reasons for inadequate investment during the 1970's has been offered: inability of American companies to manage technology, inexperience with competing in world markets, management's short-term outlook created by fear of risk and obsolescence resulting from technical innovation, high cost of capital, resurgence of inflation,

*For purposes here, high technology manufacturing industries are those that make extensive use of computer controlled mechanical devices.

inability to work with labor unions, etc. Could it be that Naisbitt's "no-tech manager" is to blame for our national economic decline [19]? Have companies relied too heavily on conventional procedures for justifying proposed capital investments because "Managers can all too easily hide behind the apparent rationality of such financial analyses while sidestepping the hard decisions ..." [12]?

At the plant level, a direct result of inadequate investment (for whatever reason!) shows itself in the age distribution of our capital assets (see Table 1). Ayres and Miller point out that "the average age of capital goods is a good indicator of the rate of diffusion of cost reducing technology" [1]. They further state that there has been a substantial slowdown in the formation of capital in the U.S. thus leading to the proposition that increased technical activity and capital formation are an important part of reviving productivity growth.

Table 1. Machine Tools in Use (Under 10 Years of Age)

	Country	%
1.	United States	31%
2.	West Germany	37%
3.	United Kingdom	39%
4.	Japan	61%
5.	France	37%
6.	Italy	42%
7.	Canada	47%

Source: Ref [1]

3. TACTICAL VERSUS STRATEGIC INVESTMENT DECISIONS*

Some of the earliest capital investment studies performed by engineers were those dealing with the replacement or retirement of obsolete equipment in industrial and government organizations [8]. The field of engineering economy as we know it today has evolved chiefly into a "bottom up" set of principles and techniques for evaluating the economic merits of proposed capital investments in view of numerous real-world complications such as nonmonetary factors and uncertainty. Engineering economy practitioners typically deal with tactical rather than strategic investment decisions, particularly in the area of replacement. Thus, we usually concern ourselves with cost reduction aspects of the profit equation rather than with the revenue creation part of the equation.

Critics of engineering economy have picked up on this orientation and have repeatedly charged that we give "the narrowest possible interpretation" to our domain [13], which is "the economic evaluation of minor investment projects" [24] and the rendering of "low-level advice for management decision making" [21]. Such criticism may have been valid in the days when replacement analyses, for example, involved a comparison between a "defender" machine tool and a "challenger" machine tool that accomplished an identical function. But this is no longer generally true in high tech areas.

Replacement studies involving increased use of technology give rise to new dimensions that upgrade such studies from a tactical to a strategic outlook: (1) innovations and flexibility in automated factories create a dynamic environment in which many traditional machine processes and manual operations are being redefined or eliminated, (2) functions to be accomplished in successive replacement intervals are time-variant and they tend to become integrated over time (see Table 2) and (3) the need to reduce unit costs through adoption of new technology tends to focus on strategic business considerations and places pressure on engineers to broaden their understanding of corporate-level policy issues. (Refer to Figures 1 and 2.) Because of these new dimensions, replacement studies in high technology settings have moved from tactical to strategic in nature and force the industrial engineer to adjust his/her analytical techniques accordingly (see Table 3). For example, it is no longer possible to determine the break points in Figures 1 and 2 from a simple analysis of projected monetary costs and production volumes. First, the strategic, often technology related, goals of the firm are usually nonmonetary in nature and, second, production volume and batch size are uncertain quantities that are highly dependent on the firm's implementation schedule for its technology strategy.

4. INTEGRATING TECHNOLOGY INTO THE BIG PICTURE

Technology will become a larger and more vital component in strategic business planning in the years ahead. Amazingly enough, few corporations acknowledge that they have an analytical approach to integrating technology in their strategic planning process. An essential prerequisite to the involvement of industrial engineers in strategic investment decisions shown in Table 3 is developing a "technology portfolio" from which priorities can be established.

In a recent paper, Skinner makes the point that the lack of an analytical approach towards integration of technology into a firm's overall business plan is a formidable barrier to market entry in high tech endeavors [22]. Its high investment costs and high demands on management are often compelling reasons to shy away from high technology ventures. Uncomfortable questions arise that are fraught with uncertainty: "What business should we be in and what technology is needed to support it?" "What competition do we face in domestic and international markets for our established and planned products?" "What are the total costs, benefits and risks of new technology and will our unit costs of production really be appreciably reduced?"

Development of a technology portfolio is a vital first step to creating a technology strategy (sometimes referred to as an automation plan) that dovetails with a firm's business plan. Matching the technology needs with the business plan and

*A strategic decision is long-term in outlook and involves "top down" issues such as technology advancement, sourcing of capital, plant modernization and competitive position of the firm. A tactical decision is concerned mainly with the short-term position of the firm in such areas as cost avoidance, reduction of inventories and rework, reaction to product changes and replacement of relatively inexpensive capital assets.

prioritizing components of the resultant strategy then leads to a plan of investment that, by its nature, considers nonmonetary and monetary factors. These steps attempt to avoid the tendency of establishing a business plan without taking full account of the scarce resource called "technology."

5. THE REPLACEMENT DECISION REVISITED

A summary of one possible decision hierarchy for addressing high technology replacement decisions is shown in Figure 3. An initial determination is made regarding the strategic or the tactical nature of the replacement decision. If the decision involves strategic issues that cannot be met due to an inadequate technology base (or other reasons), the replacement decision favors the status quo or the abandonment alternative. This is a simple accept-reject (0-1) determination. At level two where independent investment opportunities are considered, a reasonably comprehensive checklist (or score card) can be utilized to approximate how well strategic objectives are achieved in each opportunity. Allocation of investment funds occurs at level two depending on technology priorities and how well proposed undertakings mesh with the overall business plan.

Inputs to level three consist of high technology ventures selected in view of (1) strategic priority, (2) capital requirements, (3) perceived risk, (4) availability of managerial and technical know-how, etc. At this point, a detailed quantification should be made of time-phased costs and benefits associated with mutually exclusive alternatives (e.g., competing robotics systems) for implementing the technology. An analysis of cash flow differences and nonmonetary costs and benefits will lead to recommendation of a preferred system. It may be useful to utilize some form of multiple attribute decision analysis to make this determination. In this regard, a robotics application is evaluated as a "level 3" problem in Section 6. Finally, the system chosen is then implemented by project management and operations planning groups within the firm.

6. A MULTIPLE ATTRIBUTE DECISION MODEL FOR SELECTING AMONG ROBOTICS SYSTEMS

In general there are two themes that underlie the literature dealing with economic justification of robots: (1) there has been incomplete accounting of all direct and indirect costs and monetary savings and (2) there has been inadequate attention given to strategic considerations associated with manufacturing systems that incorporate robots. In this section a fairly complete listing of robotics costs/benefits is presented and a procedure for estimating their magnitude is illustrated. Additionally, a multiple attribute decision model that includes strategic factors is demonstrated for an ICAM case study dealing with a selection between a manual operation and a robot. Thus, an illustration of how monetary and nonmonetary factors can be analyzed in robotics applications is provided here.

Sophisticated and complex robotic systems are quite expensive and may require costly accessories and installation. Eshleman and Pagano state that "it is not unlikely for an installed robotic system to cost from two to three times the price of the robot. This includes engineering, peripherals, tooling and installation" [5]. They further estimate the following ranges of costs for the robot and its components.

% Cost	System Component
30 - 60%	Robot
30 - 50	Engineering
30 - 50	Peripheral Equipment
10 - 15	Installation and Training

Direct costs and benefits necessary for economic evaluation of industrial robots are listed in Table 4. Indirect costs and benefits are included in Table 5. Fleischer [6] has identified additional factors that are necessary to conduct an economic justification of industrial robots. These factors include tax rates (federal, state, local), the firm's cost of capital, the number of shifts, learning curve effects, and changes in product design.

ICAM Case Study

The ICAM (Integrated Computer Aided Manufacturing) project of the U.S. Air Force included an economic analysis of a robotic system [25]. Two alternatives are considered: a robot or a manual operation. The savings and costs from this case study are the basis for discussion of various methods to assess the merits of robotics systems. Additional assumptions will be required for data that are not included in the original case study.

To provide structure to the quantification of factors in Tables 4 and 5 as they relate to the ICAM case study, the following terms are defined [11]:

T_1 = time required to manufacture one unit of a product,

T_2 = time associated with manufacturing a product in batches,

T_3 = time needed to design the product, prepare drawings and process plans, etc.

B = number of batches manufactured throughout the product's life cycle

Q = number of units produced (assume to be constant) in each batch of product.

The total time devoted to a product over its life cycle, TT_{LC}, then equals $QBT_1 + BT_2 + T_3$.

The average time spent on each unit, T_{LC}, is then

$TT_{LC}/QB = T_1 + T_2/Q + T_3/QB$

For repetitive operations of manufacturing in batches, robot applications typically reduce T_1 (through increased availability of the robot), reduce T_2 (through computer-assisted setups) and increase T_3 (until learning curve effects become favorable). By considering the product life cycle and the anticipated annual production plans for each product, it is possible to address fundamental questions regarding changes in T_1, T_2 and T_3 that produce savings from which an initial investment in a robot may be justified. Such savings must be estimated in view of (1) fixed production quotas versus unlimited production, (2) availability of production capacity with (and without) the robot application, (3) the production line balance achieved and its reliability and variability, and (4) other supply-side issues. Also demand-side matters such as timing of market penetration and eventual market share, desired inventory position, and customer backlogs must be included.

For instance, consider monetary savings in direct labor afforded by a robot in year 1 of an N-year planning horizon. The value of unit savings in T_1 for direct labor is computed as factor A in Table 6, and a similar procedure would be followed to estimate factors B through E. These unit cost factors and the associated cash flow estimates for the ICAM case study are given in Figure 4. Other factors appearing at the bottom of Figure 4 (training and supplies) and those to be added in Table 9 relate to changes in T_2 and T_3 time requirements which are affected by a robot application.

Accordingly, cost changes in "T_2 factors" occur because of different batch sizes and involve materials requirements, number of setups, amount of inventory carried, cost of a setup and so on. Cost changes in "T_3 factors" result from differences in product design, process planning, floor space requirements, safety precautions, operator training, quality assurance, etc. In a relative sense, much attention has been paid to the T_1 factors and their incremental costs and savings in robot justification studies. This is reflected in Figure 4 and discussed in the next section. However, an effort is made to include T_2 and T_3 factors for the sake of completeness in the section, Consideration of Indirect Benefits. Strategic concerns that affect robot investments are also included later. An abbreviated version of the economic justification from the ICAM case study is shown in Figure 5. Various measures of economic merit for the robot application relative to the manual operation are discussed below.

Consideration of Direct Costs and Benefits

The Payback Period. The payback calculation provides a simple form of project evaluation, which is easy to compute and to understand. The payback period is the number of years required to recover the net investment:

$$\text{Payback Period} = \frac{\text{Net Investment}}{\text{Average annual cash flows}} = \frac{\$252,000}{62,339} = 4 \text{ yrs.}$$

A payback of 4 years would probably lead to rejection of the robotics system.

The main disadvantages of the payback method are its failure to recognize cash flows that occur after the payback period and its failure to consider the time value of money. Payback does provide implicit consideration of the risk of the investment because it reflects the liquidity of the project.

Return on Investment (ROI). This method is commonly used when only accounting data provides the basis for the analysis.

$$\text{Average Return on Investment} = \frac{\text{Average net income after taxes}}{\text{Average investment}}$$

For the ICAM example, the net income after taxes is calculated in Table 7.

Using the net income determinations, the average return on investment is calculated as follows.

$$\text{ROI} = \frac{[(28728) *7 + 26028]/8}{252000} = 0.113$$

This ROI does not indicate a favorable robotics application.

The main disadvantages of the ROI method are the lack of consideration of the time value of money and the use of accounting data rather than cash flows. Also, the method is not useful to analyze the risk of the investment.

Net Present Value (NPV). After-tax cash flows from the project are discounted at the firm's minimum attractive rate of return (MARR) to determine the net present value [3]. If the net present value is greater than zero, the project will earn a return equal to at least the firm's MARR.

The Net Present Value (NPV) for the ICAM case study is calculated in Table 8 when the firm's MARR (after-taxes) is 12%. Because the NPV is greater than zero, a favorable project is indicated from the foregoing analysis of only direct costs and savings.

The NPV approach assumes that all intermediate cash flows will be reinvested at the firm's MARR - an assumption which management should consider in the analysis. Competing projects may be ranked on the basis of the net present value, and NPV is theoretically sound for prioritizing projects. Inflation and learning curve effects can be considered in the cash flows. Data gathering costs are reasonable, and the calculations may be done on a computer.

Consideration of Indirect Benefits

The consideration of indirect benefits in the economic evaluation of industrial robots may provide a more complete analysis of the investment decision. The non-quantitative benefits should be estimated, using historical data if available.

Although the indirect benefits for the ICAM project are not described, some benefits will probably accrue if the robot is installed. The benefits shown in Table 9 are assumed.

The Net Present Value of the ICAM project calculated using both direct and indirect savings is shown in Table 10. The incorporation of indirect benefits increases the NPV to $135697 from $62189. The internal rate of return is 26% while the payback period (after taxes) is 4 years.

Consideration of Nonmonetary Attributes

Because of the difficulty of evaluating and quantifying indirect costs and benefits of industrial robots, multiple attribute decision analysis (MADA) may provide a means of assisting management in the investment decision. MADA attempts to identify important attributes that are relevant to the comparison of alternatives and focuses on the required trade-offs. In addition, the treatment of qualitative factors is made explicit.

Cosineau [2] has developed a robot application matrix, illustrated in Table 11. This matrix is a means to evaluate potential robotic applications, with different factors weighted for importance. In the matrix, "must" criteria are required for any application to be feasible. The "want" criteria consist of benefits that are desirable but not necessary for the use of an industrial robot.

Cosineau's matrix has been expanded in Table 11 to include many attributes necessary to evaluate nonmonetary costs and benefits. Management must then rank arbitrarily assigned 1.0. Other attributes receive a rank that is a percentage of importance compared to the highest attribute. Next, each attribute (i) is ranked in importance across the alternatives (j) being considered so that a score, X_{ij}, reflecting how well each alternative satisfies each attribute can be estimated. An X_{ij} score of one means that alternative j maximizes performance on attribute i; a score of zero signifies minimum performance. The last step in the procedure is to multiply the ranking and scores of each alternative across all attributes. The products are then summed to give a total measure of "value" for each alternative.

In Table 12 the ICAM case study is evaluated using indirect benefits which have been ranked and weighted [18]. The two alternatives considered are the proposed robotic method and the present manual method. The robotic method has a higher overall score which indicates that indirect benefits can be included in the decision analysis and used to supplement information obtained in Table 8. It would appear that investment in the robotics system is justified. This same procedure could be easily extended to the analysis of mutually exclusive (competing) robotics systems.

The main advantage of a MADA model such as the above is that a firm's strategic objectives can be explicitly taken into account.

7. CONCLUSIONS

In this paper it has been suggested that industrial engineers who specialize in the practice of engineering economy will play an increasingly important role in replacement decisions where improved technology is involved. The conventional models that are utilized in like-for-like replacements of existing assets must be placed in a broader context where the strategic aims of a firm are considered as an integral part of the replacement decision in high technology areas. A modified ICAM case study of a robotic system was used for the purpose of demonstrating how strategic issues can be taken into account through multiple attribute decision analysis.

8. BIBLIOGRAPHY

[1] Ayres, Robert U. and Steven Miller, "Robotics, CAM, and Industrial Productivity," National Productivity Review, pg. 42-60, Winter 1981-82.

[2] Cosineau, David T., "Robots Are Easy...It's Everything Else That's Hard," The Industrial Robot, pg. 50-56, March, 1981.

[3] DeGarmo, E. Paul, Sullivan, William G. and John R. Canada, Engineering Economy (Seventh Edition). New York: MacMillan Publishing Company, 1984.

[4] English, J. Morley, "Engineering Economics and Technological Change," Engineering Education, pg. 390-393, February, 1983.

[5] Eshleman, R. L. and F. S. Pagano, "What Managers Need to Know When Choosing Robots," National Productivity Review, pg. 242-256, Summer 1983.

[6] Fleischer, G. A., "A Generalized Methodology for Assessing the Economic Consequences of Acquiring Robots for Repetitive Operations," Proceedings of 1982 Annual Industrial Engineering Conference, pg. 130-139, May, 1982.

[7] Gerstenfeld, Arthur, "A Model for Economic and Social Evaluation of Industrial Robots," Proceedings of 12th Industrial Symposium on Industrial Robots and Robots 6, pg. 341-348, 1982.

[8] Grant, Eugene L., Principles of Engineering Economy (Third Edition). New York: The Ronald Press Company, 1950, Chapters 16 and 20.

[9] Grieve, R. J., P. H. Lowe and M. P. Kelly, "Robots: The Economic Justification," Autofact Europe Conference Proceedings, Geneva, Switzerland, pg. 2-1 to 2-12, September 13-15, 1983.

[10] Groover, Mikell P., "Meeting the Challenges of CAD/CAM," National Productivity Review, pg. 29-35, Winter 1982-83.

[11] Groover, M. P. and J. E. Hughes, "Job Shop Automation Strategy Can Add Efficiency to Small Operation Flexibility," Industrial Engineering, pg. 67-76, November, 1981.

[12] Hayes, Robert H. and David A. Garvin, "Managing As If Tomorrow Mattered," Harvard Business Review, pg. 71-79, May - June, 1982.

[13] Horowitz, Ira, "Engineering Economy: An Economist's Perspective," AIIE Transactions, pg. 430-437, Vol. 8, No. 4, December, 1976.

[14] Hudson, C. A., "Computers in Manufacturing," Science, pg. 818-825, Vol. 215, No. 4534, February 12, 1982.

[15] Kutcher, M., "Automating It All," IEEE Spectrum, pg. 40-43, May, 1983.

[16] Logue, Dennis E. and Richard R. West, "Discounted Cash Flow Analysis: A Response to the Critics," National Productivity Review, pg. 233-241, Summer, 1983.

[17] Meyer, Ronald J., "A Cookbook Approach to Robotics and Automation Justification," Society of Manufacturing Engineers, Dearborn, MI, Report No. MS82-192, 1982.

[18] Morris, William T., Engineering Economic Analysis. Reston, Virginia: Reston Publishing Company, 1976, Chapter 7.

[19] Naisbitt, John, Megatrends--Ten New Directions Transforming Our Lives. New York: Warner Books, 1982.

[20] Potter, Ronald D., "Analyze Indirect Savings in Justifying Robots," Industrial Engineering, pg. 28-29, November, 1983.

[21] Radnor, Michael, "A Critical Evaluation of the Field of Engineering Economy," The Journal of Industrial Engineering, pg. 133-141, May-June, 1964.

[22] Skinner, Charles S., "The Strategic Management of Technology," Autofact 4 Conference Proceedings, Philadelphia, PA, pg. 1-6 to 1-16, November 30-December 2, 1982.

[23] Stauffer, Robert N., "Equipment Acquisition for the Automatic Factory," Robotics Today, pg. 37-40, April, 1983.

[24] Terborgh, George, Business Investment Policy - A M.A.P.I. Study and Manual, Machine and Allied Products Institute, Washington, D.C., 1958.

[25] Toepperwein, L. L., et al., ICAM Robotics Application Guide, Wright-Patterson Air Force Base, OH: Materials Laboratory (AFWAL/MLTC), Report No. AFWAL-TR-80-4042, Volume 2, 1980.

[26] Van Blois, John P., "Economic Models: The Future of Robotic Justification," Proceedings of the 13th Industrial Symposium on Industrial Robots and Robots 7, Chicago, IL, April 17-21, 1983, pg. 4-24 to 4-31.

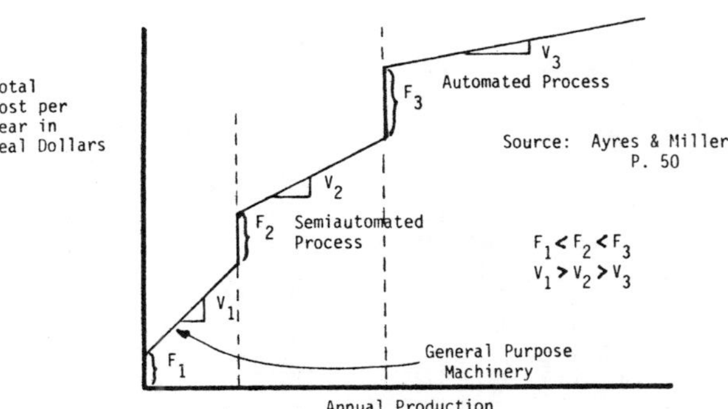

Figure 1. Total Cost Versus Annual Production

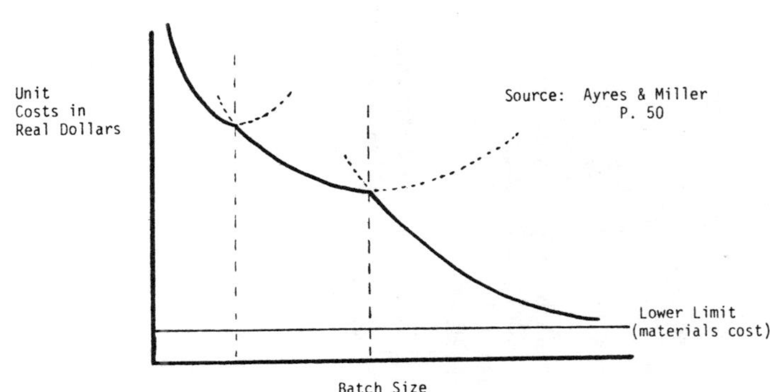

Figure 2. Unit Cost Versus Batch Size

Table 2

Current Trend Toward Integration Of Machine Tool Functions

1984	1986	1994
Tactical Approach to Replacement (Use Conventional Models)	Transition Period (Use Conventional Models, Executive Overrides to follow the Competition or "Wait and See")	Integrated Approach to Replacement (Adopt a Systems View of Entire Flexible Manufacturing Centers)
Machine Tools	Functions	Manufacturing Centers
M102 G113 T491 . . . B717	ABC . . . XYZ	α
Basic Question: Keep Asset 1 more year or replace Now?	Basic Question: Does the Function Fit into our Automation Plan?	Basic Question: Has the Full Potential of Factory Automation Been Met?

"Automate, emigrate or evaporate"

James A. Baker
General Electric Co.

Table 3

Strategic Level Investment Decisions in Which Engineers Will Play a Greater Role

Current Tactical Decisions	Future Strategic Decisions
Replacement of existing technology (like-for-like)	Replacement with improved technology (different functions)
Improvements to existing products/processes	Development of new products/processes
Expansion/relocation of departments in existing plants	Expansion/relocation of entire plants
Equipment modernization projects	Plant modernization projects
Abandonment of existing products/processes	Acquisitions/mergers/divestitures

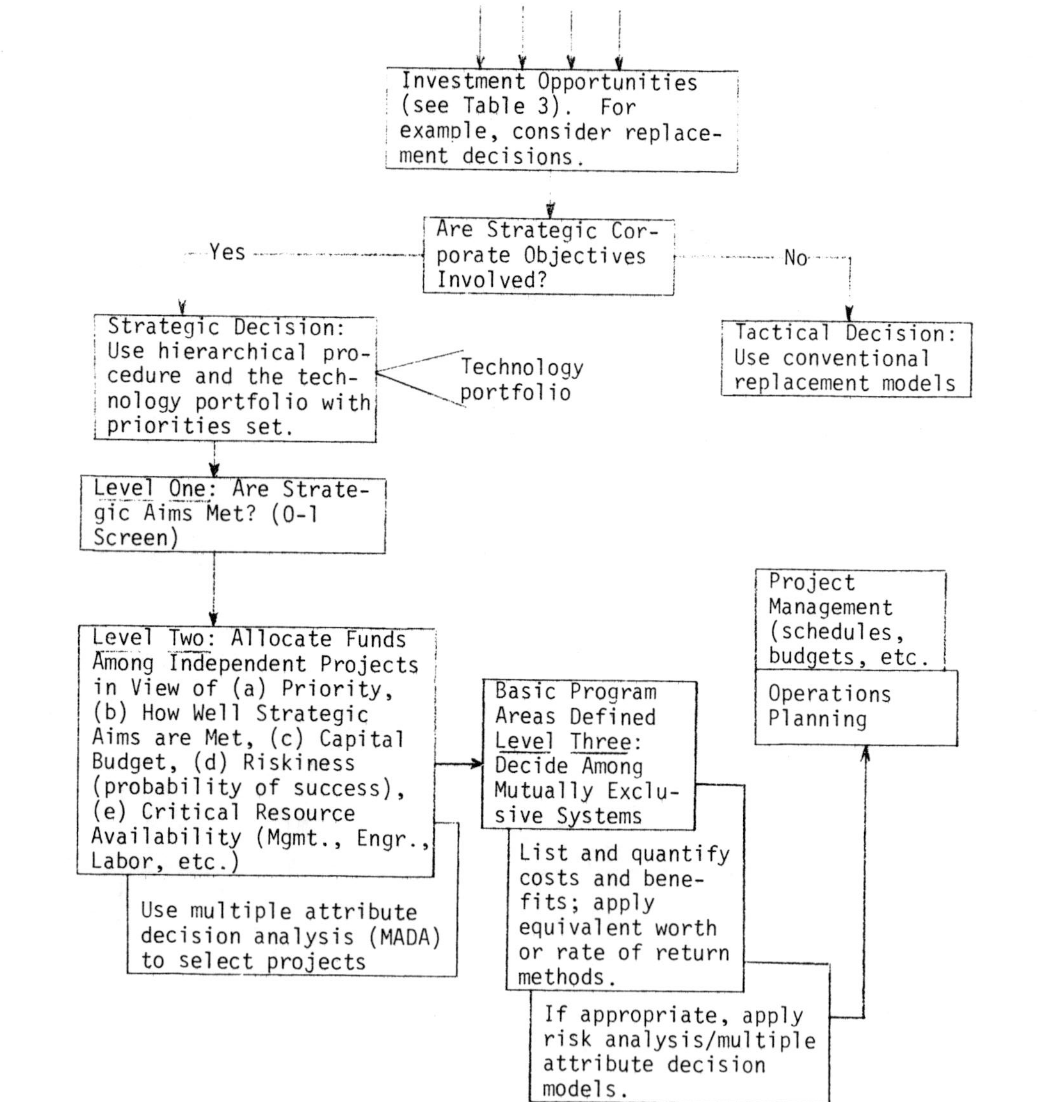

Figure 3. A decision hierarchy for replacement investments involving high technology.

Table 4

Direct Costs and Savings for Industrial Robots
[From 5,6,25]

Robot Cost - Basic cost of robot, operational equipment, maintenance and test equipment.

Accessories Costs - Additional equipment, hardware, recorders, testers, computers, tools.

Related Expense - Additional hardware costs and expenses for application, conveyors, guard rails, component cabinets, interface hardware, insurance.

Engineering Costs - Estimated cost of planning and design in support of project development, research and laboratory expenses, programming, modification of existing equipment.

Machinery and Equipment - Avoidance of new equipment or tooling to continue in base condition.

Installation Costs - Labor and materials for site preparation, floor or foundation work, utility drops (air, water, electricity), and set-up costs.

Tooling Costs - Labor and material for special tooling, end-effectors, interface devices between controller and tooling, fabrication of part positioners, fixtures and tool controllers.

Depreciation - Yearly depreciation using A.C.R.S. provisions of E.R.T.A. (1981).

Other Costs - Supplies, utilities, training, documentation, planning, warranty costs.

Product Change Costs - Product design changes, material requirements.

Direct Labor Savings - Net direct labor savings: direct labor, benefits, allowances, shift premiums, overhead, setup time, labor rate changes.

Indirect Labor Savings - Net indirect labor savings: maintenance, repair, labor support costs.

Maintenance Savings - Estimate of net maintenance savings: maintenance supplies, replacement parts, spare parts, lubricants, service contract charges.

Other Savings - Net savings or cost reductions: material savings, reduced scrap, reduced downtime, reduced nonproductive time.

Table 5

Indirect Costs and Associated Sources
[From 5,15,17,20]

In-Process Inventory - Faster process cycle, reduced/combined operations.

Finished Goods Inventory - Smaller lot sizes, faster material/ process cycle, better forecasting.

Scrap and Rework - Less operator error, "proved" program after first run, reduced scrap and rework.

Inspection - Automatic in-line 100% parts check, higher quality parts.

Operator Cost - Less training time to reach efficiency, less turnover and absenteeism.

Trucking - Fewer process operations, fewer moves, no inspection area, robotic conveyors.

Floor Space - Fewer machines, less in-process material, less room for incoming-outgoing jobs, less waiting for inspection, fewer operations, better work flow.

Efficiency - Renewed operator and foreman enthusiasm, new look at methods.

Profitability - Better delivery to customer, better quality parts, more flexibility to alter standard designs.

Utilization - Reduction in production lead time, consistent floor-to-floor times.

Production Quantity - Increase in parts produced by robot in consistent, non-stop mode, accuracy, repeatability; no fatigue, shortened attention span, boredom, vacations, illness.

Machine Obsolescence - More universal and flexible than special or hard automation.

Safety Factors - Reduced costs from not having to buffer people from adverse working conditions: safety devices, glasses, gloves, masks, ventilation systems, temperature.

Energy - Cost of utilities may be reduced.

Table 7

ICAM Net Income

Year	Before Tax Cash Flow (a)	Depr (b)	Taxable Income (c)	Income Tax (d)	Net Income (e)
1	+83200	-35000	+48200	-22172	+26028
2-7	+88200	-35000	+53200	-24472	+28728

1 ATCF = 26028 + 35000

2-7 ATCF = 28728 + 35000

(a) Total operating savings/costs
(b) Straight line depreciation
(c) a + b
(d) c x(tax rate of 0.46)
(e) c + d

Table 6

Estimation of Unit Savings (T_1) Associated with a Robot System [See 9].

Factor	Year 1		2-N
A. Savings in Direct Labor	a. Robot Output: $a_r(60/t_t)$	a. Manual Output: $a_m(60/t_m)$	. . . etc.
	b. Robot Work Hrs: $P_1/a_r\cdot(60/t_r)$	b. Manual Work Hrs: $P_1/a_m\cdot(60/t_m)$	
P_1 = planned production units/year	c. Cost of Robot "Labor": $P_1C_r/a_r(60/t_r)$	c. Cost of Manual Labor: $P_1C_m/a_m(60/t_m)$	
t_r = cycle time, min./pc., for robot			
t_m = cycle time, min./pc., for manual operator	Annual Savings in Direct Labor =		
a_r = availability of robot	$\frac{P_1C_m}{a_m(60/t_m)} - \frac{P_1C_r}{a_r(60/t_r)}$		
a_m = availability of manual operator			
C_r = out-of-pocket cost per hour for robot			
C_m = out-of-pocket cost per hour for manual operator			
B. Savings in Indirect Labor	etc.		
C. Savings in Maintenance	. .		
D. Savings in Material			
E. Other Savings			

OPERATING SAVINGS/COST ANALYSIS

Labor			
Direct			
Current Method	$140,000		
Proposed Method	$ 50,000		
Net Direct Labor Savings		$ 90,000	
Indirect			
Current Method	$ 500		
Proposed Method	$ 4,500		
Net Indirect Labor Savings		$ (4,000)	
Maintenance			
Current Method	$ 1,680		
Proposed Method	$ 3,980		
Net Maintenance Savings		$ (2,300)	
Other Savings			
Reduced Scrap (Material)	$ 7,000		
Other	0		
Total Other Savings		$ 7,000	

Other Costs	1st yr.	after 1st yr.
Training	$(5,000)	0
Supplies	$(2,500)	(2,500)
	$ (7,500)	$ (2,500)
Net Total Operating Savings	$ 83,200	$ 88,200

Figure 4

Savings Distribution Form

(From Toepperwein et al., 1980, p. 130)

Project Title: Robot Economic Analysis Example

		Year 0	Year 1	Year 2-8
Investment				
Capital Facilities				
1. Robot Cost		$ 65,000		
2. Accessories Cost		5,000		
3. Related Expense		20,000		
Development				
4. Engineering		170,000		
5. Installation		5,000		
6. Tooling		15,000		
7. TOTAL		280,000	0	0
Operating Savings/Costs				
Labor Savings				
8. Direct			90,000	90,000
9. Indirect			(4,000)	(4,000)
10. Maintenance Savings			(2,300)	(2,300)
11. Other Savings			7,000	7,000
12. Other Costs			(7,500)	(2,500)
13. TOTAL		0	83,200	88,200
Analysis				
14. Total Investment (less 10% ITC)		252,000	0	0
15. Total Savings	Line 13		83,200	88,200
16. Net Depreciation			35,000	35,000
17. Net Savings	Line 15-Line 16		48,200	53,200
18. Tax Rate 46%				
19. Net Savings After Tax	Line 17 x 0.54		26,028	28,728
20. Total Cash Return	Line 19 + Line 16		61,028	63,728
21. Net Cash Flow	Line 20 - Line 14		61,028	63,728
22. Internal Rate of Return = 18.7%				
23. Payback Period = 4 years				

Figure 5

Economic Analysis Form

(From Toepperwein et al., 1980, p. 125)

Table 8

Net Present Value for ICAM Case Study Based on Direct Benefits

Year	Before Tax Cash Flow	Depr	Taxable Income	Cash Flow for Inc. Tax[a]	After Tax Cash Flow	PV Factor[c]	Present Value
0	-280000			+28000[b]	-252000	1.000	-252000
1	+ 83200	-35000	+48200	-22172	+ 61028	.893	+ 54498
2	+ 88200	-35000	+53200	-24472	+ 63728	.797	+ 50791
3	"	"	"	"	"	.712	+ 45374
4	"	"	"	"	"	.636	+ 40531
5	"	"	"	"	"	.567	+ 36134
6	"	"	"	"	"	.507	+ 32310
7	"	"	"	"	"	.452	+ 28805
8	"	"	"	"	"	.404	+ 25746
			Net Present Value =				$ 62167

(a) Income tax rate of 0.46
(b) Investment tax credit of 10%
(c) (P/F,i,n) with i = 12%

Table 9

Direct and Indirect Savings Associated with the ICAM Project

	SAVINGS	Year 1	Year 2-7
"T_1" Factors	Labor		
	Direct	$ 90,000	$ 90,000
	Indirect	(4,000)	(4,000)
	Maintenance	(2,300)	(2,300)
	Materials	7,000	7,000
	Other Direct Savings	(7,500)	(2,500)
	Sub-Total	$ 83,200	$ 88,200
"T_2" and "T_3" factors	In-process Inventory	$ 11,800	$ 11,800
	Rework	6,600	6,600
	Floor Space	2,000	2,000
	Operator Improvement	5,000	5,000
	Safety and Energy	2,000	2,000
	Sub-Total	$ 27,400	$ 27,400
	TOTAL	$110,600	$115,600

Table 10

Net Present Value of ICAM Project Including Both Direct and Indirect Benefits

Year	Before Tax Cash Flow	Depr	Taxable Income	Cash Flow for Inc. Tax[a]	After Tax Cash Flow	PV Factor[c]	Present Value $
0	-280000			+28000[b]	-252000	1.0	-252000
1	+110600	-35000	+75600	-34776	- 75824	.893	+ 67711
2	+115600	-35000	+80600	-37076	+ 78524	.797	+ 62584
3	"	"	"	"	"	.712	+ 55909
4	"	"	"	"	"	.636	+ 49941
5	"	"	"	"	"	.567	+ 44523
6	"	"	"	"	"	.507	+ 39812
7	"	"	"	"	"	.452	+ 35493
8	"	"	"	"	"	.404	+ 31724
				Net Present Value =			$135668

(a) Income tax rate of 0.46
(b) Investment tax credit of 10%
(c) (P/F,i,n) with i = 12%

Table 11

Industrial Robot Application Matrix [From 2]

	Objectives	Weight
Must		
	Healthy Business	10
	Available Funding	10
	Large Volume of Parts	10
	Highly Repetitive Routine	10
	Labor Intensive	10
Want		
	Operational Simplicity	8
	Few Parent Equipment Modifications	9
	Minor Rearrangement Needs	4
	Inexpensive Ancillary Equipment	7
	Diminish Hazardous Situations	3
	Improve Hostile Environment	2
	Eliminate Strenuous Work	2
	Shop Management Support	5
	Labor Support	6
	High Return on Investment	9
	Low Funding Requirements	1

Table 12

Multiple Attribute Decision Analysis of The Robot and Manual Alternatives

Weight w_i	Rank	Nonmonetary Attribute i		Alternative j Robot	Manual
1.0	1	Healthy business	Rank	1	1
			x_{ij}	1.0	1.0
0.95	2	Available funding	Rank	1	1
			x_{ij}	1.0	1.0
0.95	3	Management/Technical Know How	Rank	1	2
			x_{ij}	1.0	0
0.8	4	Management commitment	Rank	1	2
			x_{ij}	1.0	0
0.8	5	Attitude toward risk	Rank	2	1
			x_{ij}	0	1.0
0.5	6	Cost savings	Rank	1	2
			x_{ij}	1.0	.7
0.5	7	Flexibility	Rank	2	1
			x_{ij}	0	1.0
0.5	8	Labor intensive	Rank	1	1
			x_{ij}	1.0	1.0
0.5	9	Repetitive motions	Rank	1	1
			x_{ij}	1.0	1.0
0.5	10	Shop management support	Rank	2	1
			x_{ij}	.5	1.0
0.5	11	Labor support	Rank	2	1
			x_{ij}	.7	1.0
0.5	12	Return on investment	Rank	1	2
			x_{ij}	1.0	0
0.3	13	Operating simplicity	Rank	1	1
			x_{ij}	1.0	1.0
0.2	14	Hazardous environment	Rank	1	2
			x_{ij}	1.0	.2
0.2	15	Strenuous task	Rank	1	2
			x_{ij}	1.0	.5
0.2	16	Tedious, boring task	Rank	1	2
			x_{ij}	1.0	.1
0.1	17	Morale improvement	Rank	1	2
			x_{ij}	1.0	.4
			Score $_j$	7.3	6.1
$Score_j = \sum_i w_i x_{ij}$			Normalized Score $_j$	1.0	0.84

William G. Sullivan is Professor of Industrial Engineering at Arizona State University. He is a Senior Member of AIIE, a director and past chairman of the Engineering Economy Division of the American Society for Engineering Education, and a member of the Editorial Board of The Engineering Economist. Dr. Sullivan received the Ph.D. in Industrial Systems Engineering from the Georgia Institute of Technology. He is a coauthor of three books and is a Registered Professional Engineer.

*Reprinted from **1981 Spring Annual Industrial Engineering Conference Proceedings.***

RISK ANALYSIS APPLICATION AND CAPITAL INVESTMENT EVALUATION

William R. Barr, P.E.
The Upjohn Company

ABSTRACT

The author briefly discusses the discounted cash flow techniques used for economic analysis and identifies shortcomings in their traditional application. He then shows how risk can be quantified and integrated into a standard evaluation to overcome existing limitations and improve analysis. The article contains a comprehensive example of the method for reference.

INTRODUCTION

The ultimate role of capital investment evaluation is more than the determination of simple profitability. In a broader context, the evaluation process provides management with the means to select investments which optimize the firm's return on capital assets and determine its future earnings potential and subsequent value to the shareholders [1].

In order to be used successfully, the evaluation process must address both the return and risk associated with an investment. Traditional analysis techniques provide only the former and usually leave the risk assessment up to management intuition. This article shows how traditional discounted cash flow techniques can be extended to provide useful, quantitative information on investment risk.

STARTING WITH A GOOD EVALUATION TECHNIQUE

All methods of capital evaluation generate an indicator of investment profitability which is used to rate projects for capital allocation.

Knowledgeable analysts favor the discounted cash flow techniques of net present value (NPV) and internal rate of return (IRR) to determine return on investment [1,2,3]. Both are relatively easy to perform once the investment cash flows have been identified, but each has its own advantages and drawbacks. IRR generates an intuitively comfortable number but cannot handle net cash flows of changing sign. It also suffers from a debate over the reinvestment assumption [4]. NPV is theoretically superior [1,2] but relies on a discount rate which is arguable and can discriminate against highly profitable investments which are small.

Analyzing a project with both techniques is the best approach because in practice they will usually indicate the same ranking among competing projects. Special cases involving questions of capital availability or discount rate have to be looked at more closely, but they usually deserve special analysis. Some firms use other investment evaluation methods such as payback and profitability index, but these techniques are inferior and should be avoided.

However, the quality and utility of an analysis are just as important as the return on investment indicator used. Although this point is basic, it deserves to be repeated. A capital evaluation method is useless if management does not understand the method employed. Worse yet, it is misleading if poorly done. It is critical to use uniform assumptions for factors such as inflation rate and recognize all pertinent cash flows. The computer installation which replaces one clerk and needs two programmers and a technician for support must be avoided.

A serious word about computers: they greatly facilitate the computations involved in capital evaluation. Many firms have in-house or rented programs which incorporate the investment analysis technique of their choice. This power enhances the analyst's ability to do his job by emphasizing data collection and is important if the evaluation is to be expanded into risk analysis.

THE TROUBLE WITH TRADITION

The basic problem a decision-maker encounters with the traditional economic analysis is that it gives him just one piece of information--the expected return on investment as measured by estimated NPV and IRR. He would also like to know how this expected value responds to changes in significant factors such as labor cost and sales volume. In general, he would like to know how reliable the expected value is.

Questions of reliability lead into the first phase of risk analysis. In order to accommodate the uncertainty associated with major project parameters, a good analyst will perform "sensitivity analysis" on the project. For example, the decision to invest in the design and production of a new product must be justified by the profits to be earned on sales. However, the sales forecast for a new product is notoriously unreliable. Sensitivity analysis in this case would involve the examination of profits at different sales levels to bracket the return on investment between high and low values. In a similar fashion, the project's sensitivity to cost overruns could be established.

The information gained from sensitivity analysis is often displayed in a format similar to Figure #1, which is a representation of the example above. Note how the parameter limits were chosen on the pessimistic side, reflecting the risk averse nature of most decision-makers.

Sensitivity analysis clearly provides a better description of potential project outcome than the expected return on investment alone. Unfortunately, it proves inadequate in two areas. It is cumbersome to perform and display for more than two parameters and becomes even more difficult as the number of outcomes examined within each parameter increases. Even more important, it gives no indication of the probability of occurrence for each event shown.

In the new product example, suppose the firm had a 30% IRR hurdle rate [5] for new product projects. When management examines the sensitivity analysis, it can see that three outcomes satisfy the criteria and three do not. This situation effectively places management in the role of second guessing the marketing and engineering specialists when it tries to determine the likelihood of each possible outcome.

<u>Project</u>: New Product Example

<u>Investment</u>: $100,000

<u>Expected IRR</u>: 35%

<u>Expected NPV</u>: $250,000

<u>Sensitivity Analysis</u>

	Project Cost	
Sales Volume	On Budget	10% Over Budget
Forecast + 10%	40%	33%
On Forecast	35%	28%
Forecast - 20%	20%	13%

Fig. 1 Sensitivity Analysis Display

A QUANTUM LEAP FORWARD

Since the discussion is getting into new ground, it's time to review nomenclature. The article will use return on investment (ROI) to describe the net benefits realized from an investment and use NPV and IRR as measures of these benefits. The project outcome is a set of physical events which will generate a financial ROI (measured by NPV and IRR), so the best estimate of these events yields an expected ROI.

The actual project outcome is affected by (or sensitive to) a number of factors which the investors cannot control. This lack of control causes uncertainty. Risk is defined as the quantification of this uncertainty, and the keystone of risk analysis is that it is possible to describe risk in a meaningful and useful way by examining the variability of major project parameters [6].

We can illustrate the utility of risk analysis by enlarging the new product example given earlier. Let the firm have the opportunity to develop one of two new products, but not both because of the cost. The two proposals are said to be mutually exclusive. Suppose that both products have an expected IRR of 35%, and the firm's hurdle rate is an IRR of 30%. At this point the firm would be indifferent to which of the two investment alternatives it chooses, and management would make its decision on qualitative factors.

However, if a risk profile of the two products shows that the outcome of A is far less likely to generate an IRR below the hurdle rate than B, we know a great deal more about the desirability of both products.

Such a comparison is shown in Figure #2 where area "1" is the probability that product A will yield an IRR below 30%, and area "2" is the probability that product B's IRR will fall below the hurdle rate. Given this information, most risk averse management will choose product A, even though both products have the same expected IRR.

This is the power of risk analysis. It actually allows the financial analyst to quantify the reliability of his ROI estimates and provide management with important information regarding the riskiness of a capital investment project.

A RISK ANALYSIS APPLICATION

The concept of risk analysis is easy enough to illustrate and understand, but performing an actual analysis is quite another matter. The sheer volume of computation requires an investment in analyst or computer time in addition to project cost. For this reason, it is best suited to large projects where the added expense is only a small portion of total cost.

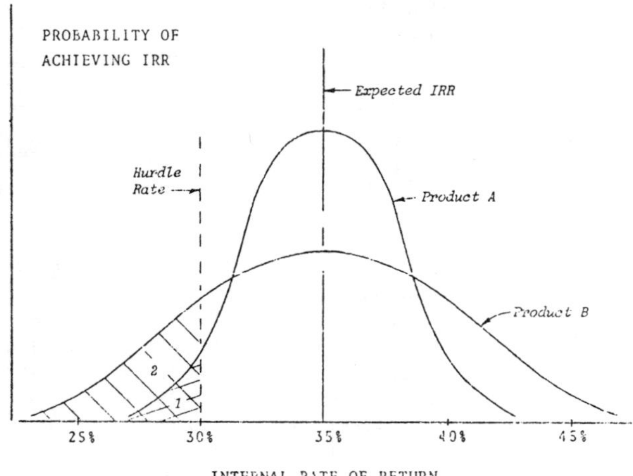

Fig. 2 Risk Profile of Products A & B

Is the added information worth the extra cost? Usually, although the answer is subjective. The larger a project investment, the more parameters there are that will affect eventual outcome. In short, more can go wrong and the harder it is to intuitively assess the reliability of the expected ROI. As so often happens in life, it becomes increasingly more important to do the job right as the job gets harder to do.

The author used risk analysis to evaluate a pharmaceutical process renovation where the project ROI was highly sensitive to five major parameters:

1. Production (sales) volume. Sales for the product, which was manufactured at a dedicated facility, had recently declined, and the forecasters did not know if this reflected a new trend or a seasonal slump.

2. Estimated cost savings. The dollar savings estimated depended in turn on labor time, production yield and materials use, maintenance support, and energy consumption. Since the process being considered was new, these estimates were prone to error.

3. Investment cost. In addition to the usual chance of a cost overrun, the project was subject to regulatory activity that might have increased cost through the imposition of new air quality standards.

4. Inflation rate. The analysis was conducted at a time when the magnitude of cost increases for labor, materials, and energy was highly uncertain.

5. Tax rate. Because of a unique organizational situation the operation was subject to fluctuating tax rates.

Under these conditions of uncertainty, management had to develop a quantitative likelihood of success or forego tremendous benefits to be gained in operating savings, new technology, and output capacity.

To do so, the project team identified the significant factors above and developed a probability profile for each, which is illustrated in Figure #3. In general, profile development is most easily done as shown here, by selecting several events within a range of possible outcomes and assigning probabilities to each that sum to one [6].

Production Volume		Project Savings	
Volume	Probability	Savings	Probability
800 lots/yr.	.10	1200/lot	.10
700 lots/yr.	.15	1100/lot	.40
600 lots/yr.	.40	1000/lot	.30
500 lots/yr.	.25	900/lot	.10
400 lots/yr.	.10	800/lot	.10

Project Cost		Inflation Rate	
Cost	Probability	Rate	Probability
$2.75 million	.20	15%	.20
$2.50 million	.60	12%	.50
$2.25 million	.20	9%	.30

Tax Rate	
Rate	Probability
40%	.80
30%	.20

Total number of possible outcomes =

5x5x3x3x2 = 450

Fig. 3 Risk Profiles of Significant Project Factors

This process is effective for two reasons. First, it forces the project team to explicitly identify the significant factors which will eventually affect project outcome. Second, it places risk assessment in the hands of those most qualified to measure it. For this project, the sales forecasting group estimated production volume and associated probability. Engineering compiled the project cost profile and so on. In most cases, the concept of assigning probabilities to a range of events was resisted at first but ultimately embraced when the individuals understood that their estimates were the best available and certainly better than none at all.

It is important that the risk factors be chosen so that they are independent of each other. For example, labor cost cannot be defined with a rate which changes with inflation, then inflation itself selected as another risk factor. Interdependent risk factors will skew the analysis and render it meaningless unless advanced probability theory is used to correct it. There is no point in making this evaluation tougher than it has to be.

With independent risk factors, the total number of possible outcomes is the product of the number of events examined in each risk factor profile, as illustrated in Figure #3. If each outcome is computed, the analyst can build a discrete histogram to describe the project's probability distribution, similar to the type of profile shown in Figure #2. (The Figure #2 profile is continuous, and the only way to develop such a curve in practice is to specify a mathematical probability distribution for each risk factor and combine them--again the theory becomes very sophisticated and unnecessary).

At this point the computer is needed unless the number of outcomes is trivially small. For this project, the team spent about $1,000 for programming and computer time to generate a probability distribution histogram. This can be done either by a Monte Carlo simulation or decision tree-type analysis. The simulation approach is well suited for very large projects with greater than about 5,000 outcomes, but this project used the decision tree approach and computed IRR, NPV, and probability for each outcome by following every branch of the tree to conclusion. This technique is illustrated in Figure #4 with a simple example.

Since the project programmer was blessed with common sense, his program rank ordered and counted the computation results, giving the project team a completed risk profile. Such a study provides an overwhelming amount of information, but the team limited its report to three critical items. With a hurdle rate of 15% IRR, it showed management that:

Median IRR = 19%

Probability of exceeding 15% IRR = .95

All outcomes fell between IRR of 12% and 28%

The project was approved and funded on the basis of its high probability of exceeding hurdle rate and excellent downside risk protection.

RISK ANALYSIS EXTENSIONS

Before we look at this technique's future applications it is worthwhile to review basics for a moment. Just as there is a bit of art in correctly identifying the proper cash flows for NPV and IRR calculations, there is judgment involved in the selection of risk factor profiles. Management must be made intensely aware that the project outcome probabilities shown for evaluation are a direct result of the risk factors used, and even encouraged to review the individual risk factor profiles. The project team in the previous example also found that such a review was quite valuable in selling this "new" evaluation technique to management.

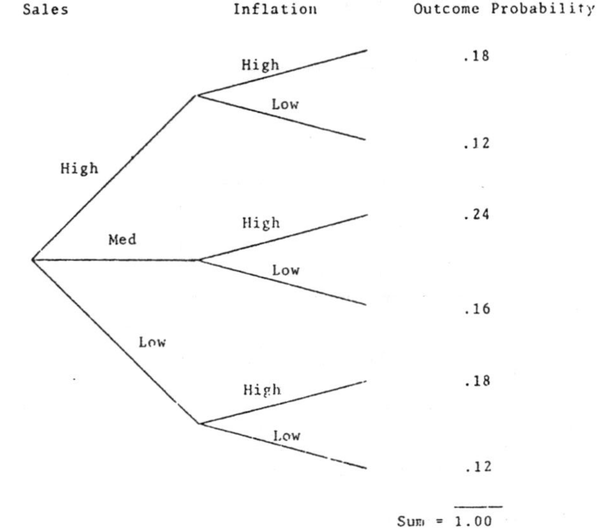

Fig. 4 The Decision Tree

What else can be done with risk analysis? It is useful as described above but could be developed further. In probabilistic approaches to investment portfolio analysis, the standard deviation is a commonly used and well accepted measure of risk [1,3]. It is a simple task to compute the standard deviation of a project IRR once the project probability distribution has been developed.

Given such a complete and quantitative investment description, it would be possible to explicitly address the risk return trade-off in investment evaluation [1]. The capital allocation process could then utilize a moving hurdle rate to demand higher expected project ROIs as project risk increased, a move one step closer to truly maximizing the aggregate return on invested capital.

CONCLUSION

This article has painted a picture of what can be done to enhance capital evaluation rather than present a stepwise algorithm for risk analysis. The analyst who is comfortable with discounted cash flow and has a nodding acquaintance with probability can master risk analysis with little trouble. The manager who is only vaguely familiar with either will need help to successfully apply the technique.

Many investments are made based on a feeling of intuitive need with little formal justification. There is a hierarchy of capital evaluation techniques illustrated in Figure #5, where each level provides more information than the one below. As one moves upward the quality of the evaluation improves, but the cost of analysis increases at the same time. While small investments can be evaluated at lower levels with a small penalty cost for error, the large capital projects on whose success the firm's future depends are well worth the added effort of a proper evaluation.

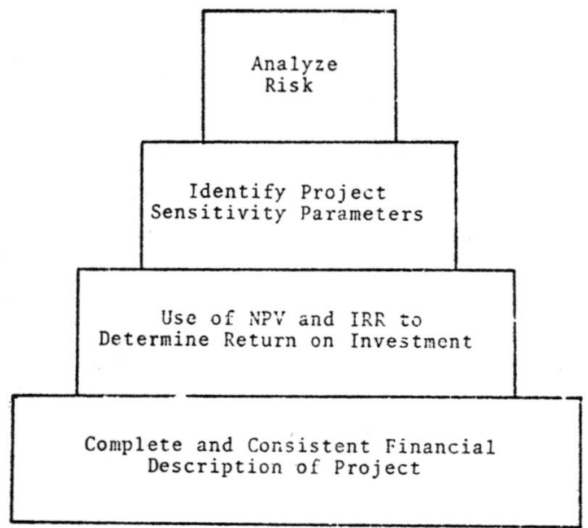

Fig. 5 Hierarchy of Capital Evaluation

ACKNOWLEDGEMENTS

The author would like to acknowledge the work of D.B. Magerlein on the analysis described here and extend his thanks for Dave's many contributions in the probability theory and computer programming needed.

We would also like to thank R.H. Paulson for his editorial guidance and acknowledge D.B. Hertz's 1964 article on which so much of this work is based.

REFERENCES

[1] VanHorne, J.C., Financial Management and Policy, Prentice-Hall, Inc., Englewood Cliffs, New Jersey, 1974.

[2] Baumol, W.J., Economic Theory and Operations Analysis, Third Ed., Prentice-Hall, Inc., Englewood Cliffs, New Jersey, 1972.

[3] Smith, G.W., Engineering Economy: Analysis of Capital Expenditures, Third Ed., Iowa State University Press, Ames, Iowa, 1979.

[4] Blankenship, J.E., "Discounted Cash Flow and and Reinvestment Assumption," Managerial Planning, May/June, 1978, pp. 16-20.

[5] Brighan, E.F., "Hurdle Rates for Screening Capital Expenditure Proposals," Financial Management, Autumn, 1975, pp. 17-25.

[6] Hertz, D.B., "Risk Analysis in Capital Investment," Harvard Business Review, Jan.-Feb., 1964, pp. 95-106.

BIOGRAPHICAL SKETCH

William Barr is a staff Industrial Engineer for the Pharmaceutical Production Division of The Upjohn Company in Kalamazoo, Michigan. He has been with Upjohn since 1974, and also held positions as a financial analyst for the Consumer Products Marketing group and liaison for Puerto Rico Pharmaceutical Operations. In addition to his experience in the private sector, he has taught evening classes at Western Michigan University and served in the U.S. Army for two years.

Mr. Barr holds degrees in Industrial Engineering from the University of Michigan, receiving his B.S. in 1970 and M.S. in 1974. In 1977 he completed an MBA in Management at Western Michigan University. He is a registered Professional Engineer in Michigan and a member of the American Institute of Industrial Engineers and the Tau Beta Pi engineering society.

Reprinted from ***1985 Annual International Industrial Engineering Conference Proceedings.***

EQUIPMENT REQUIREMENTS PLANNING USING A SIMULATION SUPPORT SYSTEM

Laurie J. Rolston
Pritsker & Associates, Inc.
P.O. Box 2413
West Lafayette, IN 47906

ABSTRACT

The use of a manufacturing simulation language in conjunction with a simulation support system is well illustrated by a consulting project performed recently. This project involved determination of equipment requirements in terms of machine tools, number of fixtures, and number of material handling devices.

INTRODUCTION

Computer simulation has proven itself to be an invaluable technique for analyzing manufacturing systems. Equipment requirements planning, bottleneck identification, capital investment evaluation, and constrained resource scheduling are examples of common manufacturing problems which can be addressed using simulation.

State-of-the-art simulation software now makes it easier than ever to perform simulation projects. Simulation languages tailored specifically to manufacturing systems have been developed and successfully applied. Integrated simulation support software is available to perform the model building, data management, data presentation, and data analyses tasks inherent in a simulation project.

The use of a manufacturing simulation language in conjunction with a simulation support system is well illustrated by a consulting project performed recently. This project involved determination of equipment requirements in terms of number of machine tools, number of fixtures, and number of material handling devices.

SYSTEM DESCRIPTION

Figure 1. contains a layout of the FMS under study. Part flow is illustrated in Figure 2. This FMS performs machining operations on castings. Castings are initially loaded onto pallets (carrying 16 parts each) and sent to one of two lathes via conveyor. Upon completion of the turning operation, the parts are transported, again by conveyor, to a wash load area before being sent to the machining center via a wire-guided vehicle.

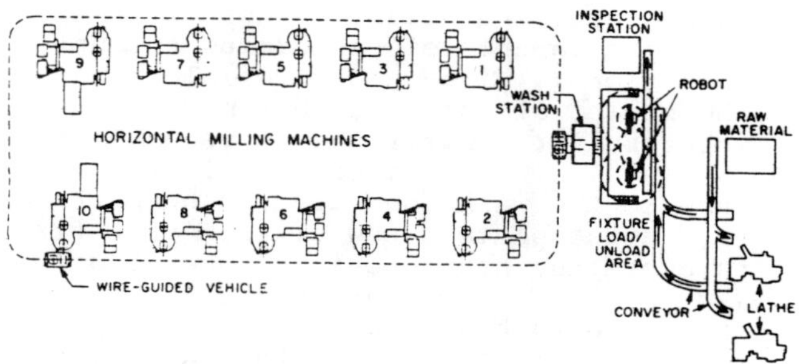

Fig. 1. System Layout

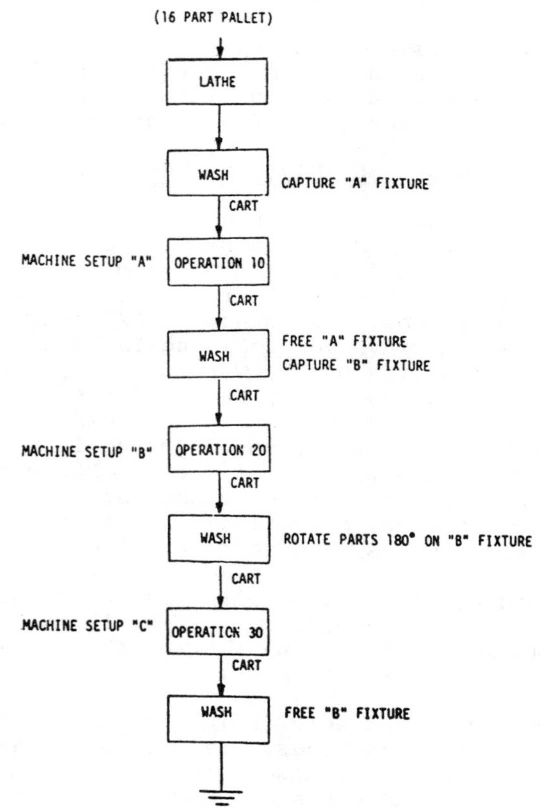

Fig. 2. Part Flow

The machining center consists of ten identical horizontal milling machines which can perform any one of three operations. The machines are set up to be either dedicated (i.e., assigned to only one operation) or flexible. The three operations, referred to as 10, 20, and 30, require 7.5, 2.5, and 3.5 minutes per part, respectively.

Two types of fixtures (A and B) are used in this system. Fixture A is used for operation 10 and Fixture B is used for operations 20 and 30.

Before a pallet is sent to a machining operation, each part in that pallet is attached to a sixteen part fixture and sent through the wash station. When buffer space becomes available at one of the machines capable of performing the next operation, a wire-guided vehicle transports the pallet of parts to that machine's input buffer. After the parts have been machined, a wire-guided vehicle returns them to the wash/load area. The parts are then attached to a new sixteen part fixture to await the next machining operation. (In the case of operation 30, the parts are not attached to a new fixture; instead, they are rotated 180° on the same fixture (i.e., Fixture B).) After all three machining operations have been completed, the parts are sent to final inspection for subsequent departure from the system.

The purpose of the simulation study was to determine the optimal level of milling machines, carts, and fixtures to achieve a production goal of 3140 parts per week.

MODEL BUILDING AND EXECUTION

The simulation language chosen for the project was MAP/1, a simulation-based modeling and analysis program developed specifically for discrete part manufacturing systems. Building a MAP/1 model of this FMS involved defining the system in terms of its parts, workstations, material handling equipment, and fixtures. Parts were defined by specifying their arrival pattern and their routing (the list of operations to be performed). For each operation performed on the parts, the workstation, operation time, fixturing requirements, and next operation were specified. In the case where the next operation could be performed at either a dedicated machine or a flexible machine, rules for selecting between the two were specified in the routing.

Workstations were defined in terms of the number of machines, size of the inventory storage areas, associated material handling device, and operation mode (either a regular machining process or a fixturing operation). The material handling devices were defined in terms of their capacities, transport and response times, and, in the case of conveyors, the path to be followed. Fixtures were defined in terms of their capacity and the operations during which they were required.

Since MAP/1 contains constructs to represent the generic physical and control components of discrete part manufacturing systems, the tasks of model conceptualization and model development were significantly reduced. Abstraction from the system description into network or block diagram formulations, or into FORTRAN programming of discrete event logic, (required when using general purpose simulation languages) was eliminated. The facility layout, equipment specifications, and part routing sheets were directly translated into the corresponding MAP/1 constructs.

A MAP/1 interactive input system was available to facilitate the task of model building. The input system was implemented using a series of forms which contained specifications for all the manufacturing system components and their parameters. Figure 3. contains example PART, ROUTE, STATION, TRANSPORTER, CONVEYOR AND FIXTURE forms created while building a model of the FMS just described. On-line error checking, help, and documentation features significantly reduced the amount of time spent on the model building task.

MAP/1 automatically reported system performance in terms of throughput, work-in-process levels, equipment and personnel utilizations, and part time-in-system. These performance measures facilitated the comparison of various system configurations against established system performance requirements.

The use of specialized simulation software and model building support software can significantly reduce the time required to create simulation models of manufacturing systems. The utility of simulation analysis in terms of project costs and timely project results is thus increased. Use of a simulation language specifically developed for manufacturing, or one which contains a significant amount of manufacturing-specific constructs, reduces model development time and eliminates costly redevelopment of modeling logic which is common across a variety of system models. Use of interactive model building support software decreases model development, model entry, and model debugging time by reducing or eliminating the task of formatting model input for computer translation by providing on-line error checking, and by offering on-line help and documentation features.

PART FORM

PART TYPE NAME: CASTING PART TYPE PRIORITY: 0

PART TYPE INTERARRIVAL: 42.67

FIRST PART ARRIVAL: 0.0000E+00

ARRIVING LOT SIZE: 2

EXPECTED FLOWTIME: 0.0000E+00

INITIAL ROUTE

STATION NAME: LATHE STATION OPERATION NUMBER: 1

SETUP TIME: 0.000000000

OPERATION TIME: 42.67

UNLOAD FIXTURE: NO LOAD FIXTURE: NO

FIXTURE NAME: NUMBER OF PARTS TO LOAD: 0

INITIAL ROUTE STATION LIST

STATION NAME	OPERATION #	CONDITION OR PROBABILITY
STORE	1	
	1	
	1	
	1	
	1	
	1	
	1	
	1	
	1	
	1	

ROUTE FORM

PART TYPE NAME: CASTING

STATION NAME: WASH STATION OPERATION NUMBER: 2

SETUP TIME: 0.000000000

OPERATION TIME: 2

UNLOAD FIXTURE: YES LOAD FIXTURE: YES

FIXTURE NAME: FIXTUREB NUMBER OF PARTS TO LOAD: 1

STATION LIST

STATION NAME	OPERATION #	CONDITION OR PROBABILITY
MILL2	1	NPRE(MILL2).NE.2
MILLU	2	
	1	
	1	
	1	
	1	
	1	
	1	
	1	
	1	

STATION REGULAR FORM

STATION NAME: MILL2 STATION SIZE: 2

PREPROCESS INVENTORY STORAGE SPACE: 2

POSTPROCESS INVENTORY STORAGE SPACE: 2

MATERIAL HANDLING CLASS: REGULAR

MATERIAL HANDLING EQUIPMENT NAME: CART TRANSPORTATION LOT SIZE: 1

EXCESS RULE: BLOCK

SHIFT SCHEDULE IDENTIFIER: 1 END OF SHIFT RULE: START

MAXIMUM OVERTIME: 0.1000E+10

OPERATION MODE: REGULAR

TRANSPORT FORM

TRANSPORTER NAME: CART

NUMBER OF TRANSPORTERS: 2

TRANSPORTATION TIME: 1

RESPONSE TIME: 0.000000000

SHIFT SCHEDULE IDENTIFIER: 1

TRANSPORTER VELOCITY: 1.000

CONVEYOR FORM

CONVEYOR NAME: CONVEYR1

CONVEYOR VELOCITY: 120.0

EXCESS RULE: ACCUMULATE

CONVEYOR TYPE: OPEN

SHIFT SCHEDULE IDENTIFIER: 1

PATH LIST

JUNCTURE TYPE	JUNCTURE NAME	CAPACITY	DISTANCE
STATION	LATHE	10	10.00
STATION	STORE	3	0.0000E+00
STATION		3	0.0000E+00
STATION		3	0.0000E+00
STATION		3	0.0000E+00
STATION		3	0.0000E+00
STATION		3	0.0000E+00
STATION		3	0.0000E+00
STATION		3	0.0000E+00
STATION		3	0.0000E+00

FIXTURE FORM

FIXTURE NAME: FIXTUREB

NUMBER OF FIXTURES: 12

SHIFT SCHEDULE IDENTIFIER: 1

FIXTURE ALLOCATION RULE: SEQUENCE

STATIONS:

WASH

ANALYSIS

Using conventional analysis techniques, the number of milling machines required could be calculated based on throughput requirements, machine efficiencies, part processing times, and production hours available. For a throughput goal of 3140 parts per week, assuming a 90% machine efficiency, and given part processing times of 7.5 minutes per part for operation 10, 2.5 minutes per part for operation 20, and 3.5 minutes per part for operation 30, with 4800 production hours available per week, the number of milling machines required for each operation was:

Number of parts/minute

$$\frac{3140 \text{ parts/week}}{(5 \text{ days/week} \times 16 \text{ hours/day} \times 60 \text{ min/hour}) \times .9} = .73$$

Number of machines

.73 parts/minute x minutes/part

Operation 10	.73 x 7.5 =	5.5
Operation 20	.73 x 2.5 =	1.8
Operation 30	.73 x 3.5 =	2.6
		9.9

Clearly, 10 machines were required to achieve the desired throughput. However, the decision as to how to allocate the machines to the various operations was not so clear, since non-integer values were obtained from the calculations. Given that the number of machines required for operation 20 was between 5 and 6, the number of machines required for operation 20 was between 1 and 2, the number of machines required for operation 30 was between 2 and 3, and a total of 10 machines were required, the scenarios listed in Table 1. are feasible:

Scenario #	Operation 10	Operation 20	Operation 30	Flexible
	NUMBER OF MACHINES			
1	5	2	3	0
2	6	2	2	0
3	6	1	3	0
4	5	1	3	1
5	6	1	2	1
6	5	2	2	1
7	5	1	2	2

Table 1. Number of Machines

Assuming that 12 fixtures of each type would be sufficient, along with 2 wire-guided vehicles, the evaluation of alternative equipment levels consisted of first varying the number of machines at each workstation according to Table 1.

In order to handle the data management and data presentation tasks inherent in the simulation project, the MAP/1 simulator was used in conjunction with the TESS software. TESS offers database and graphics capabilities for storage, analysis, and display of MAP/1 and SLAM II simulation data. Data from the simulation can be automatically stored in the database under user-specified scenario, data definition, and variable names. This data can then be summarized, analyzed, and presented in the form of plots, pie charts, histograms, bar charts, range charts, and animations.

For this analysis, data on the system throughput and equipment utilizations were stored in the database for each scenario listed in Table 1. This data was used to prepare the throughput comparison shown in Figure 4. Clearly, the only scenario which could meet the throughput goal is scenario #6, where 5 machines were dedicated to operation 10, 2 machines were dedicated to operation 20, 2 machines were dedicated to operation 30, and 1 machine was set up to accommodate all three operations.

In order to obtain a better understanding of system behavior and the interactions between the four sets of machines, plots of the number of busy machines over time were prepared for each machine grouping. These plots illustrate how the equipment levels in one group can affect the utilization of subsequent groups, and aid in understanding why certain alternatives proved more productive than others.

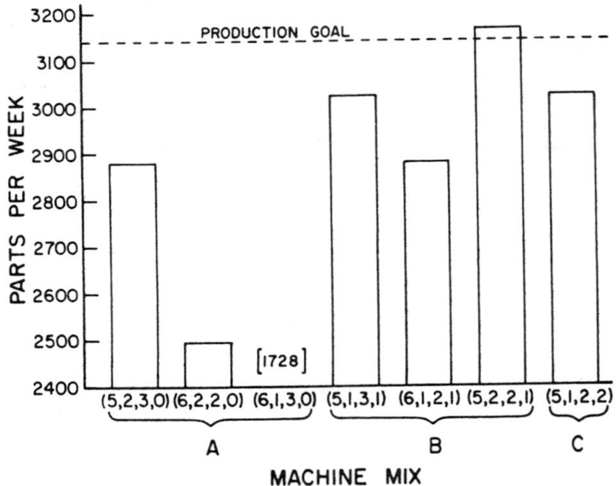

Figure 4. Throughput Versus Machine Mix

For instance consider the plots of machine utilization over time shown in Figure 5. These plots correspond to Scenario 1. After an initial startup period, all five machines dedicated to operation 10 are busy throughout the simulation. However, the plot of the utilization of the two machines dedicated to operation 20 exhibits a cyclical pattern indicating that these machines are not being fed fast enough from the previous operation. This lag is intensified in the case of those machines dedicated to operation 30, to the point where the machine group is rarely used to capacity.

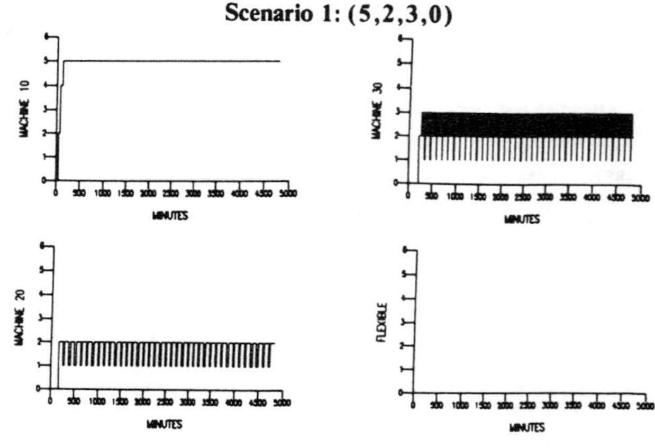

Figure 5. Utilization Plots

Figure 6. contains utilization plots corresponding to Scenario 4. Again, after an initial startup period, all five machines dedicated to operation 10 are busy throughout the simulation. Similarly the machine dedicated to operation 20 is continuously used after the startup period. However, the plot of the utilization of the machines dedicated to operation 30 definitely exhibits a cyclical pattern, indicating delays in receiving material from the previous operations. Although the flexible machine was continuously busy, it could not handle the overflow from operation 10 and operation 20 in a timely manner.

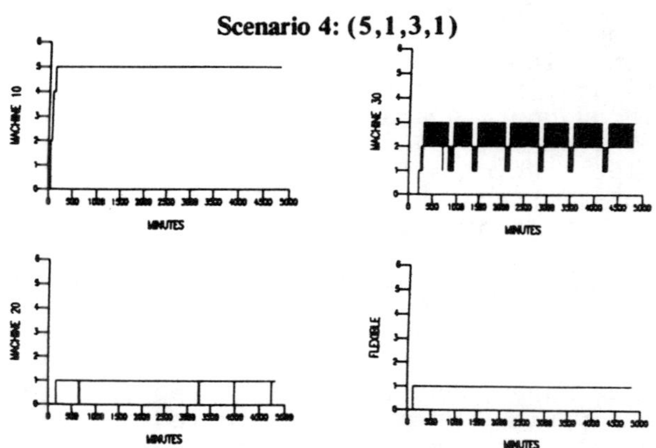

Fig. 6. Utilization Plots

Figure 7. contains the utilization plots from Scenario 6. All machine groups were used to capacity continuously, after an intital startup period, except for the machine group dedicated to operation 20. Even though the use of this machine group was cyclical between 1 and 2, the average use appeared to be close to 1.8, which was confirmed by the output statistics. Apparently the additional .5 capacity required by operation 10, and the additional .6 capacity required by operation 20, were sufficiently supplied by the one flexible machine.

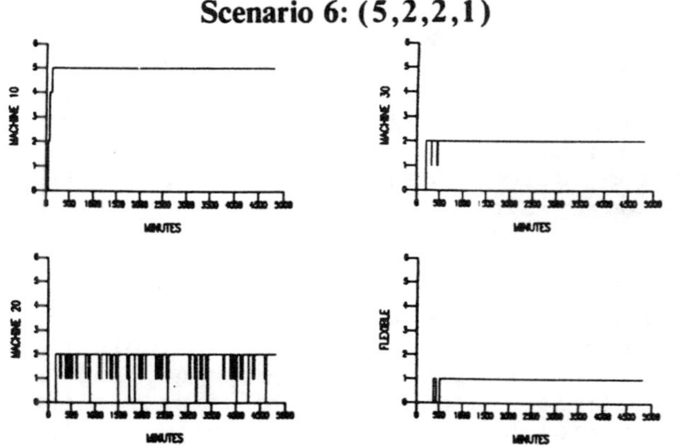

Fig. 7. Utilization Plots

Once the number of machines to be allocated to each operation was determined, equipment levels for the wire-guided vehicle and the fixtures A and B were considered. A pie chart of the utilization of the cart during Scenario 6 was prepared and is shown in Figure 8. Based on the average utilization of only 24.6%, it appeared that one cart could easily be eliminated. This was confirmed by a subsequent execution of the simulation model with only one cart available.

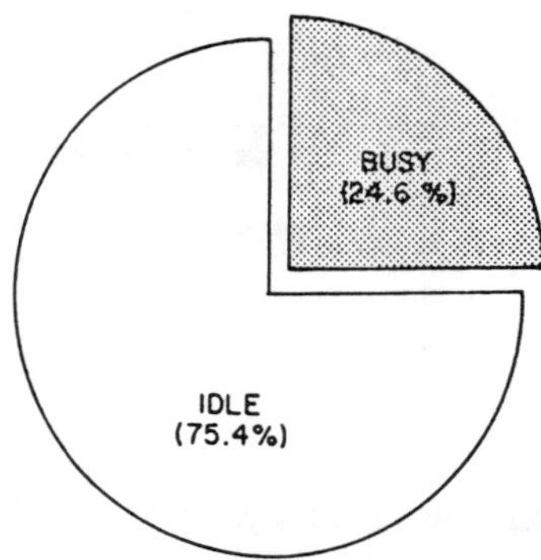

Fig. 8. Cart Utilization

Figure 9. shows a plot of the utilization of fixtures A and B over time. After an initial start up period, the number of busy A fixtures remained constant at 12. However, the number of busy B fixtures varied widely, cycling between 6 and 12, with an average utilization of around 9. Based on these plots, the next alternative evaluated consisted of decreasing the number of A fixtures to 11 and decreasing the number of B fixtures to 9. Execution of this alternative resulted in a throughput level the same as that achieved with 12 fixtures of each type. Thus, a reduction in the number of fixtures can be achieved without affecting the throughput of the system.

Based on this success, the number of fixtures of each type was again decreased in two steps. First, a combination of 10 A fixtures and 9 B fixtures was simulated, then a combination of 11 A fixtures and 8 B fixtures. Throughput results from these and the previous two alternatives were summarized on the bar chart shown in Figure 10. The plots of fixture utilization for the (10,9) scenario are shown in Figure 11. The number of busy A fixtures soon became constant at 10, but the number of busy B fixtures cycled between 4 and 7, averaging close to 6. Obviously, the use of B fixtures was closely tied to the use of A fixtures, and 10 A fixtures were not sufficient to supply the following operations utilizing B fixtures. Based on these results, fixture levels were reduced from (12,12) to (11,9), representing a substantial cost savings.

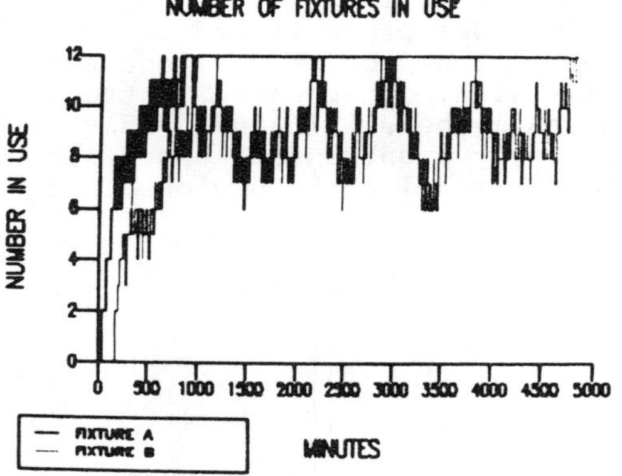

Fig. 9. Utilization Plots

Throughput Vs. Fixture Mix

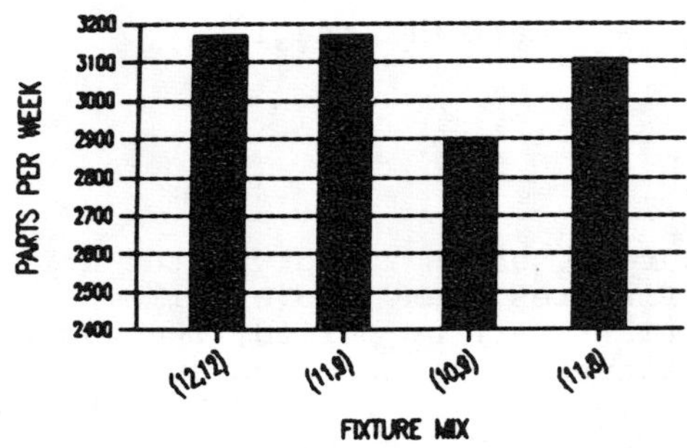

Fig. 10. Throughput Versus Fixture Mix

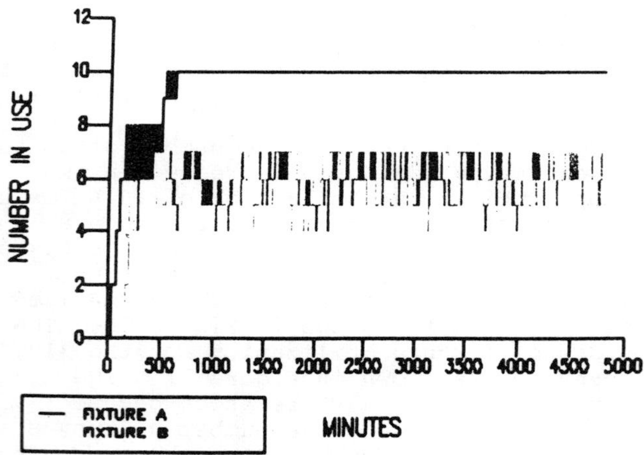

Fig. 11. Utilization Plots

Results of the simulation analysis were as follows. The number of dedicated and flexible machines which maximized throughput was determined. The levels of the cart and fixture resources required to support this throughput level were minimized. Cost savings over the initial recommendation were substantial, amounting to over $170,000.

A simulation support system was successfully used to support the analysis. Use was made of automatic data collection facilities which stored data from the simulation program directly into the support system database. Data from the various scenarios could be easily located and referenced through the user-assigned scenario name. This data was used to prepare displays showing the comparison of alternative equipment configurations and resource levels based on throughput. Detailed data on the utilization of resources over time was used to prepare plots, which offered insight into system behavior and the interrelationships of system components.

A simulation support system offering data management, data analysis, and data presentation capabilities is a definite aid in performing simulation projects. Data management facilities keep track of the various scenarios evaluated and the outputs corresponding to each. Automatic data collection frees the user from the burden of defining data storage formats which are compatible with separate data analysis or data presentation software.

Detailed data from the simulation can be manipulated to obtain various forms of statistical information using the data analysis capabilities. Statistical measures of means, standard deviations, minimums, and maximums can be calculated in total, by time intervals, or by entity or resource attributes. Frequency tables for selected cell specifications can be prepared. Confidence intervals for multiple run comparisons can be estimated.

Graphical data displays provide information in a form more easily understood and interpreted than numerical tabular reports. Detailed data which offers insight into system behavior and system component interactions can be displayed and interpreted. Animations of system diagrams provide realistic views of system operation. Data from various scenarios can be presented side-by-side for comparison purposes.

SUMMARY

Results of this simulation project included optimal allocation of machine tools to workstations to maximize throughput, and reduction in the number of fixtures and material handling devices while maintaining throughput. Equipment cost savings were substantial, paying back many times the costs of the simulation study.

REFERENCES

1. Miner, Robin J. and Laurie J. Rolston, MAP/1 User's Manual, Pritsker & Associates, Inc., West Lafayette, Indiana, 1983.

2. Standridge, Charles R., et al, TESS User's Manual, Pritsker & Associates, Inc., West Lafayette, Indiana, 1984.

3. Musselman, Kenneth J., "Computer Simulation: A Design Tool for FMS", Manufacturing Engineering, September 1984, pp. 117-120.

4. Wortman, David B., James R. Wilson, "Optimizing a Manufacturing Plant", Computer-Aided Engineering, September 1984, pp. 48-54.

5. Standridge, C.R., J.R. Hoffman, and S.A. Walker, "Presenting Simulation Results with TESS Graphics", Proceedings of the 1984 Winter Simulation Conference.

BIOGRAPHICAL SKETCH

LAURIE J. ROLSTON is a Systems Consultant at Pritsker & Associates, Inc. She holds a Bachelor of Science in Industrial Engineering from Purdue University. Since joining P&A in 1980, she has had experience applying SLAM to industrial problems and in the development and support of P&A's commercial software. Ms. Rolston was involved in the development of MAP/1 and is currently responsible for user support, maintenance and software enhancements. She is a member of Alpha Pi Mi, Tau Beta Pi, and AIIE.

ERRATA

This section replaces pages 331-339 in the volume, **Justifying New Manufacturing Technology**

Control and Evaluation in the Innovation Process: The Case of Flexible Manufacturing Systems

DONALD GERWIN

Control and Evaluation in the Innovation Process: The Case of Flexible Manufacturing Systems

DONALD GERWIN

Abstract--Control and evaluation mechanisms are a critical part of the innovation process. They provide information on whether or not an innovation should be adopted, and the extent to which implementation is a success. Yet, we have very little information on how they operate in practice.

This paper explores the problems in controlling and evaluating a radical process innovation in a batch manufacturing firm. Data come from interviews with individuals in various hierarchical levels and functions. The decision to adopt tended to be qualitative in nature and essentially made by two representatives of the organization's technostructure. Their major objective was to minimize the chance of disaster rather than to maximize efficiency.

During implementation, quality control faced frequent, complex technical problems at the same time that there was pressure for immediate solutions. Cost accounting procedures were affected by the difficulty in developing completely objective standards. The strategies employed to solve these problems are discussed along with implications for the design of control procedures.

One intriguing implication is that the characateristics of a radical new technology, along with other factors, influence the validity of the control procedures designed to evaluate it.

I. INTRODUCTION

For some reason, the vast literature on organizational control systems emphasizes conceptual development over empirical findings. We know very little about how control systems actually operate [15]. This paper contributes to closing the knowledge gap by discussing some issues raised in the control and evaluation of a computer-aided manufacturing system in a division of a large manufacturing company. A broad interpretation of the meaning of control will be taken. The term will apply to the division's evaluation of its selection of an alternative machining system, of which there are several to choose from, and to the division's attempts to insure that the system chosen produces the desired results. Hence, the paper deals not only with the theory of organizational control and evaluation, but also with decision making, capital budgeting, technology, and innovation. It should also have practical benefits for firms either contemplating the purchase of advanced manufacturing systems, or dealing with the implementation problems of these systems. However, control is a multifaceted concept. The definition used here does not embrace all of its important dimensions. Ettlie [7], for example, has argued that organizations control manufacturing technology by modifying it to fit their needs.

The use of highly automated computer-controlled machinery in manufacturing has had a significant impact on business firms and society. The greatest effects have occurred in process and mass production technologies where the closest approximations to the completely automatic factory have been reached. Firms engaged in batch production have not shared in these developments to the same extent. Typically, they have had two types of machining alternatives. One choice involves single-purpose or dedicated machinery such as transfer lines, best suited for very large batch or mass production of a specific part. The second choice involves general purpose machine tools, which are most economical for one of a kind or very small batch production of many different parts. For the large number of firms producing several parts each in lot sizes between these two extremes, neither alternative is very appropriate. Recently, however, the development of flexible manufacturing systems (FMS's), sometimes called advanced batch manufacturing systems, has provided the overwhelming majority of batch producers with an automated alternative expressly suited to their needs.

An FMS is a highly customized manufacturing system which generally consists of several machine tools operated by a central computer, and automated material handling which is also computer controlled. One of its most significant features is its capability for substituting random processing of parts for batch production. Parts that are capable of being machined by the system can be run in any sequence at any time without the need for costly setups. For further details see Cook [4], and Hughes, Hutchinson, and Gross [11].

This paper is part of a larger study of the organizational impacts of an FMS in a farm tractor division of a large multidivisional corporation. During 1978-1979, about twenty-five semistructured interviews were held with individuals representing various levels and functions within the division. Respondents were asked such questions as: "Explain your role in the decision to purchase the FMS," "What major changes has the FMS produced in your job (department?),"

"What major changes would you like to see occur?" These probes further stimulated discussion which led to ad hoc questions and further discussion. From the outset, it was clear that control and evaluation had been, and is currently, one of the most significant issues for the unit. Preliminary results from the larger study appear in Gerwin and Leung [9].

A. Background for the Study

The corporation, a large multidivisional firm with sales of $2 000 000 000 and a net income of $81 400 000 in 1979, is a diversified international leader in the manufacture and distribution of special machinery. Organizationally, it is divided into four operational sectors, including industrial process equipment for fluids and solids; agricultural equipment including tractors, combines, and other farm implements; material handling and outdoor power equipment; and electrical motors and generators. Since 1968, when new corporate management took over, the organizational structure has undergone continuous transformation under the influence of strategically planned changes in the mix of businesses. The principal feature has been planned growth in process equipment, which was 16 percent of total sales in 1969 and 49 percent in 1979. It was not possible to obtain further information on organizational structure due to its confidential nature.

Weber [18] has documented the situation facing the corporation in the late 1960's. In 1967, it was targeted by conglomerates as a prime candidate for takeover, partially due to a turnaround in performance that had begun a few years earlier. The resulting uncertainty as to its future direction had an impact on morale and effort. At the same time, strategic problems were arising due to the changing nature of markets, competition, and technology. There were also many obsolete and inefficient plants with high labor costs. However, key personnel were diverted from formulating strategic responses in order to help preserve the company's independence. In 1966, sales had been $857 200 000 and earnings a record $26 000 000. By 1968, sales were down to $767 300 000 and there was a loss of $54 600 000.

In 1968, the Board brought in an outsider as company president who had a reputation for turning operations around. He vigorously pursued a policy of remaining independent, while making top management changes, streamlining staff, extending international markets, acquiring new firms, engaging in joint ventures, developing new products, raising badly needed equity capital, and selling off noncompetitive areas. Total employment was reduced by 8000 from 1968 to 1971. From 1969 to 1975, about $240 000 000 was spent on capital improvements.

These efforts paid off. By the spring of 1970, the corporation had fought off its last takeover bid. From 1971 through 1979, there were nine years of uninterrupted growth. Sales and earnings which had been $853 500 000 and $5 200 000 in 1971, grew to $1 972 700 000 and $81 400 000 in 1979.

The problems facing the corporation in the late 1960's were mirrored in the tractor division. Basic changes in the nature of farming were occurring, including moves toward fewer larger sized farms, and the accelerated substitution of capital for labor. Consequently, farmers began demanding larger, more sophisticated, and more customized tractors. For tractor manufacturers this meant an opportunity for larger unit profits but also a wider range of models and smaller volume per model. However, the division was still producing only standardized small and medium tractors, using many old, dedicated manufacturing processes. The division's declining sales and profits necessitated that the corporation make a choice between either getting out of the tractor business entirely, or regenerating the subunit so that it could compete more effectively. Ultimately, the latter alternative was selected by the new corporate management team. A decision was made to produce and market a new line of high-horsepower tractors, and to replace the obsolete single-purpose machinery. The capabilities of the entire shop were transformed by the purchase of about 50 machine tools from over 20 suppliers at a cost of $15 000 000.

The FMS, which went on line in 1972, was only the second one installed in the United States. It is one of the largest in the world, consisting of ten machines and two computers at a capital cost of about $6 000 000. It plays the most critical role in the division's manufacturing process by machining almost all of the major housings for the new product line.

A history of the division's market share, sales, and profits is considered proprietary information. The overall share of the tractor market is not large. However, due to the new product line, the division is well positioned in the larger horsepower, higher profit segment to which farmers have been turning. The most relevant sales and profit figures which are reported are for the corporation's agricultural equipment sector of which the division is a part. In 1972, the year of the FMS's installation, sales of agricultural equipment were $204 000 000 or 23 percent of the corporate total. Sales increased in every succeeding year but one, until they reached $679 000 000 in 1979 or 34 percent of the corporate total. Income before taxes for the agricultural sector, which was negligible in 1972, rose to $66 000 000 by 1976 (52 percent of before tax income for all company sectors), but by 1979 was down to $38 700 000 (21 percent). These data suggest that the corporation's investment in the division was well founded.

II. EVALUATION OF CAPITAL ALTERNATIVES

The preparation and evaluation of the new product line proposal were made under conditions of extreme uncertainty with a distinct likelihood of a highly negative payoff occurring. Consider that this was one of the first transactions of its kind in the United States. For the vendor, it was the first sale of this type of machining system. Of course, for the company it was the first purchase. The system was highly customized in nature so that numerous complex technical decisions had to be made in order to develop it. In addition, the FMS was purchased to machine parts for a completely new product line. The high degree of risk and novelty associated with the manufacturing system indicates that it was a radical innovation for the firm.

A. Considerations Within Manufacturing Engineering

Proposals for a new product line had existed for several years before the new corporate management took over. When that event occurred, the head of manufacturing engineering for the division appointed a five person innovation group from his department to determine whether it was feasible to build the new line. The group worked closely with a liaison from purchasing who helped in locating a source of supply, and a member of accounting who prepared financial analyses.

The innovation group's task was exceedingly difficult in that it was faced with a large number of highly technical decisions that had to be made using little reliable information and under considerable pressure for correct choices. The same situation existed for the managers who had to accept or reject the proposal. Consequently, it was necessary to employ some simple strategies to reduce the problem to manageable proportions. Very comprehensive financial calculations at considerable effort were made only for the program as a whole. The calculations produced figures for the payback period and discounted cash flow. The exact values of these numbers could not be considered very reliable because of the extremely uncertain future. However, they did serve two useful functions. First, they satisfied the corporate guidelines for rationality in capital budgeting. Second, they provided some indication of whether the minimum limits on acceptability were likely to be exceeded. In making the calculations, every legitimate opportunity was taken to have the payback period as low as possible, and the discounted cash flow as high as possible. For example, cash flow items, which under normal circumstances might have been considered intangible, were quantified. At the same time, there existed strong pressures in the minds of the proposal generators to be as accurate as possible. It was considered critical to preserve the credibility of those individuals who were selling the proposal, in order not to jeopardize its chances of approval.

Since the program was treated as a whole, the numerous capital decisions that had to be made were considered to be a single investment. Sophisticated financial analyses were not made on an individual machine or machining system basis even for the manufacture of major housings. At the plant level some cursory analyses were made. For example, the leading alternatives for the production of a part might be ranked on various machine specifications using a numerical scale. Then it would be asked if the additional cost of the alternative with the most points was worth the added expense. These questions were answered intuitively. Beyond the plant level, laterally and vertically, there was little attempt to inquire into machine tool decisions. These were considered as technical choices which were best left to manufacturing engineering. If, on the other hand, a single machine tool is to be requested for an existing product line, justification would have to include sophisticated financial analyses which would be intensively reviewed. Thus it can be inferred that the major housing decision, in spite of its magnitude and uniqueness, lost visibility by being one among many choices which were highly technical in nature.

Some information was available concerning the details of the major housing decision. The search began with six or seven alternatives which were quickly narrowed down to three major contenders; two transfer lines and the FMS. This conforms to what is known about industrial purchasing in other contexts. An initial superficial search screens out alternatives until there remain a very few, which are reviewed intensively [5]. During most of the evaluation period the innovation group favored a transfer line, but late in 1971 a change in thinking occurred. At least three factors were responsible; acquisition cost, projected sales volume, and vendor relationships. These will be discussed in turn.

Corporate financial theory tells us to consider the relationship between acquisition cost and projected cash flows in choosing among machine tool alternatives. However, acquisition cost assumed a disproportionate role in the thinking of requesters and decision makers in this situation. For one reason, there was the difficulty in making accurate forecasts of future net revenues. For another, corporate headquarters was unable to invest a large sum in the division's resurgence, and the division for its part did not want its proposal rejected on this basis. In effect, dealing with the certain here and now was preferred to dealing with a hard-to-anticipate future. The FMS had a relatively small acquisition cost compared to its rivals and this represented a useful selling point to corporate management.

The merchandising department's sales forecasts

for the new product line were steadily decreasing during the evaluation period under pressure from a hard-to-convince management. This had repercussions on the innovation group. It began to be felt that estimated production quantities were falling out of the realm of economical transfer line operations. In addition, the changes in the forecasts implied that merchandising was not particularly confident in its projections. While there seemed to be almost unanimous agreement that high-horsepower tractors were the wave of the future, it was much less clear at what horsepower levels the new tractors should be positioned, and what would be the sales volume at each level.

The result was a change in objectives within the innovation group, and at higher levels, from developing an optimal program to developing one which was most likely to prevent disaster. The FMS fit into this thinking. If the new product line, or some segment of it, failed to appeal to farmers, the system could machine redesigned models or be put to other uses with little added cost. There might still be another chance to keep the division going. The salience of minimizing disaster is perhaps best exemplified by the preparation of a "Panic Plan" which calculated discounted cash flow for the program, assuming sales turned out to be very low. The result was a rate of return almost half of that which was expected, but one that turned out to be just at the floor of corporate acceptability at that time.

The company had a close relationship with the vendor of the FMS. They are located nearby each other in the same city and have a history of dealing with each other. Here was one instance in which some reliable information existed. As a result, the company was convinced that the vendor could develop a system along the lines it was promising. The vendor's professional sales effort, which projected a capable image, reinforced the company's thinking. Geographical proximity would also facilitate testing and servicing of the new equipment.

In spite of these three factors, the innovation group was not unanimous in its enthusiasm for the FMS. Some thought that the system represented too many unknown, complex variables and, therefore, it would be difficult to operate and to diagnose problems. It was maintained, for example, that the system would not be able to adhere to as close tolerances as a transfer line. Further, the vendor who had installed the only American FMS in operation at the time, preferred not to build one for the company. It believed the company's requirements made for a system which would require very large engineering development costs. The group's confidence in the FMS alternative faltered at this point.

Within manufacturing engineering the decision to adopt the system was made by the department head in conjunction with the head of the innovation group. Both were particularly concerned about minimizing the chances of disaster. They received support from the head of product engineering who appreciated the FMS's ability to facilitate design changes. Little formal analysis was involved. An evaluation sheet was prepared which rated four machining systems on each of 26 dimensions including capital cost, labor requirements, flexibility to add parts, time to deliver and install, and estimated amount of running time per week. The FMS came out far ahead of its closest competitor. In part, the evaluation served to help crystallize the decision makers' thinking. However, the fact that their minds were already pretty well made up probably had some effect on the evaluation's outcome. In part, the evaluation also served as a documented rationale for their choice.

B. Selling the Program

The task of selling the program went on as it was being developed. At least four hierarchical levels above manufacturing engineering had to be taken into consideration. The head of that department took the lead in dealing with the first three. He was characterized in the interviews as being either totally for or against an issue. Undoubtedly, his solid commitment in such an uncertain situation gave him a great deal of influence. At the very least, it prompted his superiors to look elsewhere for possible flaws in the program.

The first step was the plant manager who was very supportive of the proposal all along. His leadership style included relying upon manufacturing engineering for technical decisions, and being particularly concerned that they keep him informed of what they were doing. The next level, operations management, was more difficult to convince. Here, reluctance existed to refer the program upward until it had been thoroughly checked for flaws. The sales forecasts were of particular concern. At the next level, the division manager was not difficult to convince. He had been brought into the company by the new corporate president, and was committed to turning the division around.

The division manager took the lead in selling the program to the corporate office. Resistance at this level was strong. There were not a great deal of funds available for investing in the divisions, and it was thought that the tractor division did not provide a good opportunity for generating cash flows. It was believed that in the near future there would be just two competitors in the tractor business with the division not being one of them. The division manager argued that getting out of the tractor business was not a realistic alternative to regenerating the division. It was proving difficult to find a company that would purchase the division's assets, and the dealer network would be adversely affected if a full line of agricultural equipment including tractors was not offered. Not only the substance of his arguments but also his style of delivery proved to be compelling. He was able to articulate the case for the proposal in language the corporate personnel could understand.

Ultimately, the new corporate president made the decision to approve the program. By this time, almost three years of program development had elapsed, and over 30 revised proposals had been made. As Gold *et al.* [10] pointed out, major innovation decisions seldom take the form of one-time commitments. Rather they involve successive reviews of past estimates, and utilize developing information and experience.

III. MANAGERIAL CONTROLS FOR THE FMS

Evaluation took on a different meaning for the division after the decision to adopt the FMS was made. It became important to investigate whether the new manufacturing system produced the desired results, and to take corrective action if appropriate. According to Kiresuk and Lund [12] post-decision evaluation falls into three major categories: outcome, efficiency, and impact. Briefly, outcome evaluation focuses on the short-run outputs produced, efficiency evaluation on the ratio of short-run inputs to these outputs, and impact evaluation upon long-run unanticipated consequences.

In manufacturing, it is usually possible to develop reasonably precise outcome and efficiency measures. Pressures are thereby created to concentrate on short-run accountability and to neglect intangible long-run considerations; a classic example being the emphasis upon daily production at the expense of preventive maintenance. When a computer-aided manufacturing system is adopted it becomes much harder to develop, implement, and maintain effective short-run controls while the chances increase for long-run unanticipated consequences. The main reasons stem from the uniqueness and complexity of the technology.

In the case of the firm studied, additional factors had implications for the managerial control systems. The corporate office expected returns on its investment in the short run, which meant that the vendor had to work within a shorter delivery time than is usually expected, and that daily production was expected soon after installation. The vendor found it impractical to conduct in-house capability studies of such a large system, hence, tests had to be conducted after installation. Sales of the new product line turned out to be much greater than anticipated, compelling use of the FMS on a round-the-clock basis. In addition, operating, staff, and service personnel had little experience in how to deal with such a complex system, and had to be concerned with all the other new machines that were part of the capital improvement program.

A. Outcome Evaluations

The purpose of outcome evaluations is to appraise the short-run results produced by the entity under study. Perhaps the most critical outcome of FMS operations is the quality of the machined castings. Farmers are concerned mainly that equipment does not break down when it is in use, and are concerned secondly about price and other factors.

Divisional quality control became involved with the FMS sometime after the adoption decision but before installation. They found a completely different level of complexity than they had previously dealt with. Defects may occur due to any number of factors in the machine tools, computers, loading, material handling systems, the part being processed, or some combination.

At the same time it is more difficult to find opportunities for making inspections. Quality checks can be made readily at the end of the machining process or between machining sequences because at these points natural pauses exist in the flow of work [14]. However, unlike the situation for conventional machines, these off-line checks are not sufficient to maintain quality standards. For one reason, a complete check requires that a part be removed from its fixture. If a defect is found, it is difficult to realign the part in the same position in the fixture so that reworking can proceed. More important, waiting until a defective part reaches a natural pause increases the likelihood that other parts being processed or even the machine tools will be damaged. The time span from the occurrence of a defect until its discovery can be as much as 2 h, while the value of a part is in the hundreds of dollars.

It is also more difficult to identify the problem sources which lead to defects; a critical issue because of the pressure for immediate corrective action. If problems are not readily discovered and corrected, the system or part of it must be shut down to avoid further damage to the parts being processed or to the machine tools. Technical factors are part of the reason it is hard to find the sources of defects. Not only are there many more things that can go wrong, but different problems can produce the same symptoms. For example, it is difficult to tell whether a defect is due to loading or to a machine tool. Once more, a misalignment affecting one operation can throw off subsequent related operations, making it more difficult to relate symptoms and causes. Consequently, it took a long time for inspectors to learn how to use symptoms heuristically to narrow down the number of potential problem sources.

There is also an interpersonal aspect to the difficulty in locating causes. Problem solving forces quality control personnel to interact a great deal with individuals from other areas such as shop management, the FMS department, purchasing, manufacturing engineering, product design, maintenance, and the foundry. People bring to these interactions differences in objectives, backgrounds, and statuses which could impede task accomplishment. Therefore, it is necessary for them to develop communication, negotiating, and social maintenance skills to offset these factors.

Given these problems, scrap and rework levels turned out to be higher than originally anticipated. Attempts by quality control and other departments to reduce these levels produced several coping strategies of which three in particular will be discussed.

One strategy was to lower the original theoretical objectives for tolerances when they proved to be difficult to attain in practice. During the design phase, the vendor, manufacturing engineering, and product design believed that the FMS could achieve rather tight tolerances. This objective was considered to be important in meeting farmers' demands for serviceable equipment. Tight tolerances reduce the chances of breakdowns caused by loose fitting parts, and also facilitate prompt servicing by increasing interchangeability of parts. When the FMS was implemented, the desired tolerance levels could not be attained. Exhaustive testing indicated that more sophisticated inspection and greater machine capability would be needed. Operating personnel and quality control argued for bringing tolerances to a more realistic level, and ultimately, the theoreticical specifications were changed. However, the FMS is still able to produce tolerances comparable with the shop's conventional machines.

The second strategy involved trying to gain more control over environmental factors, which were making it difficult to achieve desired quality levels. Originally, the two largest castings machined by the FMS were obtained from external suppliers, making it difficult to exert control over the quality of the castings, the speed with which quality problems were corrected, and costs. The new manufacturing system, however, required castings with more precise specifications than are needed for conventional machines. One reason is that there are less opportunities for humans to make adjustments that could compensate for variations in the quality of raw materials.

The solution was to have the division's foundry supply virtually all of the castings for the FMS, a decision which required a considerable investment in upgrading facilities. The head of quality control was transferred to manage the effort. Currently, the FMS receives more preferential treatment than could be obtained from external suppliers as 30 percent of the foundry's business is with this one unit. A liaison person from the foundry checks the FMS's production at least twice a week to see if quality problems exist and to take corrective action if possible. The geographical propinquity between the two subunits also facilitates interaction.

The third strategy involved adapting quality-control procedures to the new technological reality. In order to create additional opportunities for locating defects, a procedure involving programmed stops and operator control was developed. First, artificial pauses are introduced during machining operations by programming the DNC computer to automatically shut off machines at certain times. Then, the operators perform a series of inspections to check generated characteristics. They are also responsible for taking corrective action such as changing tools. This procedure locates about 50 percent of all the defects that are found. However, it decreases machine utilization, a critical issue with such expensive equipment, and cannot guarantee that discrepancies will be detected at the first part in which they occur. The long-run solution is to install automated continuous monitoring. A limited amount of monitoring exists now, but a complete system must await state-of-the-art advances in automating sophisticated geometry checks.

B. Efficiency Evaluations

Efficiency evaluation is concerned with the amount or cost of resources expended relative to outcomes produced. In the division studied, this kind of control is exercised through a standard cost accounting system which compares expected output:input ratios to actual output:input ratios. The accounting system's main tool is a monthly budget which reports on total planned net department costs for planned production and total actual net department costs for actual production. Planned and actual cost components are also presented. A weekly machine and labor utilization report examines items that manufacturing has some control over in the short run in its attempts to improve budgetary performance.

Implementation of the FMS had a profound impact on the conduct of efficiency evaluations. It was soon realized that control over the new machining system could not be exercised using the same procedures as for the rest of the shop. The principal issue involved determination of the standards against which actual performance was to be judged. What units were to be used in expressing standards? Where were data for the computation of standards to come from? How flexible were standards to be? Attempts to answer these questions have led to conflicts between accounting and manufacturing which still have not been completely resolved.

The most fundamental problem concerned the units to be used in expressing standards, because the answer to this question determines the nature of the control system. The standard cost of machining a part on a shop facility had always been expressed in terms of direct labor hours, the most commonly used basis in accounting. Multiplication by the standard direct labor hours expended in machining the part yields the facility's share of the part's cost. Due to the automated nature of FMS operations, almost all labor hours are part of the burden; they cannot be attributed to the machining of a particular part. Therefore, it was decided to express FMS standards in terms of machining hours, the second most commonly used basis in accounting. However, machining hours could be defined in different ways. Manufacturing argued that cutting time, transportation time, and loading and unloading time should all be included, but accounting settled on just the first. Use of cutting time only facilitates the calculation of machine loads.

The switch to a machine hour basis for FMS accounting had an impact on departments which perform services for the new machining system. Quality control, for example, normally justified requests for more inspectors in terms of additional direct labor hours. It was soon clear that a large amount of inspection activity was going to be necessary for FMS production, and that direct labor hours was not an adequate basis for justifying requirements. New criteria for justifying requests were developed in order to solve this problem.

The fundamental changes in the nature of accounting procedures caused some unanticipated behavioral problems. Accountants, whose training and experience had been primarily in terms of direct labor hours, had some initial difficulty in working with the new concepts. The fact that there were now two accounting systems, one for the FMS and one for the rest of the shop, made cost comparisons between facilities more difficult. Determining whether a part should be removed from some area of the shop and machined on the FMS is an example.

Manufacturing managers found it difficult to interpret the meaning and significance of the new concepts. They had always relied upon informal procedures for controlling operations that were based upon direct labor hours. Managers could detect various behavioral clues that indicated direct labor hours were being lost, as when a worker was relaxing in the job. They could also quickly translate specific amounts of lost hours into corresponding amounts of lost production. Within limits, they knew how to take corrective action to get workers to increase their efforts. These internalized procedures were of little use in controlling the FMS because of the indirect relationship between how hard people work and the amount of production. Supervisors could not readily develop informal control procedures based on machine hours because the FMS's complexity forces reliance upon technical specialists from quality control, manufacturing engineering, and maintenance.

Where were data for the computation of standards to come from? None of the conventional sources could be used. There was, of course, no relevant information within the division. The FMS was radically different from all other areas of the shop, and completely new parts, which could not be handled elsewhere in the shop, were going to be machined. It was also impossible to obtain information on similar facilities elsewhere in the country. When the FMS was adopted only one other system of this kind was in operation. Therefore, it was necessary to base standards on intuitive estimates, a hazardous undertaking in view of the technology's uniqueness and complexity. Currently, a completely reliable source of data for computing standards has still not been found. About seven years of data are available on the FMS's performance, but most data cover starting up conditions or a period of abnormally high demand. There are now only about eight similar systems in operation nationally and they are highly customized.

Consequently, variance analysis, the fundamental tool of a standard cost accounting system, has limited utility in controlling FMS operations. Total planned costs usually are a reliable benchmark, but planned values of some cost components will either be too high or too low, making it difficult to determine whether the actual costs are within acceptable limits of control. Other criteria must be employed for evaluation in order to avoid situations such as a foreman being reprimanded for an unfavorable rework variance when he cannot recall there being many defectives in the recent past. Accounting will attend to FMS costs over which some control can be exerted and which represent large portions of the budget; maintenance and rework being the primary examples.

How flexible are standards to be? If their values change infrequently they can be used to detect long-run trends but many problems will be discovered only after they have occurred. If values change in the short run, problems can be detected before they become serious but it will be hard to identify long-run trends. The FMS's technology creates a need for relatively quick detection of problems and rapid corrective action, but the costs of gathering and computing the information needed to change the values of standards are currently excessive. Consider machine utilization. Its computation is based on theoretical capacity, equal to the number of machine hours per week the FMS is used. This standard can only change if the number of machines, the number of daily shifts, or the number of active days per week change. Practical capacity, which reflects the existence of downtime, machine loading requirements, and differing cycle times, is too difficult to calculate on a weekly basis. A second example is that cost standards do not give credit to the FMS department for machining a part up to the point a defect is found due to, say, the foundry.

The relative inflexibility of standards is also reflected in the timeliness of evaluation reports. Budgets appear only at the end of each month and may be delayed because closing the books takes precedence. Further, since the FMS budget requires special preparation due to its being based on machine hours, it is sometimes prepared after the other budgets. The weekly utilization report is prepared with about a five-day lead time.

All of these problems with cost accounting standards have put manufacturing management in a difficult position. They are being evaluated with benchmarks expressed in units, which are difficult to understand, that are computed from data, which are not completely reliable, and which are not as timely as desired. A dual coping strategy has emerged. Attempts are being made to bring FMS costs into acceptable limits of control. At the same time, the assumptions behind the control system are being questioned. The questioning of assumptions has emerged only recently since a great deal of effort had to be invested in understanding the intricacies of the accounting procedures.

A good example of the dual strategy concerns the weekly utilization report. Machine utilization, the ratio of hours actually used to capacity, is a critical quantity because it is one of the few variables that can be at least partially controlled in the short run to reduce FMS costs. Currently, the FMS's machine utilization is below that of other areas of the shop. Attempts to increase it have included machining more units of existing parts and adding new parts to be machined.

Concurrently, manufacturing has questioned the appropriateness of using theoretical capacity in the denominator of utilization, on the grounds that practical capacity is more realistic and is employed everywhere else in the shop. Use of practical capacity would of course increase the computed values of utilization. It has also been argued that practical capacity tends to be lower for the FMS than for other shop facilities because the greater technical complexity creates more frequent breakdowns.

Manufacturing has also tried to develop its own supplemental FMS evaluations. The foreman keeps his own daily record of production on the basis of the type of part, to augment information found in the weekly utilization report on an aggregate basis. When time permits, the process engineer assigned to the FMS produces his own machine utilization report based on practical capacity. Various intuitive evaluations are performed by manufacturing managers. Each day, the superintendent for the area which includes the FMS checks the system five or six times to see if it is operating, if breakdowns had occurred previously, if the daily production schedule is being met, if any defectives have been produced, and if it is being manned adequately.

C. Impact Evaluation

This type of evaluation examines the often unanticipated consequences on the larger environment within which the item being studied is embedded. For our purposes, the division housing the FMS will be considered its environment, and we will be concerned with the new system's impact on other subunits. Impact studies are difficult to conduct and few successful ones have ever been reported. The reasons include difficulties in measuring variables, in controlling variables, and in finding the time for a study.

Once more, the division found it difficult to establish a reference point against which to judge whether presumed FMS benefits had in fact occurred. Of course, data exist on cost conditions after adoption, but there is no relevant data source for cost conditions prior to adoption because of the new product line and revised accounting procedures. It was expected, for example, that the new system's capability for randomly processing parts would decrease in-process inventory. All the major housings needed for a single tractor can be machined and then assembled. It is not necessary to machine different batches of housings and store all finished parts until the last batch is completed. However, there are no prior inventory costs against which comparisons can be made. The same situation exists for other presumed benefits such as reduced lead times.

It is little wonder that no systematic impact investigations have been made in the division. However, interviews with division personnel indicated that the FMS has had both obvious and subtle effects on its environment. The sample to be discussed here indicates that the new technology's impacts extend as far as the development of manufacturing strategy.

While the FMS presented quality control with some novel problems, it provided this department with an opportunity to introduce new methods and procedures into operations in order to solve these problems. Prior to the FMS's purchase, the department had been stymied in its efforts to extend the concept of operator control, the use of machine operators to check measurable characateristics, to highly automated machining operations. When the FMS arrived, it became apparent that this complex system required some degree of operator control. After the benefits of the new procedure became evident, the concept was extended to the transfer lines. Quality control was also able to switch its precision measurement machine from manual to computer-assisted control. Thus the FMS possessed gateway capacity [20]. It created opportunities for certain departments to attain their objectives, and in the process facilitated the adoption of more technological changes.

The FMS requires high-quality raw castings within narrow tolerance limits. When it was realized outside suppliers could not meet this need consistently, a decision was made to have castings supplied internally by the foundry. The impact on the foundry, in terms of upgrading operations, was large. Core room facilities were modernized, control over sand quality was improved, and inspection activities were increased. The changes, which were made to accommodate the FMS, had the side effect of upgrading the quality of castings provided to the rest of the division. On the other hand, the foundry became more dependent upon it for the disposition of castings. A shutdown or slowdown in the division's operations, particularly in the FMS, could affect the foundry's activities within a week.

An FMS's ability to machine a variety of parts gives it the potential for assuming the dominating role in a manufacturing process. Then, if this facility's functioning is impaired, the entire factory can soon be affected. The division studied finds itself in this situation. Virtually all the major castings for its new product line are machined on the FMS, and no production alternatives exist in the shop. Once more, these parts are critical if assembly is to proceed, since smaller parts from elsewhere in the shop are assembled into them. While the chances are small that the FMS will completely shut down for an extended period of time, due to its semi-automatic backup capabilities, employing the backup options will slow down machining to varying degrees. Assembly activities are particularly vulnerable, but eventually all new product line activities would be affected. This suggests that one criterion for FMS parts selection should be not to make this single facility more indespensible than is necessary.

IV. CONCLUSIONS

Innovation is a process with two major steps, initiation, in which the decision to adopt is made, and implementation, in which the new item is brought to commercial readiness [20]. This paper has explored the interrelationships between an organization's control system and a radical manufacturing innovation as the process unfolds in a batch production firm. The evaluation scheme employed during initiation influences whether or not an innovation will be accepted by an organization. During implementation, the control system determines the degree of success of the innovation thereby influencing further commitments to it by the adopting firm. The diffusion process is also affected. Evaluative information spreads to firms considering adoption, thereby becoming a factor in their justification analyses. We have also seen that an innovation's characteristics, along with other factors, have an impact on the nature and results of the control system designed to evaluate it during both stages.

A. Initiation Phase

Bower's [1] conclusion that capital investment evaluation is a process of study, bargaining, persuasion, and choice, operating at many levels of the organization and over long periods of time, seems especially pertinent. Motivation for the innovation was a severe performance gap that threatened the division's viability. The problem's scope called for a radical solution of which the FMS was an integral part. Evaluation of the proposed solution was a sequential process corresponding to the steps in the organizational hierarchy. Consideration of the FMS at all levels was influenced by a lack of available information as to its probable consequences, its highly technical nature, and its being part of a larger proposal. These factors dictated a qualitative rather than a quantitative review which, beyond the proposal generation stage, would have been difficult to make very intensive. Nasbeth and Ray [16], who studied the diffusion of new industrial processes in the United States and Europe, found that companies had difficulty in assessing the anticipated profitability of numerically controlled machine tools and that calculations tended to be subjective. Gold *et al.* [10] in studying process innovations in the U.S. steel industry concluded that estimating the profitability of adoption is hampered by a lack of information. There seems to be mounting evidence that only a veneer of objectivity surrounds the adoption decision for major technological innovations.

Effectively, the decision to adopt the new machining system was made deep within the corporation's technostructure [8]. Beyond this point, acceptance hinged upon management's reaction to the entire proposal. Accordingly, the three criteria employed by the technical experts were particularly critical in the adoption decision. Minimizing the chances of disaster as oppposed to maximizing efficiency is the basis for the sociotechnical approach to the design of production systems. Miller [13] explicity made this point in his followup study of Rice's Indian Textile mill experiments. We have seen that a similar choice was made by the division's manufacturing engineers. Having a system with the flexibility to [add] machine altered versions of the new product line was more important than maximizing net cash flows. One must agree with Miller that "the goal of designing systems to minimize the chances of disaster may be more appropriate to industrial organizations than is generally recognized."

However, there is at least one critical difference

between the minimax criterion's position in sociotechnical theory and its use in the situation studied here. Miller would see the selection of minimizing disaster over maximizing efficiency as designing for long-run viability instead of short-run needs. To the proposal generators, the former objective met a more pressing short-run need than the latter. Less consideration was given to weighing long-run consequences versus immediate returns. The situation a couple of years after the adoption decision illustrates their thinking. Demand for the new product line unexpectedly surged upward, eliminating the possibility that the new line would would have to be redesigned to gain market position. Under these circumstances, it was necessary to augment machining capacity, and a transfer line was selected rather than an addition to the FMS. Maximizing immediate returns, by taking advantage of the transfer line's more economical operation at high production volumes, was given higher priority than improving long-run viability by maintaining the flexibility to deal with any future environmental contingencies.

The degree of confidence in a potential vendor's ability to meet mutually agreed upon specifications also played an important role in the decision. In my personal communications with researchers conducting an ongoing study of German FMS's, the salience of this criterion was verified. One way in which confidence is developed is through satisfactory prior relationships with a vendor. A supplier with whom a company has no experience represents more of a risk in terms of delivering what was promised. Supporting evidence comes from Cunningham and White [5] who found that 60 percent of the machine tool purchases they studied involved suppliers with whom the buying firm had previous transactions. Cardozo and Cagley [2] asked industrial buyers to participate in a controlled experiment using a buying game. Over 60 percent of bid selections and purchase decisions were characterized by suppliers having a prior relationship with the purchaser.

The discovery and subsequent use of acquisition cost, the third criterion, illustrates some key political aspects of the capital budgeting process. Apparently, requesters will try to discover which criteria are important to decision makers and then attempt to develop their proposals to fit them, thereby increasing the proposals' chances of being accepted. Carter's [3] study of capital projects in a small computing firm testifies to the generality of this phenomenon.

The choice process beyond manufacturing engineering was characterized by the strong commitment of two individuals; the head of manufacturing engineering, who sold the proposal to the division, and the division manager, who convinced the corporate office. Both people clearly served as idea champions [6]; individuals who genuinely believe in an innovation and are willing to invest effort in persuading key officials. Apparently, technical experts who are idea champions must convince a middle manager who then seeks to persuade top management. Differences in outlook between technical people and top managers often prevent them from being able to communicate well with each other. A middle level manager is sometimes needed to play the critical role of translator. One respondent specifically stated that the division manager spoke to corporate personnel in language they could understand.

Bower [1] and Carter [3] have observed that managers who must choose between competing capital projects are influenced by the quality of information sources as well as the content of information. Doubts concerning the quality of sources may lead to intensive reviews of details, requests for rechecking the accuracy of information, and ultimately rejection of a proposal. The people responsible for selling the proposal appeared to be well aware of this point. Throughout the long process of seeking approval they were concerned about maintaining the credibility of the information they forwarded to decision makers. This necessity limits any tendency to bias supporting data, a point not discussed enough in current research literature. Maintaining a confident image and preserving the appearance of rationality also contribute to the quality of an information source.

B. Implementation Phase

The design of evaluation mechanisms during implementation was affected by a number of conflicting factors. First, the FMS's characteristics created pressures for new control procedures; it was too radical to fit within the existing shop routines. For example, it was hard to make off-line quality control checks, and direct labor could not be used as the basis for cost controls. There was also a need for more immediate evaluatory feedback on quality and costs than could be supplied by existing procedures.

Second, the cognitive and affective needs of individuals who designed, operated, and were judged by the control systems created pressures to maintain the existing routines. Accounting people were concerned about the difficulties in operating two different sets of cost controls. The ability of manufacturing people to understand shop operations was rooted in the existing procedures. Further, the shift to more impersonal evaluation meant tht control was passing out of manufacturing's hands into those of technical experts and the FMS itself. Reinforcing pressures came from the lack of time to develop new procedures due to environmental and policy constraints, and the lack of availability of procedures expressly suited for integrated manufacturing systems.

The evaluation systems that emerged were a

compromise between both sets of forces. The new quality control procedure, emphasizing operator control, was essentially an adaptation of methods already used on the shop's manual machines. It was already familiar to line and staff personnel. The machine hour basis for accounting was new to the shop people but not to the accountants. Woodward's [19] findings on the nature of control systems in manufacturing support the notion of a compromise. She found that in unit and small batch production, and process production, the nature of controls is determined by the technology. However, in large batch and mass production, the category pertinent for the division, management had more freedom to influence the control structure.

The design problem was further complicated by the new manufacturing process being associated with a new product line. Consequently, there was no totally reliable reference system against which to judge the FMS's performance and its impact upon the rest of the division. The results were lack of clarity as to the FMS's costs and benefits, disagreement between line and staff people over the control system's assumptions, and a dearth of evaluative information for potential purchasers of the new technology. Moreover, this was not a unique problem. Major investments in new manufacturing processes are often made in order to develop a new product line. The critical lesson for the design of control systems is that the characteristics of a radical innovation are likely to affect the validity of the procedures developed to evaluate it.

Because there could be no completely adequate solution to the design problem, supplemental strategies had to be found. Thompson [17] helps us to understand the significance of the switch from outside suppliers to the company's foundry. Inputs must be adapted to fit within the limits of the technology processing them. If their source is the environment they cannot be regulated easily. Vertical integration is a strategy for circumventing dependence upon the environment. In addition, manufacturing was forced to develop its own personal short-run controls because the staff's impersonal procedures could not provide immediate feedback.

A case study does not settle issues very often; its main utility is in raising them. Hopefully, this research has raised some intriguing issues concerning the role of control and evaluation in the innovation process.

ACKNOWLEDGEMENT

The author would like to thank T.K. Leung who conducted some of the interviews on which this paper is based, and J. Ettlie who made comments on an earlier draft.

REFERENCES

[1] J. Bower, *Managing the Resource Allocation Process*. Boston, MA: Div. Res., Graduate School of Business, Harvard Univ., 1970.

[2] R.N. Cardozo and J.W. Cagley, "Experimental study of industrial buyer behavior," *J. Marketing Res.*, vol. 8 pp. 329-334, 1971.

[3] E.E. Carter, "The behavioral theory of the firm and top level corporate decisions," *Admin. Sci. Quar.*, vol. 16, no. 4, pp. 413-429, 1971.

[4] N.H. Cook, "Computer managed parts manufacture," *Scien. Amer.*, vol. 232, pp. 22-29, 1975.

[5] M.T. Cunningham and J.G. White, "The behavior of industrial buyers in their search for suppliers of machine tools," *J. Manage. Stud.*, pp. 115-128, May 1974.

[6] R.L. Daft and S.W. Becker, *The Innovative Organization*. New York: Elsevier, 1978.

[7] J.E. Ettlie, "A real-time case study of organization and innovation," Technol. Inst., Northwestern Univ., Evanston, IL. 1971.

[8] J.K. Galbraith, *The New Industrial State*, 3rd ed. Boston, MA: Houghton-Mifflin, 1978.

[9] D. Gerwin and T.K. Leung, "The organizational impacts of flexible manufacturing systems," *Human Syst. Manag.*, vol. 1, no. 3, pp. 237-246, 1980.

[10] B. Gold, G. Rosseger, and M.G. Boylan, *Evaluating Technological Innovations*. Lexington, MA: Lexington Books, 1980.

[11] J.J. Hughes, G.K. Hutchinson, and K.E. Gross, "Flexible manufacturing systems for improved mid volume productivity," *Understanding Manufacturing Systems*. Milwaukee, WI: Kearney & Trecker, 1976.

[12] T.J. Kiresuk and S.H. Lund, "Program evaluation and the management of organizations," in *Managing Human Services*, W.F. Anderson, B.J. Frieden, and M.J. Murphy, Eds. Washington, DC: International City Management Assoc., 1977, pp. 280-317.

[13] E.J. Miller, "Socio-technical systems in weaving, 1953-1970: A followup study," *Human Relations*, vol. 28, no. 4, pp. 349-386, 1975.

[14] E.J. Miller and A.K. Rice, *Systems of Organization*. London, England: Tavistock, 1967.

[15] H. Mintzberg, *The Structuring of Organizations*. Englewood Cliffs, NJ: Prentice-Hall, 1979.

[16] L. Nasbeth and G.F. Ray, *The Diffusion of New Industrial Processes*. London, England: Cambridge Univ. Press. 1974.

[17] J.D. Thompson, *Organizations in Action*. New York: McGraw-Hill, 1967.

[18] C.E. Weber, "Epilogue," in W.F. Peterson, *An Industrial Heritage*. Milwaukee, WI: Milwaukee County Historical Society, 1976.

[19] J. Woodward, *Industrial Organization: Behavior and Control*. London, England: Oxford Univ. Press, 1970.

[20] G. Zaltman, R.B. Duncan, and J. Holbeck, *Innovations and Organizations*. New York: Wiley, 1973.

Manuscript\ received May 8, 1980; revised February 12, 1981. This work was supported in part by CERESSEC, Cergy, France, and the the Management Research Center, University of Wisconsin-Milwaukee.

The author is with the School of Business Administration, University of Wisconsin-Milwaukee, Milwaukee, WI 53201.

ABOUT THE EDITOR

Jack R. Meredith is associate professor of production/operations management and director of operations management at the University of Cincinnati. He is former editor of *Operations Management Review* and past associate editor of the *Journal of Operations Management.* Meredith's writings have appeared in IIE's *Industrial Engineering* magazine as well as other journals.

In addition, Meredith has coauthored *The Management of Operations* (2nd ed., Wiley, 1984), *Fundamentals of Management Science* (3rd ed., Business Publications Inc., 1985) and *Project Management: A Managerial Approach* (Wiley, 1985). He has served as consulting editor for John Wiley & Sons, Inc., and has been a member of the technical staff of TRW, Inc. Other professional responsibilities include chairing the publications committee of the Society of Manufacturing Engineers' Manufacturing Management Council, and serving as a member of the board of directors of the Operations Management Association.

Meredith's research and consulting activities are in the area of managing advanced manufacturing technology, such as computer-integrated manufacturing. He is working on a grant funded by the Cleveland Foundation concerning the management of high technology.